Marx's Scientific Dialectics

Studies in Critical Social Sciences Book Series

Haymarket Books is proud to be working with Brill Academic Publishers (www.brill.nl) to republish the *Studies in Critical Social Sciences* book series in paperback editions. This peer-reviewed book series offers insights into our current reality by exploring the content and consequences of power relationships under capitalism, and by considering the spaces of opposition and resistance to these changes that have been defining our new age. Our full catalog of *SCSS* volumes can be viewed at www.haymarketbooks.org/category/scss-series.

Series Editor
David Fasenfest, Wayne State University

Editorial Board
Chris Chase-Dunn, University of California–Riverside
G. William Domhoff, University of California–Santa Cruz
Colette Fagan, Manchester University
Martha Gimenez, University of Colorado, Boulder
Heidi Gottfried, Wayne State University
Karin Gottschall, University of Bremen
Bob Jessop, Lancaster University
Rhonda Levine, Colgate University
Jacqueline O'Reilly, University of Brighton
Mary Romero, Arizona State University
Chizuko Ueno, University of Tokyo

Marx's Scientific Dialectics
A Methodological Treatise for a New Century

Paul Paolucci

Haymarket Books
Chicago, IL

First published in 2007 by Brill Academic Publishers, The Netherlands.
© 2007 Koninklijke Brill NV, Leiden, The Netherlands

Published in paperback in 2009 by
Haymarket Books
P.O. Box 180165
Chicago, IL 60618
www.haymarketbooks.org

ISBN: 978-1-60846-039-7

Trade distribution through Consortium Book Sales, www.cbsd.com

Cover design by Ragina Johnson.

This book was published with the generous support of Lannan Foundation and the Wallace Global Fund.

Printed in the United States.

Library of Congress Cataloging-in-Publication Data is available.

Contents

Figures, Lists, and Tables vii

Preface ix

I. *Reconsidering Marx*

Chapter One   Problems Reading Marx 3

Chapter Two   Marx and Classical Sociology 45

II. *Marx and His Scientific Dialectic*

Chapter Three   Marx's Onto-Epistemological Assumptions 69

Chapter Four   Marx's Analytical Procedures 103

Chapter Five   Marx's Conceptual Doublets 147

Chapter Six   Marx's Models 189

III. *From Marx's Science to His Politics*

Chapter Seven   From Political Economy to the Communist Project ... 227

Chapter Eight   Recovering Marx 259

References 283

Index 305

Figures, Lists, and Tables

Figures

Figure 4.1	Marx's Levels of Generality and their Associated Conceptual Strategies	114
Figure 4.2	Marx's Laws as Scatter Plots	141
Figure 4.3	The Phases of Inquiry Leading to Marx's Presentation of Political Economy	143
Figure 5.1	The Movement of Successive Abstractions	171
Figure 5.2	From Modes of Production to Class Systems in General to their Submodes	172
Figure 5.3	Conceptualizing Capitalism out of All Class Systems	172
Figure 5.4	Conceptualizing Submodes of Capitalism	172
Figure 6.1	Model of Social Change within & between Systems	195
Figure 6.2	Abstract Model of Expansion & Contraction in Early Capitalist Development	212
Figure 6.3	Model of Change within the Capitalist Mode of Production	217
Figure 6.4	The Relationship between Models of Production and Historical Change	219
Figure 8.1	Scientific Dialectics and the Communist Project	275

Lists

List 1.1	Marx's Work Published and Unpublished in His Lifetime	17
List 5.1	Marx's Epistemological Moments	157
List 5.2	Marx's Moments of Conceptualization	174

List 6.1	Summary of Abstractions, Forms, and Modes of Determination	193
List 6.2	Structural Components of the Capitalist Mode of Production and its Central Tendencies	198

Tables

Table 4.1	Marx's Relationship to Selected Contemporaries	110
Table 4.2	The Phases of Maturation of Marx's Research	110
Table 5.1	Synthetic Unities within Successive Abstractions	164
Table 5.2	Abstract Model of the Relations between Synthetic Unities	167
Table 5.3	Determinant Abstractions Beginning with Production in General	169
Table 5.4	Determinant Abstractions Beginning with Class Systems in General	169
Table 5.5	Determinant Abstractions Beginning with Capitalism in General	170
Table 5.6	Determinant Abstractions Beginning with Industrial Capitalism in General	170
Table 5.7	Social Systems in General at Moments of Abstraction	183
Table 5.8	Production in General and Moments of Abstraction	183
Table 5.9	Feudalism as a Specific Abstract and General Concrete Mode of Production	184
Table 5.10	Capitalism as a Specific Abstract and General Concrete Mode of Production	185
Table 5.11	Capitalism in the Specific Concrete	186
Table 6.1	The Abstract and Concrete in Historical Materialism and Political Economy	220
Table 8.1	Interconnections between Marx's Moments of Inquiry & Levels of Generality	273

Preface

The realities of our time press on us with urgency. Military conflict, economic instability, persistent social inequality, racism, sexism, and ongoing political strife remain endemic to modern life. Escalating ecological crises hang over our collective heads and demands that their solution be "business friendly" reveal much about their source.

Since its birth about 500 years ago, capitalism – a system of expanding production – has enveloped the world. To assume our social, political, and ecological problems are unrelated to this economic system strains reason. Simple observation shows us capitalism's extensive wealth and poverty, its withering effect on democratic potential, and ecological destructiveness. Scientific disciplines, which strive to demonstrate the interconnections between various realities, have contributed to these problems but also must help solve them. This is less a logical paradox than it is a contradiction in our social relationships.

In turning our knowledge into solutions for our collective problems, we must understand the realities towards which we have to act. To inform our action, science must help answer several questions: What is our moment? How does it work? From where did it come and where is it taking us? How can we shape the future being prepared for us?

Karl Marx's ideas about how to study capitalism give us the best way to address these questions. Variously depicted as a misguided idealist, a romantic revolutionary, a democratic dictator, a consummate scholar, a second-rate mind, the father of communism, a neglectful father, a delightful and loving father, a dangerous visionary, a heroic visionary, a monster, a religious figure, and a threat to religious tolerance, his supporters, always a minority, confront an unhappy family of interpretations. Predictably perhaps, skeptics remain unmoved. Three criticisms especially have been leveled. First, many observers have viewed "really existing Communism" as representative of Marx's political program. For critics, Marx's ideas are irretrievably bound with the baggage of these authoritarian regimes. Second, many social researchers remain unconvinced about Marx's methodological principles. For them, little of value exists in his writings that can assist scientific inquiry. Finally, many sociologists view Marx's work as a narrow economic analysis, one unable to inform wider sociological questions. Refuting these criticisms would go far in defending what remains valuable in his work.

If we can harness his methodological ideas, and if his analysis of capitalism remains incisive, then Marx's work is a powerful tool for sociological inquiry and a source of insight on the predicaments we confront in modern society. To do this work, we must take seriously Marx's commitment to both dialectical and scientific reason. I hope to show how Marx was cognizant of the standards of science in general and incorporated its methods within his work. In this effort, I start with the supposition that all sciences, whether social or natural, adopt assumptions, contain conceptual language, and employ analytical procedures. With this as my beginning point, I investigate how Marx incorporated these moments of science into his dialectic. This approach contrasts with two traditions in Marxist scholarship. First, Engels's approach was to apply dialectics to natural science, which, I believe, reverses relationship between dialectics and science found in Marx's work. Second, although other scholars often rightfully mobilize "the dialectical method," "historical materialism," "political economy," and "communism" as Marx's core concerns, these moments of inquiry are not coveralls nor are any of them alone a "science." Each is one moment in a whole, each has specific relationships with the others, and all of them incorporate scientific principles. If we are to read Marx in a more satisfactory way, how dialectical and scientific reason informed these inquiries and how these inquiries inform and relate to one another must be understood. In approaching Marx's work in this way, we have, I believe, a standpoint that more closely approximates his stated intentions.

Overview of the Book

The analysis in the following pages is not dissimilar to what Foucault (1972) described as an "archaeology," though it does not unwaveringly follow his conceptual framework. Rather, it is animated by its general spirit and tack. Foucault (1972: 5, 23) notes that "historical descriptions are necessarily ordered by the present state of knowledge," though any one author's work exists as a "node within a network." Foucault (1972: 26) makes use of traditional disciplinary boundaries to discover "what unities they form; by what right they can claim a field that specifies them in space and a continuity that individualizes them in time." Taking the method of analysis that Marx arrives at in his mature work as my starting point, my conclusions are informed by the historical network of discourse from which his ideas emerged and how they were interpreted and used after his death. My aim is to uncover the unity of inquiry in what Marx termed "scientific dialectics." This is only possible by using current views about Marx and comparing them to the evolution of his thought, and comparing these to his most mature form of methodological development.

This study is divided into three parts. Part I clears a path of interpretative obstacles. Chapter One, "Problems Reading Marx," elaborates Marx's moments of inquiry, acknowledges textual and interpretive problems in reading his work, and examines several traditions in Marxist scholarship in light of this framework. Chapter Two, "Marx and Classical Sociology," compares his approach with classical sociology on several important ontological and epistemological issues. Showing how many of Marx's principles are squarely within the discipline's central assumptions forces us to re-think their relationship to each other. These chapters place Marx's ideas in a context wherein his methodology can be better reconstructed.

Part II elaborates Marx's distinctive brand of inquiry. Chapter Three, "Marx's Onto-Epistemological Assumptions," outlines how Marx moves from an assumption about labor as central to social relations to a method of studying the relationships between history and social structures. Chapter Four, "Marx's Analytical Procedures," looks at how Marx's central analytical strategies were a product of his onto-epistemological assumptions and a synthesis of dialectical reason and scientific method. Chapter Five, "Marx's Conceptual Doublets," examines how Marx's terms are rooted in these onto-logical-epistemological assumptions and analytical procedures. Chapter Six, "Marx's Models," puts this previous framework together by explaining the analytical, interpretive, and explanatory models Marx used.

Part III rounds out my defense of the Marxist research agenda. Chapter Seven, "From Political Economy to the Communist Project," examines how dialectical, historical materialist, and political-economic inquiries informed Marx's political program and how it was transformed – and distorted – in

Soviet society. Showing how this happened provides a defense against critics who hail the "death of Communism" as evidence of his thought's obsolescence. Chapter Eight, "Recovering Marx," reflects on the implications of the previous chapters. Ultimately, I hope to present a more satisfying picture of how Marx went about his work and what we might take from it as the modernity with which he struggled continues on its path through our lives and into an unfolding future.

Revolutionary activity, political-intrigue, social upheaval, health problems, and familial obligations prevented Marx from putting his methodological outlook into publishable form. One senses, when reading his letters, the singular importance he afforded scientific dialectics as *the* appropriate mode of study for a capitalism he feared would transform the world in destructive ways if left unimpeded. If Marx did his job as a scientist, then he must have had an identifiable logic of inquiry as well as a useful set of conceptual terms, analytical procedures, and explanatory models. If he developed this while uniting dialectical reason with scientific method, then we must read him in these terms if we are to understand, evaluate, and use his work. We are still struggling with the question: What, if anything, can Marx's way of thinking about and studying the world offer us? I attempt to help answer this question in the pages that follow.

Acknowledgements

This book would not have been possible were it not for the efforts of friends, family, and colleagues. I would like to thank David Fasenfest for suggesting this project and for his advice in seeing it to its end. I owe much gratitude to Bertell Ollman for reading previous versions of several chapters and demanding greater precision in their presentation. Maya Becker offered me patient encouragement as I struggled to ever-improve this presentation. Her efforts were invaluable. Mary Paolucci, my mother, and David Estrin each made important contributions to the copy editing process. Two of my colleagues at Eastern Kentucky University – Elizabeth Throop and Alan Banks – each provided valuable assistance in the project's final stages. Another colleague – Dionne Smith – I owe many thanks for providing an ear as I worked through many ideas during the investigative process. A much deserved and hearty "Thank you!" to all.

Part I: Reconsidering Marx

Rare is it that a reader comes to Marx's texts without any foreknowledge about the man and his ideas. The foreknowledge with which one is armed can play a decisive role in what one takes away from their encounter. However, the quality of this foreknowledge may or may not serve the reader well. There are myths and misconceptions within popular knowledge and the literature, as well as problems internal to Marx's texts. Knowing what these problems are puts us in a better position to grasp what Marx is trying to tell us and to understand how his approach compares to the outlook of sociological inquiry in general.

Chapter One
Problems Reading Marx

Discussions of Marx vary widely in their quality and utility.[1] Some commentaries amount to little more than eulogies or attacks. Others disagree on his use of dialectics and science. When Marx's commitment to science is recognized, there is little agreement on what it means or how to see it in his work. Although Marx's texts and letters refer to a spectrum of experts and schools of thought, he never published a fully developed essay on method, leaving statements on epistemological principles scattered across his writings. For skeptical critics, research on what Marx called "scientific dialectics" has been tantamount to a quest for a nonexistent Holy Grail.

Even after twenty years of work on political economy, Marx (1992a: 18) confessed, "Every beginning is difficult." The introductions to *A Contribution to the Critique of Political Economy*, *Grundrisse*, and *Capital* reveal his thoughts on historical processes, problems in constructing conceptual frameworks, and his position on various intellectual traditions. They also lack comprehensiveness and easy communicability, revealing that his method developed as he used it. However, though these essays are only a rudimentary elaboration of his analytical tool kit, it does not follow that

---

[1] An earlier version of several arguments in this chapter can be found elsewhere (see Paolucci 2000).

his method was unsophisticated. Marx's presentational form often packed complex arguments on the philosophy of science into tightly circumscribed observations, making it easy to miss how he aligns himself with various schools of thought. As testaments to his developing research practice, they bear witness to the idea that a "science must be developed up to a certain point before it is capable of dialectical presentation" (Meikle 1979: 15).

There is much to be learned about Marx's method from his private letters with acquaintances, friends, and colleagues. Many of these were unavailable until years after his death. Their quality is as such that any picture of Marx's method must take them into account. Much of what he is believed to have not understood about science is in fact there. Much that has been attributed to him is absent.

Scholarship on Marx also has much to teach us. On the positive side, many Marxist scholars have not been dissuaded by the authoritarian and totalitarian societies that made use of Marx's name or by the pressures from the keepers of proper thought in their own societies. The unfinished archive that is Marx's work requires patience and careful study and we know more about Marx today because of the work of Marxist scholars. On the negative side, numerous assertions about and/or critiques of Marx's work that have subsequently proven untenable are still repeated in the literature. The result has been a carving out of parts from the whole done with varying degrees of congruence with Marx's overall thinking and/or the interpretive traditions he inspired. Economic determinism, the value of dialectical reason, whether he holds a teleological view of history, his relationship to metaphysical conceptions of social reality, the belief that he accepted positivism as a legitimate mode of inquiry, and the assumption that violent revolution is the only path to a future after capitalism have all been accepted – and rejected – as representing his ideas. It has been difficult to present a united front to skeptics in the midst of such interdisciplinary inconsistency.

The aforementioned issues concern debates deep within the halls of Marxian scholarship. What about the basics? Will a novice at least find agreement on the views that have come to be known as "Marxism"? Unfortunately, no. Carver (1982: 3), for instance, notes that the "traditional terms of interpretation, found in commentators from Engels to G.A. Cohen, are not Marx's terms and are ill-suited to clarifying what he said." If so, the concepts with which Marxists paint their shingle and hang for passers-by are as much a point of contention as the backroom debates.

Evidence supports Carver's claim. *Communism* as a political philosophy and program was developed before Marx was an adult, primarily from the ideas of Henri de Saint-Simon, Robert Owen, and Charles Fourier, each of whose views he disdained. He is not communism's father and communism was not a utopian theory he conjured up outside of real social relations. Marx saw communism as a potential future made possible by capitalism's material foundations. The terms *Marxist* and *Marxism* have been applied to Marx's work, his egalitarian goals, utopian schemes he rejected, the Social-Democrats he criticized, peasant uprisings in agrarian societies, authoritarian Leninist-Stalinist regimes, and academic discourses. Not only are these often incommensurable, but Marx neither coined nor endorsed the terms and certainly would have rejected calling anything a "Marxist" society. He also never used the term *dialectical materialism*: It was coined by a supporter and developed into a world-philosophy by Lenin and the Soviets, at which point it ceased to resemble Marx's views (more in Chapter Seven). Marx *was* a dialectician and *was* a materialist but to conjoin the two into "dialectical materialism" is over-simplification and caricature. Finally, *false consciousness* is not found in Marx's work, though this concept, like dialectical materialism, is still regularly attributed to him (e.g., see Allan 2005: 88–91).[2] With the validity/utility of Marx's ideas encumbered with such interpretive spin-offs, many critiques have often been based on theoretical constructs that are either not in his work and/or do not bear on its meaning. It is no wonder that Marx's supporters often claim he is misunderstood. For critics, if Marxist scholars cannot agree on the basics, perhaps Marx is not to be taken seriously?

Recognizing such problems helps us handle them. Interior to these problems are the conflation of Marx's moments of inquiry, problems in his presentation of his work and Engels's editorship of it, imprecise conceptualization of his use of science, and finally, grasping Marx's concerns about speculative philosophy, mystification and metaphysics, and obscurantism.

---

[2] The idea of a "false consciousness" was adopted from a letter by Engels (2004: 164): "Ideology is a process which is, it is true, carried out by what we call a thinker, but with a consciousness that is spurious. The actual motives by which he is impelled remain hidden from him, for otherwise it would not be an ideological process." Engels, however, does not use the term as is often attributed – poorly – to Marx, i.e., as the ideology of the ruling class. Ideology, for both Marx and Engels, was a mix of part-true, part-false, and often distorted ideas that clouded our thinking, but not something reducible simply to the machinations of a ruling class, who remained but one element, though a strong one, in its production.

Marx's Moments of Inquiry and their Interrelationships

Kemple (1995: 11) notes how often "readers of Marx's writing...remark upon on their fragmentary, scattered, and condensed aspects, their overall roughness and incompleteness, and thus the difficulty of making sense of them." Carver (1983: xi) reports a common reaction: "if the commentator does not understand a work by Marx or a passage in one of his works, *Marx* must be confused." Although it may be true that Marx himself was rarely confused, he certainly can be confusing. One step in overcoming this confusion is to divide his ideas conceptually (as opposed to biographically, temporally, or textually) into their major moments of inquiry, i.e., historical materialism, political economy, communism, and the dialectical method. *Historical materialism* refers to his study of various social realities across time and space, especially class systems. *Political economy* is his study of a specific class system, namely capitalism. *Communism* was a political program grounded in an extrapolation of capitalism's inherent possibilities. The *dialectical method* animates these inquires. Understanding both how these moments are interrelated and how to specify them from one another is vital. Neither arbitrarily connected nor a seamless whole, they are interconnected in a specific manner.

The Dialectical Method (DM). Marx's research method lies, in part, within dialectics. However, absent a systematic essay on the topic, digging through his texts for instruction is an arduous task. His view on others' use of dialectics is one source of insight. Marx (1976a) takes Hegel (and his followers) to task for their speculative, idealist and metaphysical philosophy, though he adopts from Hegel the study of change and a focus on contradiction (also see Marx and Engels 1956, 1976). Marx (1988a: 149) also praises Hegel for conceiving of "the self-genesis of man as a process." Though he applies Hegel's dialectical principles to his historical-materialist research, he only sporadically references them in an explicit manner. In the Second German Edition of *Capital*, Marx (1992d*) does reproduce a reviewer's summary of his method, though it is hardly a sufficient explanation.[3] This incomplete and fragmentary

---

[3] Citations with an asterisk (i.e., Marx 1992d*) are from a reviewer's comments that Marx (1992c: 28) evaluated positively, declaring, "what else is he picturing but the dialectical method?" Marx added this to the Afterword of the second edition of *Capital*, suggesting he thought it mirrored his own view. They are reproduced here on this assumption.

state has discouraged many detailed investigations of his dialectical method. By seeing how Marx mobilized dialectics in conjunction with scientific method across his other moments of inquiry we can come to grips with his overall approach.

Historical Materialism (HM). Although Marx never used the term "historical materialism," as a concept it approximates several of his views better than does "dialectical materialism" (for reasons we will later see). Historical materialism took shape out of several periods of investigation, including Marx's dissertation, his and Engels's critique of Hegel's idealism and adoption of his dialectical logic, and their agreement with Feuerbach's materialism but with a rejection of his ahistorical approach. Engels (1980: 469) first uses the phrase "the materialist conception of history" in his review of Marx's 1859 work, *A Contribution to the Critique of Political Economy*. Historical materialism later became an indexical expression for Marx's general theories, though this is slightly misleading. Historical materialism is that moment where Marx focuses on the history of class societies, widely construed (e.g., ancient/tribal, slave, Asiatic, feudal, and capitalist modes of production). This "open-ended approach to history" (McLellan 1975: 36) was guided by a "working hypothesis about social change within and between types of social systems centered on two main principles: 1) The ideological superstructures of a society are primarily conditioned by its material bases; [and] 2) The contradictions between the relations and the forces of production cause societies to both change within their own structure, and over time, to change into qualitatively new forms" (Paolucci 2000: 304). Embedded in these principles are the primacy of material relationships, the role of class relations in society and thus class analysis within research, and the emergence and practices of the state and its place in models of class history. This was an intellectual program that emerged over time (see Marx 1911, 1968a, 1973, 1978b, 1978d, 1992j: 85–86, Marx and Engels 1976, 1978a).

Political Economy (PE). Marx's political economy is an analysis of capitalism as a specific mode of production. Because capitalism is a changing and dynamic system, Marx mobilized dialectical and historical materialist principles for this political-economic research. His goal was to uncover the "economic law of motion of modern society" (Marx 1992a: 20). My goal here is to elucidate Marx's methodology and thus any detailed examination of political economy will only be in the context of showing how his overall epistemology informs his study of capitalism (see Marx 1909, 1934, 1963, 1968b, 1971, 1973, 1983a, 1992j).

The Communist Project (CP). Marx believed in advancing the communist program that had developed in Europe (especially France) at the onset of capitalist development. Not a utopian project apart from extant society, in Marx's view communism would have to be forged within the remnants of capitalism as working classes used what it provided them. After workers' revolutionary success, classes and states would increasingly dissolve into history's dustbin and humans would then be free to cultivate potentialities that capitalism makes possible but unrealizable (see Marx 1988a, 1978g, Marx and Engels 1978a, Engels 1971).

These moments of inquiry (DM, HM, PE, CP) are neither mutually exclusive nor without important interrelationships. Interaction occurs between them with varying degrees of necessity and contingency, where narrower levels are not predetermined results from the principles of more general levels. Narrower orders can thus be invalidated in specific instances without raising fundamental problems for broader ones. A series of interpretive rules are suggested: First, Marx's communist project should not be conflated with the dialectical method, historical materialism, or political economy; second, the dialectical method should not be conflated with historical materialism or political economy; finally, historical materialism should not be conflated with political economy. From these interpretive rules, several conclusions follow:

- The dialectical method – Marx's most broad level of inquiry – informs historical materialist principles.
- Historical materialist principles inform research in political economy.
- Conclusions in political economy, however, are not *necessary deductions* determined by either the dialectical method or historical materialism;.
- Therefore, although political economy can be read without foreknowledge of historical materialism and the dialectical method, it helps to know both.[4]

---

[4] "In the *preface* I then tell the 'non-dialectical' reader that he should skip pages x–y and read the appendix instead. This is not merely a question of the philistine, but of youth eager for knowledge, etc." (Marx 1936i: 222; emphasis in the original).

- Conversely, difficulties for political economy do not negate historical materialism.
- The possibilities of communism stem from the conditions created by capitalist development and are examined through the relationships between the dialectical method, historical materialism, and political economy.
- However, communism is not a *necessary* result of any of these three preceding levels and is therefore not their measuring stick.
- Thus, problems with the communist program, political economy, and/or historical materialism do not necessarily raise problems for dialectical method.
- As a result, one can use the dialectical method without being a historical materialist or a political economist.[5]

One is not required to examine all of these moments in tandem. Indeed, the dialectical method (Bhaskar 1993, Ollman 2003) or historical materialist principles (Rader 1979, McMurtry 1979) can be the central focus of analysis. Moreover, one can (and should) use the former to examine how it informed the latter (Sayer 1987). However, should this be done without grasping one moment of inquiry's inner-connection with the others, the result is a specification that is less precise than is possible. For example, the tradition of critical realism, by defining Marx's objects of analysis as "the social world" and "society (and nature)," collapses various human historical forms into a general abstract whole (Bhaskar 1989: 2). But Marx's approach urges us to differentiate social wholes and make clear the various levels of social complexity they contain (Magill 1994, Groff 2000, Ollman 2003, Collier 1989: 43–72). A more comprehensive use of Marx's work would consider how dialectics inform historical materialism and political economy and examine how all of these inform his communist project. Reconstructions of Marx's thought that do not do this – or do it only partially – meet with mixed success.

General discussions of Marx often lack specificity of his moments of inquiry and their interrelations (Bottomore 1975). Scholars might extract a particular concern – such as Marx's ontology (Lukács 1978), the Asiatic mode of production (Sawer 1977), or the class-state relationship (Miliband 1983) – but do not trace the role these play across his moments of inquiry and/or fail to address

---

[5] This can be seen in Foucault's work (Paolucci 2003a).

how scientific methods operate in his analyses. Other scholars examine Marx's thought in a temporal order (Kolakowski 1978), which, as we will see, makes mapping the interrelations between his moments of inquiry more difficult. In a similar tack, Marx's framework is presented in a linear extrapolation centered on pre-capitalist, capitalist, and communist modes of production (Gandy 1979), which fractures the relationship between historical materialism, political economy, and the communist project. Sometimes Marx's inquiries arrive in a patchwork, such as moving straight from dialectical method to political economy without mediating this relationship through historical materialism (Eldred and Roth 1978), focusing on political economy denuded of its historical materialist and dialectical relationships (Kautsky 1925, Robinson 1966, Wolff 1984), or moving from the dialectical method to historical materialism to the communist project without having political economy mediate the latter two, thus producing a metaphysical view of history (Leff 1969). Similarly, some scholars adopt "dialectical materialism" as encompassing all of Marx's ideas and approach them from the viewpoint of philosophy (Lefebvre 1968) and even push his ideas as a world-philosophy (Adoratsky 1934). Others reduce historical materialism to Marx's "system of sociology" (Bukharin 1925), which makes the complexity of his thought (and politics) harder to piece together. Combining these approaches, Mandel (1968) brings historical development into his analysis but imports a sort of universal dialectic into Marx's ideas and later allows historical materialism to swallow up political economy (1977).

This history scholarship testifies to the possibilities that result from Marx's work. We need to reconstruct Marx's approach in a way that – like the above approaches – uses a non-arbitrary system but also relates the articulation and order of one moment of his discourse with another. What Marx says about capitalism (PE) is not necessarily appropriate for a study of society in general (HM). What can be said about society in general may have little bearing on events within capitalism. When the relationships between dialectics and historical materialism, or those between class societies and capitalism and toward which Marx directs his historical materialist and political-economic analysis, are not differentiated, the moments of inquiry are collapsed together (see Reuten 2000, Sayer 1984: 85, 129, Bologh 1979).

Sometimes these moments of inquiry are not so much collapsed as reconstituted in a form that makes their interrelations harder to elucidate. For example, although Garaudy (1967) sees the dialectical method, historical materialism, and political economy as distinct orders of discourse, he only partially traces out their interrelations. Derek Sayer (1979) begins with political economy and works back to the dialectical method but skips historical materialism, only to reconnect it at the end. Finally, both Ollman (1993, 2003) and Sherman (1995) detail Marx's methodological inquiries but his dialectic's interconnection with political economy requires additional clarification. These approaches advance our understanding of Marx's thought but they need to be more consistent in connecting his moments of inquiry with another and more specific on how he mobilizes the principles of science to animate his project as a whole.

This discursive history has elicited efforts to reconstruct Marx's theories. Commentators such as Williams (1973, 1978), Habermas (1979a–b), and Giddens (1995) see in historical materialism a too-rigid construction and/or a universal-linear theory of history. Others accept the scientific features of Marx's work while denying the value provided to them by dialectics (Little 1986), or strip Marxian analysis of its dialectical bases and appeal to the principles of analytical philosophy (Levine, Sober, and Wright 1987, Elster 1982, 1985, 1986, Roemer 1986a–b). However, these evaluations either rest on problematic assumptions about what sort of theory is operative in Marx's method or fail to see how stripping his approach of its dialectical moorings eliminates exactly what makes his approach both what it is and why it is necessary for a *social* science. Moreover, it is the relationship between his dialectical, historical, and materialist thinking and his concern with political economy where we find the connections between Marx's scientific and political views.

The Political and the Analytical Within Marx's Moments of Inquiry

There have been numerous views on how to approach the relationship between the scientific and the political in Marx's thought, often revealing different sides of a similar coin. One approach sets up the revolutionary history of the communist project as the ultimate testing ground for Marx's science and remains committed to both (Burawoy 1990). However, to view proletarian revolution as a prediction – and as *the* test for Marx's science – misconceptualizes his views on science and social change. Another approach is to strip "Marxism" of its scientific aspirations and adopt its political spirit, focusing on its ability to inform anti-systemic discourse but without bringing dialectics, historical materialism, and political economy together (Laclau and

Mouffe 1982, 1985). But aligning oneself with Marx's political spirit may do little to advance his scientific understanding of society, making the politics Marx urges upon us simply a bias. What these views share is that socio-political allegiances are the basis upon which we should evaluate the combined meaning of Marx's moments of inquiry.[6]

When observers do not differentiate Marx's method of inquiry (DM, HM, PE) from his politics (CP), his social-scientific ideas are liable to be interpreted in light of societies that claim a Marxist mantle (Coates 1945, Marcuse 1954, Mandel 1968, 1977, Turner 1993b). Sidney Hook "wrote, for instance, in 1983, fifty years after the publication of *Towards the Understanding of Karl Marx*: 'history which according to Marx is the acid test of historical predictions showed that I was wrong [in this volume] – together with Marx – about capitalism, and even more so about the Soviet Union'" (E. Hook 2002: 10–11). Marx (1978f: 544), however, thought that the "economic requisites do not exist" for a socialist revolution in the Russia of his time, making it highly unlikely for "a European social revolution, resting on the economic foundation of capitalist production, to take place on the level of the Russian or Slavic agricultural and pastoral peoples and not to overstep that level." Thus, communism as it evolved in Russian practice did not contain the historical-materialist presuppositions relevant to Marx's theories.

It has been argued that the Soviet Union's decline rather than its rise makes Marx's ideas more relevant not less:

> It has only been, that is to say, by continuous state interventions to reduce the 'contradictions' Marx emphasized between a society's productive powers and its structure of capitalist ownership, overlaid by pervasive new systems of ideological conditioning and control, that capitalist social organization has developed a way of overcoming its various crises of depression and social opposition. We might say overall that Marx's general historical materialist principles have thus been confirmed, but his radical inference of

---

[6] "Although Marx sympathized profoundly with the sufferings of the working classes, it was not sentimental considerations but the study of history and political economy that led him to communist views. He maintained that any unbiased man, free from the influence of private interests and not blinded by class prejudices, must necessarily come to the same conclusions" (Lafargue 1890; cited in McLellan 1981: 67).

a coming new order, the absolute opposite of capitalism, has been disconfirmed (McMurtry 1992: 304).

This measured and sympathetic account, however, is only partially commensurable with Marx's thinking. Although Marx's historical materialist principles as applied to political economy can tell us much about Communism's decline (e.g., see Mayer 2002), the above view equates political economy and historical materialism with one another, depicts Marx's communist project as the "absolute opposite of capitalism," and presents the Soviet Union as an evaluative test case, thus reproducing many traditional readings. Such an approach assists in defending Marx's work against claims that the fall of official Communism makes it obsolete, but it is insufficient for advancing Marx's methodological principles.

Collapsing the dialectical method, historical materialism, political economy, and the communist project into an undifferentiated whole, "Marxism" is often evaluated against the lack of worker revolution within capitalist society. Although this approach implicitly connects historical materialist principles with those of political economy and both combined with the prospects of *one possible type* of structural change, it often insufficiently acknowledges the element of contingency in his thinking. Critics who adopt this approach often see Marx as "predicting" proletarian revolution and its failure disconfirms "Marxism" in general (Croce 1982, Popper 1972: 312–335, Gottheil 1966, Turner 1993a–b). Some of Marx's supporters also accept that revolution functions as the criterion for evaluating the value of "Marxism" (Wetter 1963), though they disagree on the timing (Burawoy 1990).

If one predicted rain and it did not rain relatively soon afterward, one could simply claim that it will rain eventually and still be technically correct, though not scientifically sound. Conversely, if a meteorologist predicted rain and it did not, this would not mean that she or he lacked scientific credentials. Forecasters appeal to degrees of probability. In incorporating dialectical thinking, historical materialist principles hold that contradictions often exist between forces and relations of production and this produces change, no matter the social system. Marx's political economy hits its apex at the theory of crises, which intersects with historical materialism's principle of class struggle and dialectic's principle of contradiction. Combined, dialectics and historical materialism contribute the idea of social transformation to political economy and extend it to communism's potentiality. However, a dynamic tension between necessity and contingency exists within the crises endemic to capitalism and the possibility of their transcendence (DeMartino 1993). Crises may not reach a critical mass. Forces may impede worker action and/or economic collapse. Revolution may occur, but its proletarian character is contingent on political praxis rather than predetermined in advance.

To say the value of "Marxism" (in the reading here, the totality of inter-relations between DM, HM, PE, and CP) lays in the assertion that proletarian revolution is always somewhere off in the near/middle/distant future is similar to predicting it will eventually rain. Such approaches elicit critics' skepticism: "If Marxism is to be maintained as a revolutionary ideology, Marxist theory must be corrected in a way that does nothing to undermine the authority of Marx himself" (Alexander 1982, Volume 2: 65). Historical materialist principles, as well as Marx's communist program, do not dictate outcomes to political economy nor to real history; they only help guide study and action. When the most contingent and speculative levels of Marx's political-project – i.e., proletarian revolution and long-term communist development – are used to indicate the validity of political economy, historical materialist principles, and/or dialectical method, the relevant relationships in his moments of inquiry are conflated and/or even reversed.

It would be both an overstatement and incorrect to view the aforementioned traditions in Marxist discourse as uninformative or simply wrong. There are multiple ways to approach Marx's ideas and texts and the methods of analysis and reconstruction one chooses can render the overall interpretation anywhere from partial to incomplete to distorted. When scholars recognize Marx's moments of inquiry as distinct orders of discourse but do not conceptualize their interrelations, the outcome is only a partial reconstruction. When these moments of inquiry are related to one another but without tracing the methods of science Marx used to inform these researches, the outcome is incomplete. When these moments of inquiry and Marx's scientific commitments are collapsed into an undifferentiated whole – often "Marxism" and/or "dialectical materialism" – the result is a distortion, often in ways that qualitatively transform his ideas into something they were not intended to be, i.e., a finished world-philosophy.

What is required then is, first, understanding the relationships between Marx's four moments of inquiry and, second, mapping out how he brings the principles of science in general to bear upon them. This requires that we trace Marx's assumptions (chapters Two and Three), analytical procedures (Chapter Four), conceptual development (Chapter Five), and descriptive and explanatory models across historical materialism and political economy (Chapter Six). We must, at the same time, understand how both dialectical thinking and scientific methods animate these inquiries and how the combination of all of them inform his communist project (Chapter Seven). This approach to Marx's work makes a greater level of specificity and precision possible. However, this work will not be sufficient if we do not prepare ourselves for additional problems internal to Marx's work and texts, including the manner in which he presented them to the public (or failed to), the influence Engels had on interpretations of his work, acknowledgement of the intellectual traditions that informed Marx's thinking, and, finally, his critique of speculative, mystical, metaphysical, and obscurant modes of thought. The remainder of this chapter examines these issues.

The Presentation and Interpretations of Marx's Work

The more varied an author's work, the more difficult it is to put a single cast on it, although the author's death can transform her or his writings into a seeming whole (Foucault 1984). Marx often worked on some topics until he had learned what he needed from them only to turn to new research without putting his findings in presentable order. Arnold Ruge complained about his colleague: "Marx reads an enormous amount; he works with uncommon intensity and has a critical talent that sometimes degenerates into a wanton dialectic, but he finishes nothing, is always breaking off and plunging afresh into an endless sea of books" (cited in McLellan 1975: 5). His in-progress manuscripts, projects unpublished and those never completed – e.g., *The 1844 Manuscripts*, the Theses on Feuerbach, *Grundrisse* – provide crucial insight into Marx's thought but remained long inaccessible.[7] Others remained in obscurity. After *The German Ideology* was written, for instance, "Marx never subsequently stated his materialist conception of history at such length and

---

[7] Engels (1941: 8) described Marx's Theses on Feuerbach as "notes hurriedly scribbled down for later elaboration, absolutely not intended for publication, but they are invaluable as the document in which is deposited the brilliant germ of the new world outlook."

in detail...[with such] cogency and clarity.... Yet it remained unknown for almost a century" (McLellan 1973: 151). There is no doubt that the history of debate surrounding Marx's work and ideas would have been different if such unpublished works been available. This brings up two questions: How should we relate Marx's unpublished work to his ideas? What does the work that he did not finalize for publication tell us about the work that he did? Though difficult questions, there are avenues to pursue.

One common interpretation of Marx is that he formulated his theoretical models first and only then proceeded to research (see Popper 1966, Elster 1985). Gerdes (1985: 2), for example, asserts that "much of the structure of the first volume of *Capital* was deductive." Both Marx's and Engels's testimonials and the documentary record belie such assumptions. The year before the Preface to *A Contribution to the Critique of Political Economy* was published, Marx (1983e: 270) confessed: "But the thing is proceeding very slowly, because no sooner does one set about finally disposing of subjects to which one has devoted years of study than they start revealing new aspects and demand to be thought out further." In his review of Marx's book, Engels (1980: 469) claimed that the materialist conception of history was "a scientific task requiring years of quiet research." Aware that *Capital*, also a product of long years of research, might be read through conventional deductive philosophies of science, Marx (1992c: 28) warned that if his method was successful in its research and presentation, "then it may appear as if we had before us a mere a priori construction." None of these comments argues in favor of historical materialism or political economy as the products of a preconceived theory.

Marx struggled with his method of presentation. It is not clear if what is available today is in his preferred form, equally true for work published and unpublished in his lifetime. Of his unpublished work, to what extent can we be confident what we read is what Marx would have wanted had he known it would have an audience? "It goes without saying that a writer who works continuously cannot, at the end of 6 months, publish *word for word* what he wrote 6 months earlier" (Marx 1982a: 51; emphasis in the original). Of the work he did publish, how much of it is in a form Marx found satisfying? He confessed to a having a "quirk...of finding fault with anything I have written and not looked at for a month, so that I have to revise it completely" (Marx 1985f: 356). Marx's qualifications warn us not to assume everything published under his name today involves a position he would have endorsed without qualification. Some lines of thought were retained, others were dropped, and still others were presented only once.

The tension between the need to publish with timeliness versus continuing his studies shaped significant aspects of Marx's textual legacy. His works differ in quality, emphasis, and level of completeness. Their availability when many critical appraisals of his ideas were made has been variable. For example, List 1.1 presents works published and unpublished during Marx's lifetime. McLellan's bibliography, although revealing, does not include Marx's writings on subjects including suicide, gender, anthropological issues, ethnography, and mathematics (see Marx 1972, 1983i, Harstick 1977, Plaut and Anderson 1999, Anderson 1998, Rubel 1980: 100). This archive undoubtedly contains material of great value, but without the authoritative stamp represented by Marx's decision to publish these works, they must be interpreted carefully. One thing is clear: Many appraisals have filtered into the literature and taken on a life of their own when only a partial view of his *oeuvre* was available. This has led to numerous debates that would have been less contentious and less protracted had his full archive been available earlier.

List 1.1
Marx's Work Published and Unpublished in His Lifetime

Year	Published in Marx's Lifetime / **Unpublished**
1873	**Letter to his father**
1838–1840	**Doctoral Thesis**
1842	*Articles for Rheinische Zeitung*
1843	**Critique of Hegel's Philosophy of Law**
	On the Jewish Question
1844	**Critique of Hegel's Philosophy of Law: Introduction**
1844	**Economic and Philosophical Manuscripts**
	Critical Notes on "The King of Prussia and Social Reform"
	The Holy Family
1845	**Theses on Feuerbach**
1846	**The German Ideology**
	Circular Against Kreige
	Letter to Annenkov
1847	*The Poverty of Philosophy*
1848	*Speech on Free Trade*
	The Communist Manifesto
	Demands of the Communist Party in Germany
	Articles for Neue Rheinische Zeitung
1849	*Wage, Labour and Capital*
	Articles for Neue Rheinische Zeitung
1850	*Addresses of the Central Committee to the Communist League*
	Articles for Neue Rheinische Zeitung

Table (cont.)

	The Class Struggles in France
1852	*Articles for New York Herald Tribune*
	The Eighteenth Brumaire of Louise Bonaparte
	The Great Men of Exile
1853	*The Cologne Communist Trial*
	The Knight of the Noble Conscience
1854	*Palmerston and Russia*
1855	*Articles for Neue Oder Zeitung*
1856	*Revelations About the Diplomatic History of the Eighteenth Century*
	Articles for People's Paper and Free Press
1857	**General Introduction, Outlines of a Critique of Political Economy (Grundrisse)**
1858	**General Introduction, Outlines of a Critique of Political Economy (Grundrisse)**
	Articles for New American Encyclopedia
1859	*Preface to a Critique of Political Economy*
	Critique of Political Economy
1860	*Herr Vogt*
1861	*Articles for Die Priesse*
1862/3	**Theories of Surplus-Value**
	Articles for Die Priesse
	Manuscripts on Polish Question
1863	**Capital, Volume II** (until 1877)
1864	*Inaugural Address of First International*
	Capital, Volume III
1865	**Value, Price and Profit**
	On Proudhon
1866	*Programme for First Congress of International*
1867	*Capital, Volume I*
1870	*Two Addresses on Franco-Prussian War*
1871	*The Civil War in France*
1872	*Alleged Splits in International*
	Preface to Second Edition of Communist Manifesto
	Amsterdam Speech
1873	*Preface to Second German Edition of Capital, Volume I*
1874	**Remarks on Bakunin's Statism and Anarchy**
1875	*Critique of the Gotha Programme*
	French Edition of Capital, Volume I
1877	*Contribution to Anti-Duhring*
	Letter to Mikhailovsky
1879	**Circular Letter**
1880	*Questionnaire*
	Introduction to French Workers' Program
	Remarks on Wagner
1881	**Letter to Vera Sassoulitch**
	Notes on Morgan's Primitive Society
1882	**Preface to Second Russian Edition of The Communist Manifesto**

Source: Adapted from McLellan (1973: 459–465).

Forward and Backward Readings of Marx

Too often it is assumed that Marx forged his ideas as a young man and spent forty years elaborating them. Many of the young Marx's essays display liberal romanticism, idealism, and mysticism. His "Reflections of a Young Man on the Choice of a Profession" invokes questions about the Deity's plans and devoting a life to the service of humanity as a moral stand (Marx 1975a). When these views (and Hegel's religious idealism) are read into Marx's later work, his continuing project is interpreted as a footnote to a youthful religious mission, transforming Marx's later work into something he argued against (see Tucker 1957–58, 1964, Halle 1965, Wessell 1984). How one reads Marx's works in relation to the order in which they were offered is of consequence in terms of what sort of Marx one encounters.

If one interprets the ideas in Marx's work in their chronological order, there is no way of knowing which ideas influenced him throughout his career. In his review of *A Contribution to the Critique of Political Economy*, Engels (1980: 475) discusses possible methods of analysis and presentation. His comments are equally applicable to reading Marx's work itself:

> ...the critique of political economy could still be arranged in two ways – historically or logically, since in the course of history, as in its literary reflection, development proceeds by and large from the simplest to the more complex relations, the historical development of political economy constituted a natural clue, which the critique could take as a point of departure, and then the economic categories would appear on the whole in the same order as in the logical development. This form seems to have the advantage of greater lucidity, for it traces the *actual* development, but in fact it would thus become, at most, more popular. History often moves in leaps and bounds and in zigzags, and as this would have to be followed throughout, it would mean not only that a considerable amount of material of slight importance would have to be absorbed, but also that the train of thought would frequently have to be interrupted; it would, moreover, be impossible to write the history of political economy without that of bourgeois society, and the work would thus be endless because of the absence of all preliminary studies. The logical method of approach was therefore the only suitable one. This, however, is indeed nothing but the historical method, only stripped of the historical form and of interfering contingencies. The point where this

history begins must also be the starting point of the train of thought, and its further progress will be simply the reflection, in abstract and theoretically consistent form, of the course of history, a corrected reflection, but corrected in accordance with laws provided by the actual course of history, since each moment can be examined at the stage of development where it reaches full maturity, its classical form (emphasis in the original).

This is a statement about which objects to prioritize in the research process – all earlier periods of historical development or those periods where an object's most important features can be observed? If one took the first option, research could go in an undeterminable number of directions and coherence would suffer. In terms of the second option, if the examination of an earlier stage is to inform knowledge of its later growth, study must have an adequate view of the object's mature features. Otherwise, an analysis of something cannot be adequately informed by knowledge of its history. One could make the same argument about Marx's work, i.e., one can use the results of his research as a sieve for extracting the roots of his ideas from his early work and for interpreting their historical evolution. Using the results as a prioritizing map, *Capital* serves as both the beginning and the end point in a way that allows for sifting through his work when searching for the basics of his method and its development over time (also see Chapter Four, pp. 114–115, Chapter Six, p. 196).

In a letter to his father, Marx (1975b) mentions his critique of Kant (and Fichte), his rediscovery of Hegel and his critique of him through a study of Feuerbach. He also remarks on relational truth, empiricism, materialism, and realism. These views are often justifiably read forward as pertinent to understanding Marx's later work. At the same time, some things in Marx's mature work cannot be read backward (e.g., the early Marx was neither a communist nor a political economist), just as other things cannot be read forward (e.g., the late Marx was a dialectician who had lost all religious, philosophical and mystical trappings). A discriminatory eye is essential: Marx *must* be read backward before reading him forward if we are to maximize our chances of grasping his mature method and the foundations upon which it sits.

Knowing how Marx put *Capital* together aids us in interpreting it, especially because the meanings in its volumes are not confined to solely within the pages of each. First, *Capital*'s four parts are not in the sequence Marx originally planned or initially conceptualized them, nor are they numbered in the order written: "In fact, *privatum*, I began by writing *Capital* in a sequence (starting with the 3rd, historical section) quite the reverse of that in which it was presented to the public, saving only that the first volume – the last I tackled – was got ready for the press straight away, whereas the two others remained in the rough form which all research originally assumes" (Marx 1991a: 287; emphasis in the original). Marx's mature political-economic research thus worked from the whole backward but he began its presentation with elementary parts (i.e., "you see, I always move in dialectical contradictions" [Marx 1979b: 244]). The *Grundrisse, A Contribution to the Critique of Political Economy, Theories of Surplus-Value*, Volume Two, and then Volume Three of *Capital* were started first, of which Marx only published the *Critique*. Educating himself through these works, Marx then started and completed Volume One of *Capital*, publishing it in 1867. This accounts for its condensed style. Half of Volume Two was rewritten by 1870, but was unfinished as of 1879 and Volume Three remained virtually unedited after 1864–65, taking Engels eleven years after Marx's death to publish it, given his manuscripts' confusing and unfinished state (for discussion, see McLellan 1973: 422, Brewer 1984: 87).

Second, Marx allowed *Capital*'s public presentation to go forward with its method concealed and its intended meaning hidden. Struggling with bringing his method together with his presentation of political economy, Marx (1985e: 333) admitted to Engels that his "writing is progressing, but slowly. Circumstances being what they were, there was, indeed, little possibility of bringing such theoretical matters to a rapid close. However, the thing is assuming a much more popular form, and the method is much less in evidence than in Part I."[8] In addition, Marx (1987c: 366–367) confided in a colleague:

> *Volume I* comprises the '*Process of Production of Capital*.' As well as setting out the general theory, I examine in great detail the conditions of the English – agricultural and industrial – proletariat *over the last 20 years*, ditto the condition of Ireland, basing myself on *official* sources that have never previously been used. You will immediately realise that all this serves me solely as an *argumentum ad hominem*.... I hope that a year from now the whole work will have appeared. *Volume II* contains the continuation and conclusion of the theory; *Volume III, the history of political economy from the middle of the 17th century* (emphases in the original).

---

[8] Marx here refers to *Critique of Political Economy* as Part I, as it would have been under his earlier conceptualization of the entire work.

Volume One, with its method hidden, was meant as a set up for an extended critique contained the unpublished volumes. If Volume One was written with the assumption that other volumes would round out its meaning, then it is unfinished too. One could hardly ask for a worse interpretive scenario, especially on questions of method in general, its application to political economy specifically, and the findings of Marx's political economy itself.

Capital's four parts have different levels of interpretive authority in relation to Marx's four moments of inquiry. Marx edited Volume One several times with readers' criticisms in mind and republished it in different languages. Because Marx's presentation, terminological development, and empirical demonstrations became increasingly more sophisticated over the course of his work, and if we study "each factor...examined at the stage of development where it reaches full maturity," then Volume One reveals his most advanced research practice and we can be confident that its form of inquiry and analysis met his criteria for scientific dialectics. However, the further we get from Volume One, the more cautious we must be. On the issue of political economy, as the unfinished volumes did not receive critical reviews seen by Marx nor did he put them in their final order, Volume Two should carry more weight than Volume Three. By extension, both volumes Two and Three should carry more weight in political economy than *Theories of Surplus-Value*. The unpublished volumes of *Capital* are nevertheless useful for instruction in methodology and the application of historical materialist principles, though they should not be taken as studies in historical materialism as such. In terms of the communist project, Volume One is a handbook of political education for the working class – by extension, so are all of the volumes, though none provide much elaboration on the communist project itself.

In an intellectual biography, time, space, and ideas can be presented chronologically without distortion of the story. However, if the goal is uncovering what in Marx's method can speak to research today, then we must re-order our priorities. *Capital* provides the key to what is scientifically important in work such as the *Manuscripts of 1844*, more than the *Manuscripts* contain all we need to know for interpreting *Capital*. This is not to deny the value of the *Manuscripts* but rather the need to recognize their relational-interpretive utility. Looking from *Capital* backward allows us to extract which early ideas informed Marx's mature investigations. But one must be careful in any such analysis. For example, Althusser (1971) misread the differences between the early and the late Marx when he held that concerns about alienation did not inform the later more "scientific" Marx. The *Grundrisse* – released in the early 1970s – showed that Marx never lost his concern with alienation and freedom (McLellan 1975: 61, Nicolaus 1968, 1973). The meditations in the *Grundrisse* also help us see Marx's research agenda, including his concern with the capitalism as a social system, its central trends and contradictions, problems involved in developing concepts for historical and political-economic research, and capitalism's potential to make the transition into socialism (Bottomore 1973: 12–13). Thus the *Grundrisse* shows how Marx retained his concerns with political economy, alienation, and dialectics in his journey from the *Manuscripts* to *Capital*. Its late release closed a gap in the knowledge of his thought's evolution, though much criticism had already found a home within this gap.

Engels's Influence on Interpretations of Marx's Work

It was not only the lack of access to Marx's full works that contributed to misconceptions of his ideas. Engels – Marx's closest collaborator – also influenced this history. Their intellectual partnership spanned *The Holy Family*, *The German Ideology*, "The Communist Manifesto," and *Anti-Duhring* (Marx composed a chapter). After Marx's death Engels edited volumes Two and Three of *Capital* and wrote seventeen prefaces for Marx's works and five for "The Communist Manifesto" (Rubel 1980: 109, Carver 1983: 145). In a letter of August 16, 1867, Marx credited Engels for alone being the one who could help him bring *Capital* to publication. Their correspondence reveals discussions on dialectics, social and natural sciences, working class movements, military history, and world events, each providing the other an ear for critical feedback. Engels's work on dialectics, published after Marx's death, also played a significant distorting role in interpretations of Marx's theories (see Chapter Seven). Engels thus stands at the intersection of two paths, one that provides insight on how Marx pieced his method together and another that shows us problems in applying it to substantive topics.

Some commentators have questioned the extent of "Engels's competence, as the unchallenged executor of Marx's intellectual legacy" (Rubel 1981: 18). Their long period of cooperative work can mask subtle but important differences between them. Engels (1983: 304) confessed to Marx that the "study of your abstract of the first half-installment has greatly exercised me; it is a very abstract abstract indeed – inevitably so, in view of its brevity.... I often had to search hard for the dialectical transitions, particularly since all abstract reasoning is now completely foreign to me." Marx (1936g: 205) later wrote to Engels about what it would take to publish *Capital*: "Although finished, the manuscript, gigantic in its present form, could not be prepared for publication by anyone but myself, not even by you."

Perhaps Marx's cautious attitude was warranted. As editor of volumes Two and Three, Engels did not appeal to Marx's preparatory work for his dissertation, his early critique of Hegel, the 1844 *Manuscripts*, the Kreuznach critique of Hegel's philosophy, the 1857–1858 manuscripts now known as the *Grundrisse*, nor other notebooks and correspondence (Rubel 1981: 24). This perhaps explains why, until Lukács rehabilitated it, the concept of alienation was "almost entirely absent [in Marxian discourse] from Engels onward" (McLellan 1975: 85). Others note how Engels's edit of *Capital* could have had a higher degree of fidelity in its translation of terms (Kline 1988), the relevance of which increases across its volumes because, "with the exception of the first volume, Marx's *Capital* is accessible to us only through Engels' version of it" (Riazanov 1927: 217). Understanding that Engels had an imprint on Marx's work and understanding the nature of that imprint are the first steps in coming to terms with problems this poses for us in reading Marx.

Engels's work on dialectics changed Marx's ideas in important ways, setting the tone for many interpretations of dialectics after his death. Engels, like Marx, understood that historical materialism was not a coverall.[9] However, his use of the dialectical method displayed an "ability to simplify – sometimes oversimplify – deep and complex questions" (McLellan 1973: 279). True, Engels (1985: 331) criticized Lassalle's dialectic, which he found guilty of "applying it to every single point – as though it would gain weight in the process." And Marx (1985e: 333) agreed: "Ideologism permeates everything,

---

[9] "In general the word *materialistic* serves many of the younger writers in Germany as a mere phrase with which anything and everything is labelled without further study; they stick on this label and then think the question is disposed of. But our conception of history is above all a guide to study, not a lever for construction after the manner of the Hegelians. All history must be studied afresh, the conditions of existence of the different formations of society must be individually examined before an attempt is made to deduce from them the political, civil-legal, aesthetic, philosophic, religious, etc., notions corresponding to them" (Engels 1936b: 473; emphasis in the original).

and the dialectical method is *wrongly* applied. Hegel never described as dialectics the subsumption of vast numbers of *'cases' under a general principle"* (emphases in the original). However, Engels (1934: 62–63) does just this when he says that dialectics can be "reduced" to three laws: the transformation of quantity into quality, the interpenetration of opposites, and the negation of the negation. Elsewhere, he says that "Dialectics is nothing more than the science of the general laws of motion and development of Nature, human society and thought" (Engels 1939: 155). Engels's reduction of dialectical principles to natural science-like laws, oversimplified and formalized *a priori* laws of a universalistic and even metaphysical variety, is a distortion of Marx's approach. Although it is debatable whether Marx would have forced a study of nature into a dialectical framework, there is no doubt he never reduced all of history to three general laws. Still, there also is no doubt that Marx thought tools from the natural sciences were useful for studying the social world.

Marx's Intellectual and Scientific Roots

Understanding how and why Marx positioned himself vis-à-vis other intellectuals and traditions facilitates a more thorough reading of his moments of inquiry. Although Marx repeatedly aligned his method with dialectical reason and proclaimed he was Hegel's student, this relationship is sometimes downplayed or even denied as important (Carver 1983, Little 1986). But it is undeniable that Marx referred to Hegel as a "great thinker" and a "master" and found in his work a logic and a language appropriate for incorporating dynamics and change into his core concepts and research practices. And, although it is generally agreed that reading Feuerbach further encouraged Marx to found his inquiry on material rather than ideal relations, not all observers agree what this, nor his claim that Feuerbach led him to turn Hegel on his head, means. Kant's role in all of this remains perhaps least acknowledged. Originally finding Hegel stodgy, boring, and nonsensical, Marx's (1975b) letter to his father claims that his struggle with Kant turned him back toward Hegel. Knowing what Marx rejected and accepted in Kant (and why) helps us understand this move to Hegel and its implications.

Marx and Kant. Marx found some elements of Kant's work useful, others he altered to fit his own outlook and needs, and still others he rejected. In his *Critique of Pure Reason*, Kant inquires into "whether there exists a knowledge altogether independent of experience...called *a priori*." This is the question of whether universal categories of knowledge exist. Kant (1947: 423) answers in the affirmative:

> Pure knowledge *a priori* is that which no empirical element is mixed up [and where] there is no exception to this or that rule...that is, no possible exception. Necessity and strict universality, therefore, are infallible tests for distinguishing pure from empirical knowledge, and are inseparably connected with each other.

For Kant, there are categories of knowledge whose qualities of effect, and thus their power of explanation, exist prior to concrete events. As a result, we must use several fixed orders because these originate in our cognitive processes and it is only through these that nature is knowable. Researchers may arrange these categories in special configurations for specific cases but they are restricted by the form and content of these categories. In this view, *a priori* forms of knowledge are universal and limited: universal in the sense that they are in place for all situations; limited in the sense that no other general categories exist and explanation is limited to these nonreducible categories.[10] Although not rejecting Kantian categories *per se*, it is in reaction to their *a priori* formulation that Marx took his leave of Kant.

---

[10] Appelbaum (1988: 55–56) summarizes Kant's view: "What features must the mind possess in order to enable us to think and reason as we do? Kant concluded that without certain in-born mental properties, thought would not be possible at all. First, he reasoned, some prior concepts of time and space are necessary: Without the innate ability to intuit the reality of spatio-temporal relations, human thinking could not occur. Therefore, Kant concluded, the mind possesses two innate *forms* of time and space. Second, rational decision making requires a number of basic judgments to be made (Kant used physics and biology to formulate his criteria for such judgments). These judgments, Kant believed, are 12 in number – and they therefore require 12 corresponding innate mental *categories*.... Kant posits the existence of an innate mental capability of knowing causality *prior* to any concrete experience of a cause-and-effect relation, thereby enabling us to organize our experiences under the category of causality. In this formulation, it need not trouble us that one cannot derive causation out of experience alone, since it was there (in the mind) all along. All 12 categories operate in a similar fashion: They reflect concepts that *must* be built into the mind if we are to make judgments at all, since they cannot be obtained from experience" (emphases in the original). These twelve categories are (categories of quantity) unity, plurality, totality, (categories of quality) reality, negation, limitation, (categories of relation) substance and accident, causality and dependence, community or interaction, (categories of modality) possibility-impossibility, existence-nonexistence, and necessity-contingency.

Kant accepted the existence of pure and unobservable realities that retain their essential identity throughout changes in their empirical representations (i.e., a "thing-in-itself"). Marx, on the other hand, believed many variables are deeply rooted in practice, experience, and other types of social action. This required a different approach to knowledge. For Marx's view, the nature of time, space, and causation can change and need to be specified for their historical context. Further, as Appelbaum (1988: 57) observes, if all we can possibly ever know is the world through sense perception, then why assume a "thing-in-itself" at all, why these categories, and why are they fixed? – i.e., "Kant's solution is both static and arbitrary." Despite such reservations, *a priori* concerns about time, space, and causality do find their way into Marx's work. His critique is less a denial or rejection than an observation that the quality of effect of Kant's categories should be understood as dynamic rather than static.

In a letter to his father, Marx (1975b: 15) explained that in his study of law his "mistake lay in [his] belief that matter and form can and must be developed separately from each other." He continued:

> The concept is indeed the mediating link between form and content.... At the end of the section on material private law, I saw the falsity of the whole thing, the basic plan of which borders on that of Kant, but deviates wholly from it in the execution, and again it became clear to me that there could be no headway without philosophy. So with good conscience I was able once more to throw myself into her embrace, and I drafted a new system of metaphysical principles, but at the conclusion of it I was once again more compelled to recognize that it was wrong, like all my previous efforts.... From the idealism which, by the way, I had compared and nourished with the idealism of Kant and Fichte, I arrived at the point of seeking the idea in reality itself. If previously the gods had dwelt above the earth, now they become its centre (1975b: 15, 17–18).

In this view, an object of knowledge "has to be put alongside something else, then it assumes other positions, and this diversity added to it gives it different relationships and truths. On the other hand, in the concrete expression of a living world of ideas, as exemplified by law, the state, nature, and philosophy as a whole, the object itself must be studied in its development; arbitrary divisions must not be introduced, the rational character of the object itself must develop as something imbued with contradictions in itself and find its unity in itself" (Marx 1975b: 12). Marx's epistemology is not constructed in a universal sense, but in a manner appropriate for the subject matter. Thus, "the Marxian conception of the foundations of knowledge is the negation and rejection in principle not only of Kant's solution but also of his way of putting the question" (Zeleny 1980: 195). These observations explain why Marx found Kant ultimately unsatisfying and why and how he was becoming dialectical, historical, and materialist in his thinking.

Marx and Dietzgen. It is also useful to acknowledge the views Marx shared with Joseph Dietzgen (1906). Engels (1941: 44) claimed that Dietzgen had "discovered [the materialist dialectic] independently of us." Like his critique of Kant, Dietzgen's outlook bears on how Marx thought a dialectical and materialist science could construct categories that correspond to the patterns of life. Marx, however, provided no elaboration of his agreement with Dietzgen and we can only surmise what this was. Instead of examining Dietzgen's views here, later chapters will elaborate Marx's approach to forming concepts based on material observations. For the time being it bears noting that coming to Marx's texts without some foreknowledge of this issue renders one less prepared than is possible for understanding his meaning.

Marx's Studies in Political Economy. Though editing *Die Rheinische Zeitung* exposed him to political economy, it was Engels's efforts, partly through his "Outlines of a Critique of Political Economy" (1975), that influenced Marx to pursue economic research and criticism. After his initial studies, he expanded his inquiries to Adam Smith, David Ricardo, François Quesnay, and Thomas Malthus, bringing with him what he learned through his critique of Kant:

> Do not let us go back to a fictitious primordial condition as the political economist does, when he tries to explain. Such a primordial condition explains nothing. He merely pushes the question away into a gray nebulous distance. He assumes in the form of fact, of an event, what he is supposed to deduce – namely, the necessary relationship between two things – between, for example, division of labor and exchange. Theology in the same way explains the origin of evil by the fall of man; that is, it assumes as a fact, in historical form, what has to be explained (Marx 1988a: 70–71).

In such critiques, "Marx...juxtaposed political economists' ideas to draw out contradictions, tensions, and a more synthetic conception of the division of labor" (Beamish 1992: 70). It was from such studies that he adopted his concern with class struggle and the labor theory of value.[11]

Marx and the Natural Sciences. Marx (1983i) progressively mastered mathematics and calculus, and statistics were central for his political-economic models (see chapters Four and Six).[12] He also studied scientific papers in the natural sciences. In these studies, he adopted an approach that was critical but positive. Darwin's work encouraged his belief that evolutionary models could be harnessed for social research, though with additional specification (Gerratana 1973, Krader 1982). There has been much disagreement on what this means. For example, both dialectics and historical materialism are often read as the crux of Marx's science. When Kantian-like natural science models are used as the interpretive frame (recognized or not), historical materialism and dialectics combine into a world-philosophy and/or a rigid explanatory model.

Elements in Marx's work do encourage universalistic and rigid interpretations. First, "The Communist Manifesto" declares that, in a history determined by class struggles, the proletariat's victory is "inevitable." Second, his Preface the *Critique of Political Economy* states:

> The general conclusion at which I arrived and which, once reached, continued to serve as the leading thread in my studies, may be briefly summed up as follows: In the social production which men carry on they enter into definite relations that are indispensable and independent of their will; these relations of production correspond to a definite stage of development of their material powers of production. The sum total of these relations of production constitutes the economic structure of society – the real foundation, on which rise legal and political superstructures and to which correspond definite forms of social consciousness. The mode of production in material life determines the general character of the social, political and spiritual

---

[11] "Now as for myself, I do not claim to have discovered either the existence of classes in modern society or the class struggle between them. Long before me, bourgeois historians had described the historical development of this struggle between the classes, as had bourgeois economists their economic anatomy" (Marx 1983b: 62).

[12] The editors of Marx's mathematical manuscripts present a chronology of his studies in mathematics (e.g., derived functions, differential calculus, residual analysis, as well as theorems of leading mathematicians), which, according to their records, start in the 1846 period, resuming in 1857–58, 1865–69, 1872–1882, with several having no dates (Marx 1983i: 237–238).

processes of life. It is not the consciousness of men that determines their existence, but, on the contrary, their social existence determines their consciousness. At a certain stage of their development, the material forces of production in society come in conflict with the existing relations of production or – what is but a legal expression for the same thing – with the property relations within which they have been at work before. From forms of development of the forces of production these relations turn into their fetters. Then comes the period of social revolution. With the change in the economic foundation the entire immense superstructure is more or less rapidly transformed. In considering such transformations the distinction should always be made between the material transformation of the economic conditions of production which can be determined with the precision of natural science, and the legal, political, religious, aesthetic or philosophic – in short ideological forms in which men become conscious of this conflict and fight it out. Just as our opinion of an individual is not based on what he thinks of himself, so can we not judge of such a period of transformation by its own consciousness; on the contrary, this consciousness must rather be explained from the contradictions of material life, from the existing conflict between the social forces of production and the relations of production. No social order ever disappears before all the productive forces, for which there is room for it, have been developed; and new higher relations of production never appear before the material conditions for their existence have matured in the womb of the old society. Therefore, mankind always takes up only such tasks as it can solve; since, looking at the matter more closely, we will always find that the problem itself arises only when the material conditions necessary for its solution already exist or are at least in the process of formation. In broad outlines, we can designate the Asiatic, the ancient, the feudal and the modern bourgeois methods of production as so many epochs in the progress of the economic formation of society. The bourgeois relations of production are the last antagonistic form of the social process of production – antagonistic not in the sense of individual antagonism, but of one arising from conditions surrounding the life of individuals in society; at the same time the productive forces developing in the womb of bourgeois society create the material conditions for the solution of that antagonism. This social formation constitutes, therefore, the closing chapter of the prehistoric stage of human society (Marx 1911: 11–13).

Third, in *Capital*'s Preface Marx makes an analogy to capitalism's laws of motion and later in the text he refers to several of these as "iron laws." Combined with his claims about being scientific, many conclude that Marx believed that *a priori* sociological laws with predictive certainty exist. Once these premises are accepted, what may be called the "traditional reading" of the Preface is that Marx views class struggle as a sort of gravitational law and a universal theory of social evolution, positing a linear model wherein all societies evolve towards capitalism, with communism as its inevitable successor (Labriola 1980, Böhm-Bawerk 1898, Durkheim 1982a, Croce 1982, Lindsay 1973, Gottheil 1966, Cohen 1978). Were this reading true, the inevitability Marx is supposed to have posited dissolves any notion of agency; humans are simply swept along with the action of structures following a predetermined teleological path. However, it is difficult, if not impossible, to reconcile such an interpretation with the fact that it was *against* any universal, metaphysical, and rigid application of science and materialism to social questions that brought Marx to reject Hegel's idealism, Feuerbach's ahistoricism, and Kant's *a priori* conceptualizations (more below).

Over- and Under-Precise Readings of Marx's Historical Materialist Principles

When Kantian-like natural science models are assumed, Marx's ideas may be interpreted as over-precise and rigidly determinist models pitched at the level of the universal instead of the flexible and historically specific approach he favored. Let us examine the traditional reading of the Preface as an over-precise reading in some detail. First, and most obviously, if Marx (and Engels) held the opinion attributed to them there would have been no reason to write "The Communist Manifesto" beyond prophecy and braggadocio. Universally (and correctly) interpreted as a pamphlet of political agitation and a call to action, writing it would have been unnecessary if they believed that revolution and communism would arrive without education, organization, and action. Second, Marx's Preface is neither a statement about *the* method of political economy nor a general theory of history. In his review of Marx's *Critique*, Engels (1980: 469) explained that the "principle features" of "the materialist conception of history" were "briefly outlined in the Preface." Later, Marx (1992c: 27) also says that his Preface states "the materialistic basis of my method." That is, the Preface is a statement of how historical materialist principles *inform* the study of capitalism.

To many readers the Preface is an *a priori* theory of history, a catalyst for reading into Marx's work strong claims about how social life is determined. The most common reading is that "Marx, like Hegel, believed that society evolves through a series of stages" (Walker 1989: 105). This forward-linear theory of history is an almost universal reading. Rejecting that Marx and Engels used it as "a heuristic," Habermas (1979b: 130) sees it as a "doctrine…indeed as a theory of social evolution." Giddens (1995: 235–236) holds that the Preface…

> …is not simply an empirical account of phases of development of specific human societies. It remains linked with a quite strongly affirmed, but only weakly elaborated, vision of humanity ascending through various stages of class society towards a new order that finally creates a 'truly human' society…[thus retaining] strong echoes of the Hegelian view of history as the progressive overcoming of human self-alienation through the clash of opposites.

Morrison (1995: 40) puts the generally accepted view succinctly: "the materialist perspective is, above all, a theory of historical development which explains human existence in terms of a series of economic stages…tribal, ancient, feudal and capitalist modes of production…[a] sequence of economic stages." This interpretation is hard to sustain, as we shall see.

Marx calls his Preface a "general conclusion" arrived at after lengthy study, which became a "leading thread" – i.e., a working hypothesis – that guided his inquiry and analysis. This assertion is something different than a linear, *a priori* model. In fact, Marx never claims that his Preface's list of modes of production represents a universal theory of history. Rather, he tells us these categories can be used to "designate…progress of the economic formation of society" implying that those that do not represent progress are superfluous for his ends. He designates these epochs simply as markers of analysis. His statements on theory and method elsewhere support this interpretation:

> The division of labour is, according to M. Proudhon, an eternal law, a simple, abstract category. Therefore the abstraction, the idea, the word must suffice for him to explain the division of labour at different historical epochs. Castes, corporations, manufacture, large-scale industry must be explained by the single word divide. First study carefully the meaning of 'divide,' and you will have no need to study the numerous influences which give the divi-

sion of labour a definite character in every epoch.... *History does not proceed so categorically* (Marx 1847: 127–128; emphasis added).

If this view seems to contradict the Preface, perhaps Marx changed his mind between this critique of Proudhon and writing *Critique of Political Economy*? Or is he offering a different sort of argument? The same year he wrote *The Poverty of Philosophy*, Marx (1976b: 317) wrote:

> Any development, whatever its substance may be, can be represented as a series of different stages of development that are connected in such a way that one forms the *negation* of the other. If, for example, a people develops from absolute monarchy to constitutional monarchy, it *negates* its former political being. In no sphere can one undergo a development without negating one's previous mode of existence. *Negating* translated into the language of morality means: *denying* (emphases in the original).

I believe Marx suggests a similar logic in his Preface, which offers an ordering of categories that inform his analysis of those modes of production most important for understanding the rise of the capitalist system. Marx does *not* argue for going back into the past and treating the future as a series of predictable and sequential stages. Rather, once he has grasped the present system and its core features as a progressive step over the system that was its progenitor, he can analytically treat the history that led up to it as containing epochs that also *represent progress* in *economic history*, not history in general. As Marx (1973: 106) noted in the *Grundrisse*: "The so-called historical presentation of development is founded, as a rule, on the fact that the latest form regards the previous ones as steps leading up to itself." Thus, Marx is simply positing a method of reading historical change in modes of production once they have occurred.

This view is indicative of the flexibility needed for both historical-structural analysis and dialectical thinking, though many traditional readings strip Marx's principles of their flexibility and force them into overly precise and rigid theories. This usually comes in the form of the belief that "Marxist thought is...the first and perhaps the most forceful theory emphasizing a single, determining factor in social change" (Timasheff 1957: 46–48). Such reductionism is commonly read into the type of determinism Marx offered. *Materialist determinism*, the most vulgar form, "is the view that nothing is real but what is material and that minds must therefore be forms of matter" (Acton 1967: 17). Many early interpreters often believed that "Marx certainly held, that the movement in thought is only the reflection of the movement in things" (Lindsay 1973: 17–18). If matter is all that is real and if dialectic is its universal causal force outside our will and consciousness, then analytical and/or political work are irrelevant – a view incompatible with the efforts Marx put into research and politics. *Technological determinism* holds that, for Marx, "technique and technology" (Jordan 1971: 33) cause change in modes of production. Cohen (1978: 134) similarly argues that "Marx assigned explanatory primacy to the productive forces...over the production relations, or over the economic structure the relations constitute," a reading that takes the Preface not as a guiding thread but as a finished "theory of history" for reading all of Marx's work. Finally, *economic determinism*, perhaps the most common reading, views historical materialism as "a rigid, metaphysical theory of historical causation...[whereby the] economic factor becomes the only determinant of historical development" (Jordan 1971: 37). Babbie (1995: 33), for instance, instructs students that "Marx put forward a view of economic determinism...that economic factors determined the nature of all other aspects of society." Although a widely accepted interpretation, it begs us to ask: Is it true?

After Marx's death, Engels attempted to counter such rigid determinist readings:

> According to the materialist conception of history, the determining element in history is *ultimately* the production and reproduction in real life. More than this neither Marx nor I have ever asserted.... The economic situation is the basis, but the various elements of the superstructure – political forms of the class struggle and its consequences, constitutions established by the victorious class after a successful battle, etc. – forms of law – and then even the reflexes of all these struggles in the brains of the combatants: political, legal, philosophical theories, religious ideas and their further development into systems of dogma – also exercise their influence upon the course of the historical struggles and in many cases preponderate in determining their *form*. There is an interaction of all these elements, in which, amid all the endless *host* of accidents (*i.e.*, of things and events whose inner connection is so remote or so impossible to prove that we can regard it as absent and neglect it), the economic movement finally asserts itself as necessary. Otherwise the

application of the theory to any period of history one chose would be easier than the solution of a simple equation of the first degree.... [H]istory makes itself in such a way that the final result always rises from conflicts between many individual wills, of which each again has been made what it is by a host of particular conditions of life. Thus there are innumerable intersecting forces, an infinite series of parallelograms of forces which give rise to one resultant – the historical event.... Marx and I are ourselves partly to blame for the fact that younger writers sometimes lay more stress on the economic side than is due to it. We had to emphasize the main principle in opposition to our adversaries, who denied it, and we had not always the time, the place or the opportunity to allow the other elements involved in the interaction to come into their rights (1936c: 475–477; emphases in the original).

Though Engels's warning has gone generally unheeded, the prevailing determinist reading (see McIntyre 2006) contrasts with those Marxists who accurately believe "Marx's scientific method is fiercely antireductionist" (Murray 1988: 116). Marx's materialism was not simply a mechanical-materialist inversion of Hegel's idealism hammered into an *a priori* theory. Marx's critique of dominant scientific theories, his analysis of ideology, and both "The Communist Manifesto" and *Capital*, would have been fruitless and contradictory had he assumed ideas and intentional action play no role in social change.

If determinist readings are over-precise, under-precise approaches use dialectics as incantation or formula. There are two primary interrelated types. The first is a "magic wand" approach whereby "the dialectic" becomes the explanatory variable used in place of what would otherwise better be understood as reciprocal, interactive relationships and/or cyclical processes. Though they may *describe* something real, many statements that appeal to "dialectic" or "dialectical" are so general that they do not *explain* causal relationships between variables. So common has been this usage that Analytical Marxists accuse some Marxists of using dialectics as a "yoga" in order "to justify a lazy kind of teleological reasoning" (Roemer 1986b: 191). Overlooking the imprecision of wielding "yoga" as an analytical critique, this argument sees dialectics as jargon and not a "method." A friend of Marx's anticipated such complaints:

> Certainly the style of 'Capital' is difficult to understand – but is the subject it treats of easy to grasp? The style is not only the man, it is also the subject matter – and it must be adapted to the latter.... To complain of the difficult,

obscure or clumsy style of 'Capital' is only revealing one's own slothfulness of thought or incapability of reasoning (Liebknecht 1901: 75–76).

Perhaps this retort is too strong? Whatever the case, dialecticians such as Sherman (1995: 217) also believe that "One must give up the notion that the dialectic approach provides answers or universal laws in some magical fashion. Rather, it should be viewed as providing some of the questions one needs to ask in order to understand and change society." But this view can and should be pushed further. Although dialectical reasoning provides for new types of questions, its acknowledgement of internal relations, contradictions, negativity, and dynamic change also provide an ontological and epistemological basis for concept construction as well as for piecing together the interactions of parts that make up wholes (see chapters Three though Six).

In the second type of under-precision, commentators draw analogies between Marx's ideas and religious, mythic, prophetic, and/or utopian ideologies (Tucker 1964, Wessell 1984, Harris 1950, Ling 1980). For Tucker (1957–58: 125), Marxism became "comparable to that of a new faith [and] history will remember him...as the apostle of a secular creed." Howard Parsons (1964: 52) claimed that Marx "was a prophet whose work was religious in both method and aim." Such commentators are not simply imagining evidence for the Marx-as-prophet/Marxism-as-religion thesis. In reading the mature Marx in light of the young Marx, he can be seen a soothsayer, prophesier, and/or utopian. Marx, however, did not develop a new creed or see himself as a messiah. After he died, Marx's works *were* consecrated as holy texts by some and he was enshrined as intellectual high-priest, though he was adamantly opposed to such characterizations, e.g., "Your letters to me, and they are ready for publication, prove that you had tried everything to force upon me the role of the 'democratic Dalai-Lama and Possessor of the Future.' How do you prove that I had ever accepted this silly role?" (Marx 1979a: 132).[13]

---

[13] McLellan (1975: 79) notes: "Marx had not offered any hard-and-fast conclusions – only a methodology.... Almost inevitably, the ideas were simplified, rigidified, ossified. Marxism became a matter of simple faith for its millions of adherents to whom it gave the certainty of final victory. But this entailed its transformation into a dogmatic ideology with the correlative concept of heresy – or revisionism, as it was often called. (The parallels with certain periods in the history of the Christian religion are obvious.) Only a very selective version of the sacred texts could be propagated, and if even then history did not conform – then history itself had to be rewritten."

Speculation, Mysticism and Metaphysics, and Obscurantism

In his analysis of Hegel, Marx (1988a: 14) wrote that he had the goal of "a critique of the speculative elaboration of that material." Marx (1976a: 12) felt Hegel's idealist views only produce "the appearance of real understanding. They are and remain uncomprehended, because they are not grasped in their specific essence." Similarly in *The Holy Family*, Marx and Engels (1956: 80) argue that "speculative philosophy has as many incarnations as there are things, just as it has here in every fruit an incarnation of the Substance, of Absolute Fruit." For them, "Where speculation ends, where real life starts, there consequently begins real, positive science" (Marx and Engels 1976: 37). Marx thought science should involve the positive investigation of the real, which Hegel's speculative philosophy and Kantian things-in-themselves make difficult. Politically, Marx's (1992c: 26) dismissal of speculation was expressed in his disdain for those "writing receipts for the cook-shops of the future."

Marx also criticized mysticism and metaphysics. To mystify something is to distort it through the way it is conceptualized. This is both an analytical and a political problem. Marx (1975k: 144), for instance, argued for the "reform of consciousness not though dogmas, but by analysing the mystical consciousness that is unintelligible to itself, whether it manifests itself in a religious or a political form." Proudhon, for example,

> ...falls into the error of the bourgeois economists who regard those economic categories as eternal laws and not as historical laws which are only laws for a given historical development, a specific development of the productive forces. Thus, instead of regarding the politico-economic categories as abstractions of actual social relations that are transitory and historical, [the conventional approach], by a mystic inversion, sees in the real relations only the embodiment of those abstractions. Those abstractions are themselves formulas which have been slumbering in the bosom of God the Father since the beginning of the world (Marx 1982b: 100).

Hegel's "stylistic peculiarity" was also "a product of mysticism" (Marx 1976a: 12). In the tradition Hegel spawned, "Not only in its answers, even in its questions there was a mystification" (Marx and Engels 1976: 28). If the questions are misshapen, the answers will be as well. Although "Nothing is easier than to invent mystical causes, i.e. phrases in which common sense is lacking"

(Marx 1982b: 96), Hegel's dialectic was not to be abandoned. "This dialectic is, to be sure, the ultimate word in philosophy and hence there is all the more need to divest it of the mystical aura given it by Hegel" (Marx 1983f: 316). Marx (1936a: 102) wanted to amend this situation, e.g., "If there should ever be time for such work again, I should greatly like to make accessible to the ordinary human intelligence, in two or three printer's sheets, what is *rational* in the method which Hegel discovered but at the same time enveloped in mysticism" (emphasis in the original). Whether it was against Hegel's idealist dialectic or Kant's things-in-themselves (each a form of speculative philosophy), Marx opposed any discourse he found to be mystifying its object of inquiry.

Marx was also steadfast against any metaphysic – dialectical or otherwise. Assertions about invisible, transhistorical forces are a metaphysical antithesis of real scientific understanding. Hegel's "mystical dualism" (Marx 1976a: 84) viewed dialectic as a metaphysical force in history, wrapping what was "rational" about dialectics in a "mystical shell" (Marx 1992c: 29). In a critique of Proudhon and metaphysics generally (including both Hegel and Kant), Marx (1847: 106) encapsulated his problem:

> If we abstract thus from every subject all the alleged accidents, animate or inanimate, men or things, we are right in saying that in the final abstraction, the only substance left is the logical categories. Thus the metaphysicians who, in making these abstractions, think they are making analyses, and who, the more they detach themselves from things, imagine themselves to be getting all the nearer to the point of penetrating to their core – these metaphysicians in turn are right in saying that things here below are embroideries of which the logical categories constitute the canvas.

Much of *The Poverty of Philosophy* is an attack on assumptions and theories about "immutable laws, eternal principles, ideal categories" that have existed "since the beginning of time" (1847: 116–117). Such claims as this (combined with those above) suggest that any strong arguments about determinism in general or a universal theory in Marx are hard to defend. Either Marx held contradictory and opposing views on these matters, or his approach to materialism and history have been read through frameworks not present in his writings. His ideas, expressed throughout his work, support the latter view much more so than the former.

Finally, in criticizing any form of knowledge if it remained clouded, Marx warned of two types of obscurantism. In one, a linguistic strategy can undermine the intelligibility of one's work. In another, the analytical framework in use covers over and/or distorts more than it reveals. For example, in an early self-critique, Marx (1975b: 15) complained that his studies had "arrived at a division of the material such as could be devised by its author for at most an easy and shallow classification, but in which the spirit and truth of law disappeared." This concern with clarity versus obscurantism carried over into his critique of political-economic theory, e.g., "what all this wisdom comes down to is the attempt to stick fast at the simplest economic relations, which, conceived by themselves, are pure abstractions; but these relations are, in reality, mediated by the deepest antithesis, and represent only one side, in which the full expression of the antitheses is obscured" (Marx 1973: 248). Marx shared this view with Engels (2001: 268), who, for instance, cautioned Kautsky on his history of the French Revolution: "I would say a great deal less about the modern mode of production. In every case a yawning gap divides it from the *facts* you adduce and, *thus* out of context, it appears as a *pure abstraction* which, far from throwing light on the subject, renders it still more obscure" (emphases in the original). For Marx (and Engels), forcing data into the framework of historical materialism when unwarranted obscures one's subject matter and thus is to fail scientifically.

In sum, Marx rejected philosophical speculation as guesswork, metaphysical dialectics as mystifying, and demanded that scientific inquiry should clarify things rather than obscure them.

Marxist thought has had an uneven relationship to these concerns. In theorizing the relations between political-economic institutions of power and the usurpation of culture by the logic and role of capital in dominating art, music, and knowledge, the Frankfurt School provided a point of view from which to launch criticism, though its dialectic remained within the bounds of philosophy. As a result, "critical theory was all too often profoundly antiempirical: It seemed to deal with abstract concepts rather than concrete institutions and actual people" (Appelbaum 1988: 13). In response to such concerns, Analytical Marxists have rejected dialectics and turned toward analytical philosophy (see citations above; also see Roberts 1996). To their credit, this tradition holds that "in order to be useful" categories of dialectics "have to be translated into a language of causes, mechanisms and effects, rather than be left as elusive philosophical principles" (Wright, Levine, and Sober 1992: 6). Analytical Marxism thus holds "four specific commitments" distinguish it, three of which include "conventional scientific norms in the elaboration of theory and the conduct of research," "the importance of systematic conceptualization [and] careful attention to both definitions of concepts and the logical coherence of interconnected concepts," and the "concern with a relatively fine-grained specification of the steps in the theoretical arguments linking concepts" (Wright 1995: 14). There is nothing unique about these commitments, however; Marx shares them too, as we shall see. Analytical Marxism's fourth commitment – i.e., the "importance accorded to the intentional action of individuals within both explanatory and normative theories" (Wright 1995: 14; also see Levine, Sober, and Wright 1987) – is one that Marx does indeed reject because it obscures the nature of capitalism by carving out a unit of analysis ill-fit for revealing its logic as a system.

In Marx's approach, rather than using micro-foundations for organizing macro-level explanations (as Analytical Marxists offer), "totalities" are broken down into constituent parts and the interrelationships of these parts are most important. When social structure is the unit of analysis, individuals are treated as representatives of structured social relationships, with individual level characteristics abstracted out of view. Marx (1992j: 151) thus asks how the force of capitalism's structure brings the capitalist to act "as capital personified...endowed with consciousness and a will," while "compelling him to sell goods under their social value, this same law, acting as a coercive law of competition, forces his competitors to adopt the new method" (1992j: 302). This is because "competition makes the immanent laws of capitalist production to be felt by each individual capitalist, as external coercive laws. It compels him" (1992j: 555). Marx thus rejects reductive frameworks for explaining the central tendencies of capitalism as a class system, e.g., "since sales and purchases are negotiated solely between individuals, it is not admissible to seek here for relations between whole social classes" (1992j: 550) – and even pre-capitalist forms – i.e., "for, in the beginning of civilisation, it is not private individuals but families, tribes, &c., that meet on an independent footing" (1992j: 332).

For Marx's (1985c: 232) outlook, any "attempt to present...history in terms of psychology, is bad, muddled, and amorphous." Idealism, rationalism, and utilitarianism are founded on "the very patterns of mystification that are being exposed" (Wolff 1988: 81). As a result, "To try to describe society in rational terms, when that society is not rationally organized, necessarily leads to confusion" (Wolton 1996: xvi). In Marx's view, by analyzing class relations with abstractions carved at the level of the individual actor, Analytical Marxism's methodological individualism is ill-prepared for explaining structural-causal properties on their own terms and thus mystifies and obscures capitalism's most important relationships.

Though Marx believed scientific dialectics could demystify the effects the "nightmare" of history had on the "minds of the living," he left himself open to multiple interpretations on metaphysics. An early reviewer of *Capital* complained that "in assigning to labour a unique power to produce value Marx...falls into metaphysics, in the bad and obsolete sense of the word" (Bonar 1898: 7, in Böhm-Bawerk 1898). With Hegel's and Kant's speculative and metaphysical philosophies dominating dialectical and scientific discourse, and with Engels's later version of dialectic presenting "a view of ultimate causes and fundamental processes in the universe" (Carver 1983: 108–109), metaphysics became a standard reading of Marx after his death. In one metaphysical reading, "the dialectic" is "the motive power for change" and "dialectical materialism" is a force in "control of man and society" (De Koster 1964: 57, 105). A more moderate position holds that his Preface's "guiding thread" was something that Marx "treated as a metaphysical theory" (Jordan 1967: 299). Finally, a more tentative view is that Marx possessed "a metaphysical theory of the nature of social reality," though "such an ontology is only implicit in [his] work" (Gould 1978: xi). Such views must defend themselves against the plethora of instances where Marx attacks metaphysical views. Consequently, proponents of this view must either hold that Marx secretly accepted metaphysics and/or the metaphysic implied in his work was something he failed to recognize, two questionable interpretations.

Outside of Marxism, Karl Popper (1950: 679) held that to posit Marx's dialectic "as a metaphysical system" was to "use Marxism irrationally." And, although Joan Robinson (1966: vii) complained about Marx's "nineteenth-century metaphysical habits of thought," Joseph Schumpeter (1954: 10) accurately understood that though Marx's Hegelian "background shows in all his writings whenever the opportunity presents itself," it was "a mistake and an injustice to Marx's scientific powers" to assume that Hegelianism was "the key to [his] system.... Nowhere did he betray positive science to metaphysics."

In fact, Marx never posited the existence of a universal causal force called "the dialectic" (i.e., a metaphysic), but rather argued that a scientific dialectic provides the tools necessary for analyzing the social world in a systematic way, i.e., an epistemology. Moreover, although he accepted that that labor was a central feature in human history (see Chapter Three), his analysis of labor – its conditions, products, and struggles over each – was always something based on the concrete and real, never settling for simply the abstract and the metaphysical – i.e., a general and universal theory of history.

The Thesis-Antithesis-Synthesis Myth. Metaphysical myths about Marx persist and are widely repeated in the literature. In the most common metaphysical reading, Marx's outlook is reduced to "dialectical materialism," which is, in turn, reduced to a triad of "thesis-antithesis-synthesis." This is a conceptualization of dialectic whereby a proposition or a fact (thesis) calls into being its opposite (antithesis), which creates a tension between them that transforms them through their mutual conflicts into a higher level of reality that contains elements of each (synthesis). This mythical reading is so commonly accepted that even Marx's birthplace (Karl Marx Haus, Trier, Germany) and gravesite (Highgate Cemetery, London) offer pamphlets that repeat it.[14] Such interpretations lead to the conclusion that Marx believed that communism was a metaphysical inevitability given how capitalists (thesis) will call up their opposite (proletarians) and the tension between them will result in a synthesis or change (revolution). This same argument, at the level of social structure,

---

[14] "Hegel had introduced a philosophical concept which he employed in dialectics and called negation. In the process of negation any character process or phenomenon develops its own opposite, so that an old quality sooner or later gives rise to a rival, i.e., a new quality producing a contradiction" (*Karl Marx and his Contemporaries: Guidebook to the Permanent Exhibition in the Karl Marx House, Trier*, 1994, Neu GmbH, Trier, p. 39); "The philosopher Hegel had advanced the theory that the world system was not static but was continually evolving through conflict between two divergent ideas, the thesis and the antithesis to produce a synthesis, which would then be challenged by a new anti-thesis. However he believed that this process would eventually end when the ideal was achieved. Marx extended this theory to include events in society, and said that progress could not be achieved by the limited reform of any one system but only when one system was challenged by another to result in a combination of the two, leading in turn to another challenge from a new system. He also believed that this process would eventually end in the achievement of the ideal and for Marx this meant a classless society, where private property was abolished and profit used to benefit the workers" (Judith Yuille, *Karl Marx: from Trier to Highgate*. Friends of Highgate Cemetery. Charity No. 1058392. VAT No. 544 5652 34. Highgate Cemetery. Swains Lane, London N6).

holds that capitalism brings communism into being metaphysically. But any such position is hard, if not impossible, to defend, given that "Contrary to what is alleged in many books, Marx never employed these terms – nor, for that matter, did Hegel" (McLellan 1973: 163).[15] Not only did he never employ this metaphysic, he explicitly argued against it, for example, accusing Proudhon of using these terms "to frighten the French by flinging quasi-Hegelian phrases at them" (Marx 1847: 104). Unfortunately, extracting Marx's dialectic from the framework of "thesis-antithesis-synthesis" has proven difficult and leaves Marx open to pseudo-refutations, "pseudo" because what is being refuted is not Marx's position (see Timasheff 1957: 47, Popper 1950: 682, 1966: 334, Appelbaum 1988: 59, Allan 2005: 73).

Although Marx criticized obscurantism, the logical form (dialectical reason) and the content of analysis (sophisticated but sometimes unfamiliar data) in his political economy often obscured his meaning. This was in part because "Marx was far removed from the conventional sources of news and so made much more use of official reports, statistics, and so on, than the majority of journalists" (McLellan 1973: 288). Marx also employed a linguistic strategy that produced problems in reading him. In 1885, John Swinton, a US journalist, related the following:

> Asking him why [*Capital*] had not been put in English, as it has been put in French and Russian, from the original German, he replied that a proposition for an English translation had come to him from New York.... He said that his German text was often obscure and that it would be found exceedingly difficult to turn into English. 'But look at the translation into French', he said as he presented me with a copy of the Paris edition.... 'That', he continued, 'is far clearer and the style better than the German original. It is from this that the translation into English ought to be made.'... A few days ago in

---

[15] "The categories thesis, antithesis and synthesis, so often and so erroneously attributed to the methods of Hegel and Marx, were used by the German philosopher J.G. Fichte, though not with reference to dialectic. Hegel himself rejected this triadic approach as schematic and distortive. Despite this rejection by Hegel, the legend that his dialectic is thesis-antithesis-synthesis is well established; it derives from an early commentator, Heinrich Moritz Chalybaus, who identified the triad as the key to Hegel's work. Other commentators seem to have identified this key with dialectic. Hegel, however, defined the dialectic in his *Science of Logic*: 'It is in this dialectic as it is here understood, that is, in the grasping of opposites in their unity or of the positive in the negative, that speculative thought consists'" (Carver 1982: 46).

> taking up the first chapter of Mr. Broadhouse's translation, my eye fell on a sentence so obscure as to be unintelligible, but in turning to the French version, the meaning of the sentence was plain (Swinton 1983: 266–67).

With a book whose own author declared its translation into another language improved its intelligibility, Marx's presentation threatened itself with obscurantism, a problem lasting to this day. It has been left to his supporters – from Engels onward – to undo these speculative, mystical, metaphysical, and obscure treatments.

Marx's dialectic was more than metaphors. His method articulates causal mechanisms and can be shown to be logical and systematic. Thus, the challenge is to reconstruct Marx's method without conflating his various levels of inquiry while respecting his use of scientific principles. Marx viewed his work on capitalism as an exercise in de-mystification, a revealing of the roots, rise, and reaches of this mode of production. But as a result of his method of presentation and of historical developments, Marx's work has become mystified as well. Rather than a "new dialectic" and a new Marx (Arthur 2002), we need to rediscover an older Marx without recourse to those interpretations about which he warned, while avoiding additional interpretive problems that have been handed down to us. In making this effort, we need to re-establish Marx's bona fides as one who is squarely within the sociological tradition rather than as the gadfly and outsider he is often held to be.

Chapter Two
Marx and Classical Sociology

Classical sociologists worked within a milieu where at least one horizon carved by Marx was within view – i.e., Weber's *Protestant Ethnic*, Simmel's *Philosophy of Money*, and Tönnies's *Gemeinschaft und Gesellschaft* were each "inspired directly by Marx's thought" (Bottomore 1973:14). Durkheim (1982a:173) found "the materialist conception of history" – read through Antonio Labriola's use of "The Communist Manifesto" (*The German Ideology* was unavailable) – "unproven" and "contrary to facts which appear established." Simmel (1964: 16) felt his approach to sociology could "point the way toward a conception of history which is more profound than historical materialism, and which may even supersede it." "The Communist Manifesto," however, offers only a general thesis informed by historical materialist principles but it does not elaborate them in detail. Moreover, "Marx never wrote an exposition of his methodology in the style of Durkheim's *Rules of Sociological Method* or Max Weber's long essay on 'Objectivity in Social Science and Social Policy'; nor did his work receive any widespread critical attention during his lifetime, such as would have led him to defend his theory in a systematic way" (Bottomore 1975: 11). Marx's comments on method in *Capital*'s Preface (1992a) and Afterword (1992c) should have been available to Durkheim and Simmel, though neither refers to them. Their work would influence the kind of Marx sociologists would inherit.

Comparisons of Marx to classical theory have yielded no one accepted conclusion. When the comparisons are of Marx with Weber (Antonio and Glassman 1985, Ferrarotti 1985, Löwith 1982, Wiley 1987) or of Marx, Weber and Durkheim (Hughes, Martin, and Sharrock 1995, Madan 1979, Morrison 1995), the most common approach has been "to examine some of the main points of divergence between Marx's characteristic views on the one hand, and those of the two later writers on the other" (Giddens 1971: vii). Marx and Durkheim (Poggi 1972) or Marx and Simmel (Turner 1993a) have been compared less often. Though he admits they have methodological "points of convergence," Turner (1993a: 88) holds that "Simmel...was concerned primarily with abstracting the 'forms' of social reality from ongoing processes, whereas Marx...was committed to changing social structures by altering the course of social processes." Depicting Simmel as a scientist and Marx as (only) a (perhaps naïve) revolutionary is incomplete, misleading, and even disingenuous. Marx did abstract forms of social reality from ongoing processes and it has been his method of doing so and the subjects to which he put his abstractions to work that has kept his sociological legacy from being relegated to the likes of Saint-Simon and Comte, as much as some may wish this to be so.

Although Marxists have found similarities between Marx's approach and sociology in general (Bottomore and Goode 1983, Colletti 1972, Lefebvre 1968, Burawoy 1982, Sorel 1983), such claims are tempered by a range of qualifications: "Marxist theory has nothing to do with the sociology of the 19th and 20th centuries as established by Comte and elaborated by Mill and Spencer" (Korsch 1983: 34); "Marx's sociology...constitutes the basis of all his work and is an indispensable key to understanding his thought as a whole" (Gurvitch 1983: 38); "Marxism is identical with sociology.... [I]t is an attempt to gain knowledge of the law-governed character of social life as a whole" (Adler 1983: 30). Many reasons for this variety of opinion stem from problems in reading Marx (Chapter One). If we compare Durkheim's essays, "The Method of Sociology" (1982b: 245–247), "What is a Social Fact?" (1982c: 50–59), "Rules for the Observation of Social Facts" (1982c: 60–84), "Rules for the Demonstration for Sociological Proof" (1982c: 147–158), and Simmel's (1964) essay, "The Field of Sociology," with Marx's Preface (1992a) and Afterword (1992c, 1992d*) to *Capital* (Volume One), his Introduction to the *Grundrisse* (1973), and supplemental material from letters and main texts we find that many of Marx's views are closer to Durkheim's and Simmel's than is commonly assumed.

A Commitment to Scientific Methods and Values

For Durkheim's (1982b: 247) sociology, "collective representations...have a nature of their own, and relate to a distinctive science." He believed "that the sociologist must take on the state of mind of the physicists, chemists, and biologists when they venture into a territory hitherto unexplored, that of their scientific field" (1982b: 246). Like natural scientists, sociologists "have only one way of demonstrating that one phenomenon is the cause of another."

> This is to compare the cases where they are both simultaneously present or absent, so as to discover whether the variations they display in these different combinations of circumstances provide evidence that one depends upon the other. When the phenomena can be artificially produced at will by the observer, the method is that of experimentation proper. When, on the other hand, the production of facts is something beyond our power to command, and we can only bring them together as they have been spontaneously produced, the method used in one of indirect experimentation, or the comparative method.... [S]ince social phenomena clearly rule out any control by the experimenter, the comparative method is the sole one suitable for sociology (Durkheim 1982c: 147).

Thus, Durkheim (1982c: 56) adopted from the physical sciences a model for social science "found in the organisms of those mixed phenomena of nature studied in the combined sciences such as biochemistry," including an assumption about the "close affinity of life and structure, organ and function" (1982c: 59, note 4).

For Simmel (1964: 13), too, sociology is "a science with its own subject matter that is differentiated, by division of labor, from the subject matters of all other sciences." Sociology, he explains, "proceeds like physics," a science "which could never have been developed without...certain assumptions concerning space, matter, movement, and enumerability" (1964: 24). In other words, a sociologist, *as a scientist*, searches for regularities, causal properties, and laws of behavior, just as is done in all sciences.

This commitment to science meant support for objectivity, communalism, and progress. Objectivity is both a value to be held and a characteristic of social facts, e.g., "the distinctive features of the social fact gives us sufficient reassurance about the nature of this objectivity to demonstrate that it is not illusory" (Durkheim 1982c: 70). If we accept that "Feeling is an object for scientific study, not the criterion of scientific truth" (1982c: 74), then "how facts are classified does not depend on [the scientist], or on his own particular cast of mind, but on the nature of things" (1982c: 76) and thus "his research must be as objective as possible" (1982c: 82). If "the appropriate method...must be historical and objective," then one can focus on those social facts that "statistics afford us a means of isolating" (1982c: 52, 55). In this process, "One must systematically discard all preconceptions...the basis of all scientific method" (1982c: 72).

For Simmel's (1964: 9) view of objectivity, "the interactions we have in mind when we talk about 'society' are crystallized as definable, consistent structures such as the state and the family, the guild and the church, social classes and organizations based on common interests," i.e., observable things rather than subjective states or hypotheticals.

Demonstrating communal and progressive values, Durkheim (1982c: 76, 83) argued for a science whose "criterion...can be demonstrated and generally accepted by everybody, and the observer's statements can be verified by others.... Only later will it be feasible to carry our research further and by progressive approaches gradually capture that fleeting reality which the human mind will perhaps never grasp completely." In this work, sociology can provide answers to questions posed by psychology and theories of knowledge (Durkheim 1982b: 247).

Simmel (1964: 12–14) also believed that sociology can "lead to a new viewpoint that must make itself felt in all so-called human studies" as it "adapts itself to each specific discipline – economics, history of culture, ethics, theology, or what not." Thus, social science promises progress to both the human sciences and society as a whole.

Marx shared these commitments. For him, political economy should go forward with its inquiry "as impartially as possible," while "only formulating, in a strictly scientific manner, the aim that every accurate investigation...must have...[i.e.,] the disclosing of the special laws that regulate the origin, existence, development, death of a given social organism" (Marx 1992d*: 27, 28), pursuits that demand both "free scientific inquiry" (Marx 1992a: 21) and "greater scientific strictness" in the research process (Marx 1992c: 22). Natural science tools available for sociological research include the experimental model (Marx 1992a: 19) and evolutionary frameworks (Marx 1992c: 28). Working for the public good, social scientists "should openly, in the face of the whole world, publish their views, their aims, their tendencies" (Marx and Engels 1978a: 473). In order to make this work accessible, anyone with average access to public information should be able to check on data sources, such as "social statistics...governments and parliaments appointed...commissions of inquiry...reporters on public health, [and] the history, the details, and the results of...legislation" (Marx 1992a: 20). In endorsing publicly available and cross-paradigmatic work, Marx (1992a: 21) welcomed his readers, lay and professional, to forward him "Every opinion based on scientific criticism." Though Durkheim and Simmel shared these values, Marx (1992c: 23) was perhaps more concerned with nonacademically trained audiences, claiming that the "appreciation which 'Das Kapital' rapidly gained in wide circles of the German working-class is the best reward of [his] labors."

Internal Relations in Society and the Disjuncture of Appearance and Essence

The study of society and its elementary parts requires acknowledgement of and an approach to the relationships between wholes and parts. In Durkheim's (1982c: 56) understanding, "It is in each part because it is in the whole." He argues that "external characteristics" are "linked to the basic properties of things," where there is a "connection between the surface and the depths" that accounts for characteristics "found identically and without exception in all phenomena of a certain order" because they are "closely linked to the nature of...phenomena and are joined indissolubly to them" (1982c: 80). Interiorizing objects within interconnected whole-part relationships is an internal relations ontology.

Simmel (1964: 6–7) formulated a similar vision. For him, "Color molecules, letters, particles of water indeed 'exist'; but the painting, the book, the river are synthesis: they are units that do not exist in objective reality but only in the consciousness which constitutes them." These levels of reality are "organically fused" (1964: 15), or, what amounts to an identical expression, internally related. "In a similar way, when we look at human life from a certain distance, we see each individual in his precise differentiation from all others. But if we increase our distance, the single individual disappears, and there emerges, instead, the picture of a 'society' with its own forms and colors – a picture which has its own possibilities of being recognized or missed" (1964: 8). In this depiction – one that corresponds to Durkheim's terminology of "connection between the surface and the depths" – social relations are irreducible and composed of internal relations between individuals, groups, and institutions.

Marx's ontology is also one of inner-connections and mutually defining characteristics. In Hegel's philosophy, and traditional political economy, the "antithesis of *propertylessness* and *property,* so long as it is not comprehended as the antithesis of *labor* and *capital,* still remains an antithesis of indifference, not grasped in its *active connection*, its *internal relation*" (Marx 1988a: 99; emphases in the original). In his *Critique of Political Economy,* Marx (1983g: 354) held that "an important view of social relationships is scientifically expounded for the first time." In search of that "which governs...phenomena [within their] definite form and mutual connexion within a given historical period" (Marx 1992d*: 27), he is concerned with how connections within social relationships produce laws that shape the form and type of realities in given periods, an assumption that history and social structure (both parts of wholes) have relationships that are internal to each other.

Because social facts can "exist without being applied at the time" and "do not present themselves with such extreme simplicity" (Durkheim 1982c: 55, 64), a disjuncture often exists between what is immediately observable and what we think we know. The disjuncture exists because "Man cannot live among things without forming ideas about them.... But, because these notions are closer to us and more within our mental grasp than the realities to which they correspond, we naturally tend to substitute them for the realities" (1982c: 60). Durkheim (1982c: 62–63) thus argues that preconceptions of things arrive to our consciousness "resembling ghost like creatures [that] distort the true appearance of things, but which we nevertheless mistake for the things themselves" while "the details of social life swamp consciousness from all sides." Given that "the social being has been fashioned historically" (1982c: 54), a scientist...

...must free himself from those fallacious notions which hold sway over the mind of the ordinary person, shaking off, once and for all, the yoke of those empirical categories that long habit often makes tyrannical. If necessity sometimes forces him to resort to them, let him at least do so in full cognisance of the little value they possess, so as not to assign to them in the investigation a role which they are unfit to play (1982c: 73).

However, the objects of any new science "are already represented in the mind, not only through sense perceptions, but also by some kind of crudely formed concepts...because reflective thought precedes science, which merely employs it more methodically" (1982c: 60). Durkheim (1982c: 81) thus asserts that the "starting point for science...cannot therefore be different from that for common or practical knowledge," though scientific knowledge of society tends to outpace "the idea that ordinary people have of it" (1982c: 66). If appearances can mask realities that stand behind them, then a method is needed for unmasking that which accounts for them.

For Durkheim (1982c: 62), because the "state of dissociation [between appearance and reality] does not always present itself with equal distinctiveness," we can trust neither simple sense perception nor popular prejudice. Given that the form of social facts can stay the same while their content varies, "events do not take on the same appearance each time nor from one moment to another...[and because of this] fluctuating character...social life consists of free ranging forces which are in a constant process of change" (1982c: 82). It is the task of method to fix such free ranging forces, if even temporarily, within thought so that they may be studied systematically. However, if the conditions required to articulate an object of study are lacking, then sound knowledge cannot be a simple reflex of experience because without "sufficiently strong perception of the details [it is not possible] to feel the reality behind them" (1982c: 63). Consequently, the conditions upon which a science rests must mature before it can be presented:

> When research is only just beginning and the facts have not yet been submitted to any analysis, their sole ascertainable characteristics are those sufficiently external to be immediately apparent. Those less *apparent* are doubtless more *essential*. Their explanatory value is greater, but they remain unknown at this stage of scientific knowledge and cannot be visualised save by substituting for reality some conception of mind (1982c: 75; emphases added).

Thus, because of its changing character, the principles of social reality "can only be established when the science has been worked out" (1982c: 64). Once the science has been worked out, the resulting knowledge is often out of sync with the "official" beliefs of society – where there "are so many idols, as Bacon said, from which we must free ourselves" (Durkheim 1982b: 246). A social scientist therefore must not be one for whom "Any opinion which is embarrassing is treated as hostile.... [C]old, dry analysis is repugnant to certain minds" (Durkheim 1982c: 73).

Simmel (1964: 18) also starts with surface images (i.e., appearances) as a place to begin but he does not remain there because "it is a characteristic of the human mind to be capable of erecting solid structures, while their foundation are still insecure." Social science develops with halting progress to the extent its objects exist in a rapidly changing evolutionary stage as opposed to stages that congeal and stabilize over time. Thus, we must treat appearances skeptically because "reality...is given to us as a complex of images, as a surface of contiguous phenomena...which is our only truly primary datum" (1964: 8). To remain at this level, however, is to be bound by surface appearances, a trait of conventional thought that injects "into reality, an *ex-post facto* intellectual *transformation* of the immediately given reality. Because of constant habit, we achieve this almost automatically...[and] almost think it is no transformation at all, but something given in the natural order of things" (1964: 8; emphases in the original). The sociologist, that is, is under no obligation to treat appearances as the natural order of things.

Simmel and Durkheim's positions can be summed up in four propositions: (1) There is often a disjuncture between social life's appearance and how its essential mechanisms in fact function; (2) A social science cannot be presented until it is worked out somewhat; (3) Social science should critique forms of popular and official knowledge; and, (4) A social science should be unconcerned about its political consequences. Marx agreed with all four propositions.

Recognizing that scientific inquiry is often in tension with cultural knowledge in a society, Marx (1992a: 18–19) warned readers that *Capital* does "presuppose, of course, a reader who is willing to learn something new and therefore think for himself." Approaches (e.g., Hegel's idealism) that "glorify the existing state of things" (Marx 1992c: 29) are antithetical to science's task. Thus a historically materialist dialectic is an "abomination to bourgeoisdom and its doctrinaire professors, because [its grasp of historical fluidity is] in its essence critical and revolutionary" (Marx 1992c: 29). Cold dry analysis is indeed repugnant to certain minds: "In the domain of Political Economy, free scientific inquiry meets not merely the same enemies as in all other domains. The peculiar nature of the material it deals with, summons as foes into the field of battle the most violent, mean and malignant passions of the human beast, the Furies of private interest" (Marx 1992a: 21). Through appealing to scientific research, thinking for oneself, and learning some difficult truths one can "raise the veil [of capital] just enough to let us catch a glimpse of the Medusa head behind it," though, in the face of such knowledge, Marx (1992a: 20) tells us that we must resist the urge to "draw the magic cap down over eyes and ears as a make-believe that there are no monsters." Guarding against undue optimism and being careful when drawing conclusions are the scientific orders of the day because social improvements in one region do "not signify that tomorrow a miracle will happen" (1992a: 21). Marx finds medusas and monsters, Durkheim ghosts and idols, and for both the power of each is strengthened through illusions found in the appearances social knowledge presents to consciousness.

Just as Durkheim held that observations of a physical phenomenon are the basis upon which everyday concepts are based, Marx (1992c: 29) argues that conventional knowledge should be understood as "the material world" expressed in "forms of thought." As social conditions develop and mature, a "simpler category may have existed historically before the more concrete, [but] it can achieve its full (intensive and extensive) development precisely in a combined form of society" (Marx 1973: 103). Marx (1992c: 24), however, accuses the political economists of his period of misconceiving what capitalism has brought into being for "the starting-point of...investigations." Because the study of political economy "can remain a science only so long as the class-struggle is latent or manifests itself only in isolated and sporadic phenomena," when "the capitalist regime is looked upon as the absolutely final form of social production, instead of a passing historical phase of its evolution," a researcher is guilty of "naïvely taking this antagonism for a social law of Nature" and thus their science has reached "the limits beyond which it [can] not pass" (1992c: 24). This was true of *Capital*'s reception by its professorial audience in Germany where...

> ...the soil whence Political Economy springs was wanting [and was] turned, in their hands, into a collection of dogmas, interpreted by them in terms of the petty trading world around them, and therefore misinterpreted...[with the result that in] place of disinterested inquirers, there were hired prize-fighters; in place of genuine scientific research, the bad conscience and the evil intent of apologetic (1992c: 23, 25).

For approaches that offer their services to the powers that be, it is "no longer a question, whether this theorem or that [is] true, but whether it [is] useful to capital or harmful, expedient or inexpedient, politically dangerous or not" (1992c: 25). Marx (1983h: 374) thus voiced a dedication "to pursue my object through thick and thin and not allow bourgeois society to turn me into a money-making machine." Like Simmel and Durkheim's, Marx's scientific practice was an objective and critical enterprise with implications for everyday action.

Inductive Analysis: Examining Nonmetaphysical Supra-Individual Realities

Durkheim (1982c: 63–64) complains that "up to now sociology has dealt more or less exclusively not with things, but with concepts." In Spencer's sociology, for instance, all "that is really essential in his doctrine can be directly deduced from his definition of society" (1982c: 65). The problem with such approaches is that...

> Instead of observing, describing, and comparing things, we are content to reflect upon our ideas, analysing and combining them. Instead of a science which deals with realities, we carry out no more than an ideological analysis. Certainly this analysis does not rule out all observation. We can appeal to the facts to corroborate these notions or the conclusions drawn from them. But then the facts intervene only secondarily, as examples or confirmatory proof. Thus they are not the subject matter of the science, which therefore proceeds from ideas to things, and not from things to ideas.... It is clear that this method cannot yield objective results (1982c: 60–61).

Beginning with concepts is to start with "a veil interposed between things and ourselves, concealing them from us even more effectively because we believe them to be more transparent" (1982c: 61). "We do not know *a priori* what ideas give rise to the various currents into which social life divides, nor whether they exist" (1982c: 70). Therefore, "however changeable that life may be, we have no right to postulate *a priori* its comprehensibility" (1982c: 83). Thus, "in order to be objective science must start from sense-perceptions and not from concepts that have been formed independently from it" (1982c: 81). A sociologist starts a study of social facts "by adopting the principle that he is in complete ignorance of what they are, and that the properties characteristic of them are totally unknown to him, as are the causes upon which these latter depend" (Durkheim 1982b: 246). With the "conventional character of a practice or an institution...never...assumed in advance" (Durkheim 1982c: 70), inquiry advances by "proceeding empirically...to describe and explain things" using "systematic interpretation" (1982c: 65, 74).

Simmel (1964: 21) also held that sociology's principles must be "ascertained inductively." He offers a method that "isolates [sociation] inductively...from the heterogeneity of its contents and purposes, which, in themselves, are not societal" (1964: 22), a condition that makes the sociologist similar to a linguist, both of whom "isolate the pure form" of their objects of study. For him, then, the sociological method is "essentially like induction" (1964: 13).

Reflecting similar commitments, Marx (1992c: 26) dismisses critics who accuse him of "mere critical analysis of actual facts" ("imagine!" he retorts). In his view, research needs, "by rigid scientific investigation, to establish...the facts that serve [as] fundamental starting points" (Marx 1992d*: 27). Marx (1985l: 29) asserts that Proudhon "and the utopians are hunting for a so-called '*science*' by means of which a formula for the 'solution of the social question' is to be devised *a priori*, instead of deriving science from a critical knowledge of the historical movement." Thus, Marx's method is an inductive one as well.

Durkheim (1982c: 55), who believed that the intensity of a social current "varies according to the time and country in which they occur," recognized time and space variables combining individuals, societies, and history, i.e., "forms of life at varying stages of crystallization" (1982c: 58). Social forms and their stages of crystallization raise an analytical problem. Social facts are not reducible to individuals because "the forms that these collective states may assume when they are 'refracted' through individuals are things of a different kind" (1982c: 54). Social facts, for him, "consist of representations and actions [but should not] be confused with organic phenomena, nor with psychical phenomena" (1982c: 52). Durkheim (1982c: 74) claims he "cannot protest too strongly against this mystical doctrine." For him, therefore, social facts "must be studied from the outside, as external things" (1982c: 70). A study of social facts – or that which "constitutes...beliefs, tendencies and practices of the group taken collectively," where individuals are abstracted as "only the representations and intermediaries" (1982c: 54) – should not be reduced to individuals, i.e., "Far from being a product of our will, [social facts] determine it from without" (1982c: 70).

Durkheim (1982c: 74) believed that "all mysticism...is...the negation of all science," especially appeals to "transcendental" forces. If it is true that "most of our ideas and tendencies are not developed by ourselves...[then social] phenomena must therefore be considered in themselves, detached from the conscious beings who form their own mental representations of them" (1982c: 52, 70). In studying these detached realities, a "social fact is identifiable through the power of external coercion which it exerts or is capable of exerting upon individuals" (1982c: 56). However, Durkheim (1982c: 54) felt that the coercion of social facts "in time ceases to be felt...because it gives rise to habits, to inner tendencies which render it superfluous; but they supplant the constraint only because they are derived from it." Nevertheless, the expression of social facts by the individual will vary according to their "psychical and organic constitution" and "on the particular circumstances in which he is placed" (1982c: 55–56). Ideal superstructures – not rooted in individual consciousness nor possessing a metaphysical (or "transcendental") reality – only arise from social relationships. Scientific knowledge "must therefore create new concepts and to do so must lay aside common notions and words used to express them, returning to observations, the basic material for all concepts" (1982c: 81).

Although he had a more sympathetic view of Kant than did Marx (see Simmel 1977, 1980), and maybe even Durkheim, Simmel (1964: 9–10, 12) also believed that there are "two insufficient alternatives" – the "contents of historical life" as "inventions of individuals" or the explanation of social phenomena by resorting to "transcendental forces." Simmel (1964: 4) rejects outlooks that hold human existence "is an exclusive attribute of individuals, their qualities and experiences," such as viewing "language, religion, the formation of states, material culture...as inventions of single individuals" (1964: 12). The complexity of social life "requires its special investigation [of] innumerable influences of the physical, cultural, personal environment – influences that come from everywhere and extend infinitely in time" (1964: 5–6). "Sociology asks what happens to men and by what rules they behave...insofar as they form groups and are determined by their group existence because of interaction" (1964: 11). Consequently, "if we examine 'individuals' more closely, we realize that they are by no means such ultimate elements or 'atoms' of the human world...the individual appears as a composite of single qualities, and destinies, forces and historical derivations" (1964: 5–7). Sociological variables are always *supra-individual* if they are, strictly speaking, properly sociological.

Marx shared this belief that society is made of multiple levels of complexity, that this required a study of the historically empirical concrete, a rejection of individualist-reductionist models, and an appeal to the analysis of supra-individual realities.

In a historical materialist analysis, if "a critical inquiry whose subject matter is civilization [cannot] have for its basis any form of, or any result of, consciousness," then "not the idea, but the material phenomena alone can serve as its starting point" (Marx 1992d*: 27). Marx's (1992a: 19) study of the "capitalist mode of production" (the abstract) is predicated on examining "the conditions...corresponding to that mode" (the concrete). In *Capital*, Marx abstracts the system into parts that make up its most important relations, working his way from commodities – that which might appear as "minutiæ" – to exchange, then to industry, and then the historical development of the system as a whole. Modernity's social relations include its roots in feudalism and the structure of the capitalist mode of production, prior religious traditions, and its ever-increasing multicultural, international character. Thus, Marx (1992a: 20) warns that "[a]longside of modern evils" we are also forced to confront "a whole series of inherited evils [that] oppress us," evils that arise "from the passive survival of antiquated modes of production [and] their inevitable train of social and political anachronisms." Just as Simmel (1964: 13) explains that "the production of phenomena through social life...occurs [through] the simultaneity of interacting individuals...[and] the succession of generations," Marx (1992a: 20) also reminds us that "We suffer not only from the living, but from the dead."

As opposed to Adam Smith's depiction of capitalism as something reducible to the collective actions of abstract individuals (*homo economicus*), Marx emphasizes capitalism's historical origins from a viewpoint that prioritizes structural relationships. Adopting a notion of supra-individual realities that prefaces Durkheim's, Marx (1975g: 353–354) explains that we should "recognise the powerful influence of general *conditions* on the *will* of the acting persons...the *factual embodiment* and *obvious manifestation* of...*general* conditions.... [T]hese conditions were...the *general, invisible* and *compelling* forces of that period...and were bound to be manifested in facts and *expressed* in separate actions which had the *semblence* of being arbitrary" (emphases in the original). Marx assumes that social change occurs "whether men believe or do not believe it, whether they are conscious or unconscious of it" and this "standpoint...can less than any other make the individual responsible for relations whose creature he socially remains" (Marx 1992a: 21). With the relations in which they are found as the object of study, in political economy individuals "are dealt with only is so far they are personifications of economic categories, embodiments of particular class-relations and class-interests" (1992a: 21). Reductionist analysis enters a game already started, overlooking how the system in which activity occurs existed before the individuals contained within it. Like Durkheim, Marx (1992d*: 27) examines social life *to the extent* it is "independent of human will, consciousness and intelligence" because social life "determin[es] that will, consciousness and intelligence."

The Method of Abstraction and the Search for Sociological Laws

Durkheim (1982c: 55) notes that sociological analysis can be facilitated "with the help of certain methodological devices." For example, "the sociologist...must strive to consider [social facts] from a viewpoint where they present themselves in isolation from their individual manifestation" (1982c: 82–83). Because of the interplay of history and social structures, "Some instrument of analysis is necessary in order to render [institutions] visible. It is history which plays...a role analogous to that of the microscope in the order of physical realities" (Durkheim 1982b: 245–246). In order to do this, "an effort of abstraction is necessary" (Durkheim 1982c: 71). The aim of the method of abstraction is not simply inventing concepts but "discovering laws of reality" (1982c: 61). These are sociological laws and not putative universal Truths such as those found in "the quest for the philosopher's stone" (1982c: 62). Durkheim's (1982b: 245) approach thus "opens up vistas for the future discovery of the laws of social evolution," though "the outcome cannot be anticipated" (Durkheim 1982c: 70).

Simmel (1964: 11) also held that the "topics of [sociology's] researches certainly arise in a process of abstraction." Simmel (1964: 18) assumes that because of "the totality of human existence," "Cognition cannot grasp reality in its total immediacy" (1964: 17). This assumption demands a methodological principle. "This principle is the abstraction, from a given complex of phenomena, of a number of heterogeneous objects of cognition that are nevertheless recognized as equally definitive and consistent" (1964: 7). Starting with the whole/part relationship, the totality of interest is abstracted into manageable parts. However, data "conceived under a specific category" can only refer to "that character of one-sided abstraction that no science can get rid of" (1964: 17, 16). Abstractions are thus always qualified: "Under the guidance of its particular conception, any science extracts only one group or aspect out of the totality or experienced immediacy of phenomena. Sociology does so too" (1964: 11). Simmel's (1964: 15) goal is to uncover "the general law that is valid 'on the whole'," where "large systems and the super-individual organizations...crystallize [and] attain their own existence and their own laws" (1964: 10). Thus, because of sociological laws, "No human wish or practice can take arbitrary steps, jump arbitrary distances, perform arbitrary synthesis. They must follow the intrinsic logic of things" (1964: 16–17).

Given that social inquiry is afforded possession of "neither microscopes nor chemical reagents," Marx (1992a: 19) also concluded that we must replace both with the "force of abstraction." Successfully used, abstraction helps uncover laws. In political economy, for instance, "it is not a question of the higher or lower degree of development" of capitalism as a whole, but rather it "is a question of these laws themselves" (1992a: 19). Though a system is "governed by laws" (Marx 1992d*: 27), method must account for the necessary and contingent factors which "can shorten and lessen the birth pangs" of its development (Marx 1992a: 20). In later chapters, we will look more closely at the nature of Marx's laws and his use of them.

Marx's Differences from Classical Sociological Theory

For this discussion, the divergences between these thinkers that stand out relate to the implications of abstractions carved at the level of "society" in general, historical explanation and the problem of teleology, the commitment to materialism, and the issue of historical progress.

Sociology's unit of analysis is perhaps the most important difference between these great thinkers. For Simmel (1964: 11–12), "Sociology…is founded upon an abstraction from concrete reality, performed under the guidance of the concept of society." The same could be said for Durkheim. After 1844–1848 or so, Marx uses the term "society" less and less as a coverall and increasingly understands that history (whole) is made up of a variety of social systems (parts) and the structures and the laws unique to them (parts of parts). Though his analysis of modernity is less a "general sociology" and more a "political economy," the methodological principles (DM, HM) Marx brings to political economy can animate broader sociological studies. For Marx, once established, capitalism's nature is to transform the world and therefore not to provide it a special place in analysis is to distort our understanding of what makes modernity unique. This is not to say that Marx fails to note realities occurring at the level of society in general, it is just that these should not be conflated with what which happens within specific systems. Simmel's "forms of sociation" and Durkheim's "social facts" are carved at the level of the general abstraction "society." Pitching abstractions based on observations made possible by capitalist relations threatens to obscure and transform the problems inherent to it into the seemingly intractable (and ahistorical) problems of society in general.

Teleology is a form of reasoning that places effect prior to a cause, arguing that the functional role of a structure is/was the reason for its development. Correspondence between Marx and Engels reveals an awareness of this issue, especially on the problems of historical application of evolutionary models (Gerratana 1973, Ball 1979). Durkheim does not reveal a similar awareness. For example, he at first rightly argues that "It is only after we have traced the currents back to their source that we will know from where they spring" (Durkheim 1982c: 70). He seems to believe, however, that we can use abstract categories based on empirical data gleaned from the present (e.g., division of labor, religion, crime) and read them back into history with the same explanatory validity, i.e., in the earliest representations of a social phenomenon we can find the clues of their ultimate sociological function. For example, Durkheim goes back to the earliest forms of "the division of labor" to extract their general functions (solidarity) that can then be applied to understanding the present (mechanical versus organic). Though the form may be different, the ultimate function is the same (i.e., social integration) for Durkheim.

Marx makes claims that invert this logic, making his views seem somewhat paradoxical or even counterfactual. For him, the division of labor and labor itself are products of modern society, i.e., these forms have historically developed and crystallized to such a degree that they can be observed in a mature form, at which point a concept can be constructed upon them. "It is therefore necessary to watch the present course of things until their maturity before you can 'consume' them 'productively', I mean *'theoretically'*" (Marx 1991c: 354; emphasis in the original). In applying a concept made possible by modern development it does not necessarily follow that the same concept applies equally to earlier periods when the phenomenon was in an embryonic form. Labor is one of Marx's (1973: 105) central examples: "This example of labour shows strikingly how even the most abstract categories, despite their validity – precisely because of their abstractness – for all epochs, are nevertheless, in the specific character of this abstraction, themselves likewise a product of historic relations, and possess their full validity only for and within these relations."

Although Durkheim seems to recognize that material conditions make certain concepts possible, as we saw earlier, he does not seem to see that such a concept cannot be read backward into history with the same sociological meaning. If this happens, the result is a teleology of the present. Here is a teleological Durkheim (1982b: 245–246):

> [A]ny institution being considered has been formed piecemeal. The parts which constitute it have arisen in succession. Thus it is sufficient to follow its genesis over a period of time, in the course of history, in order to perceive in isolation, and naturally, the various elements from which it results.... [History] not only distinguishes these elements for us, but is the sole means of enabling us to account for them. *This is because to explain them is to demonstrate what causes them and what are the reasons for their existence. But how can they be discovered save by going back to the time when these causes and reasons operated?* That time lies behind us. The sole means of getting to know how each of these elements arose is to wait upon their birth. But the birth occurred in the past, and can consequently only be known through the mediation of history (emphasis added).

Because his abstractions are carved at the level of society, Durkheim has little choice but to see contemporary social facts as representations sociological universals in terms of form, if not content. He believed, for instance, that we could discover the nature of modern religion by tracing it back to its "elementary forms" (Durkheim 1968). However, to account for the presence a social structure or institution by locating its ultimate function within its historical origins is not necessarily justified. Origins and current functions do not necessarily correlate with each other and we cannot assume that broad abstract similarity in form in one epoch compared to another translates into commensurable sociological realities and/or laws that can be validly generalized.

Although that Marx was a committed materialist should not need recounting, Durkheim – with an ontology about "social life, [where] everything consists of representations, ideas and sentiments" (1982b: 247) – is not seldom read as an equally committed idealist. This interpretation is confounded by Durkheim's rule that "states of consciousness can and must be studied externally and not from the perspective of the individual consciousness which experiences them" (1982c: 71). Individualistic reductionism is rejected because "By their very nature social facts tend to form outside the consciousness of individuals, since they dominate them" (1982c: 72) and, therefore, "private manifestations [and] socio-psychical phenomena ... are of interest to the sociologist without constituting the immediate content of sociology" (1982c: 56). If individual consciousness is not the agent for Durkheim, then *social* relationships must be our subject matter. For Marx, labor, production, exchange, and consumption are vital for the existence of individuals influenced by representations, ideas, and sentiments and are inherently social in nature. In fact, they are the primary – though not only – relations we must strive to understand. With economic determinism being the common interpretation of Marx at the time, it is understandable that Durkheim did not identify himself as a materialist and his essays are ambiguous on the priority of material relations for research. Still, sociologists often read idealism into Durkheim's methods, though whether he was a strict idealist is by no means certain.

Even though he allowed for an analysis of class struggle, Simmel (1964: 16) also held that historical-material phenomena could be the result of something else:

> And if economics seems to determine all the other areas of culture, the truth behind this tempting appearance would seem to be that it itself is deter-

mined – determined by sociological shifts which similarly shape all other cultural phenomena. Thus, the form of economics, too, is merely a 'superstructure' on top of the conditions and transformations in the purely sociological structure. And this sociological structure is the ultimate historical element which is bound to determine all other contents of life, even if in a certain parallelism with economics.

For Marx's project, the capitalist system directly causes the misery of the proletariat. Its origins, its structure, and its ordinary functioning can be understood, must be understood, if this oppression is to be overturned. A nonmaterialist sociology is likely to obscure and mystify the research needed for this project. For Simmel, this political program, even if laudable, is not itself an issue for sociology and he goes so far as to entertain the possibility that there are other, deeper structures that need to be understood but his essay does not venture as to what these might be.

As we saw in Chapter One, Marx did not view history as a linear evolution. His assumptions about historical progress were qualified and contingent. He did, nevertheless, hold out for the possibility of progress, given the outcome of class struggles. Further, he believed the laws internal to the capitalist structure were "tendencies working with iron necessity toward inevitable results" (Marx 1992a: 19). From his essay, it is unclear what Marx is referring to here, capitalism's dissolution or the emergence of a future communist society? Although one could argue for either or both cases, there is definitely more evidence – given the material found within *Capital* – that the inevitable results to which Marx alludes have to do with outcomes internal to the logic of capitalism, though these too have an internal relation with what is to come afterward. Moreover, Marx's acceptance of capitalism's inherent progress does not extend to a general theory of history. In any case, it is clear that the same beliefs did not hold true for Durkheim. For him, "the progress of humanity does not exist.... The succession of societies cannot be represented by a geometrical line; on the contrary, it resembles a tree whose branches grow in divergent directions" (Durkheim 1982c: 64, arguing against Comte). Although at first this seems to be a similarity with Marx's outlook – i.e., Engels's depiction of history as moving in "leaps and bounds" and "zigzags" (see Chapter One) – the difference here is that Marx held that capitalism is both inherently progressive over feudalism and contains the necessary conditions for eliminating class antagonisms, which, as Engels (1978b: 472) asserted, will result in "forever freeing the whole of society from exploitation, oppression, and class struggles." Durkheim makes no such claims. Simmel's essay is agnostic on this question.

Re-reading the classics often challenges our assumptions about the theorists and theories found there. For example, only in the late 20th century was the Marx-Weber relationship interpreted as a positive dialogue rather than as a critical engagement (Antonio and Glassman 1985, Wiley 1987). The typical stance toward the Marx-Durkheim relationship casts them as founders of incommensurable paradigms, a conclusion I challenge. Even politically, they shared a goal for sociology. Like Marx's (1978b: 145) 11th Thesis on Feuerbach, which claims "the point" is not just to interpret the world but "to *change* it" (emphasis in the original), Durkheim (1982b: 245) claims that "the purpose of sociology is to enable us to understand present-day social institutions so that we may have some perception of what they are destined to become and what we should want them to become." Either Marx was not so far outside the mainstream of sociology as is often supposed, or Durkheim was closer to Marx than is usually thought.

Conventional wisdom about Marx's relationship to the overall sociological project – i.e., as naïve radical, as reductionist, as someone outside of conventional sociological thinking – does not hold. Nor can the assumption hold that the sociology advocated by Durkheim and/or Simmel was something thoroughly different than Marx's. Finally, if these three thinkers shared so much common ground, and if the differences in Marx's thought pointed out here are valid in their concerns, then this can only mean that sociology as a whole suffers the further it moves from Marx's approach. For example, by conflating observations made at the level of modern society (something intertwined with the history of capitalism) with observations about processes occurring at the level of society in general, it becomes nearly impossible to create the sufficient and necessary concepts that capture the essential social relations that account for observations made about the present. Marx promises us a way to handle this issue, as we will later see.

Thinkers in the classical tradition had to position themselves toward Marx one way or another and they set the terms of debate for later scholars. Giddens (1971: 243–244) tells us that mainstream sociology has assumed that Durkheim and Weber "represent a more or less direct partisan defense of liberal bourgeois society in the face of the Marxist challenge," though any such outlook must ignore the fact that "Marx's writings share a good deal more in common with those of Durkheim and Weber than was apparent to either of the latter two authors." The same is true for Marx's relationship to Simmel. Durkheim, Simmel, and Marx were in agreement on the priority of scientific commitments and values, that individuals and social structures are internally related, that the use of the abstractive method was necessary to handle the relationship between appearance and essence and for the search for sociological laws, that induction is *the* method to uncover these laws, and that non-metaphysical supra-individual realities exist. If Marx's approach shares this epistemological space with classical theory, then rather than gadfly or simply (and only) as revolutionary, Marx's orbit of inquiry navigates around questions and approaches shared with Durkheim and Simmel. By extension, if one's goal is to understand life within capitalist society, then, through his focus on science, materialism, and political economy, Marx gives the sociologist tools he or she must have. To better understand these tools, an explanation of the ontological and epistemological assumptions upon which Marx's work rests is taken up in the next chapter.

Part II: Marx and His Scientific Dialectic

Marx's work is often viewed as a monolithic whole. This is misleading. He came to his ideas about society, history, and method over time, with discoveries about each shaping his work anew. This dynamic between continuity and discontinuity can raise analytical problems for the student. One strategy is to trace his ideas as they unfold across his life, which is useful only to a point. Another strategy is to see if an overall picture can be extracted from his work as a whole but using its mature form as an analytical point of view. Starting from the mature whole and looking backward, we find Marx retained his commitments to scientific and dialectical reason, that his materialist principles evolved over time, and that his studies in political economy came later but became immensely important to him. He committed to the communist project as a young man and it remained his polestar throughout his life.

To understand why Marx thought his approach to social and political questions fulfilled the requirements – and thus deserved the moniker – of science, we must break it apart. Marx's thought was both thoroughly dialectical and yet deeply entrenched in scientific principles. Rather than break it along an axis determined by dialectical terms and concepts, it is more useful to break it up through scientific categories. Scientific thinking works with ontological assumptions, epistemological principles, methods of analysis and conceptual development, and the creation of descriptive and explanatory models. Grasping Marx's work in this manner leaves his dialectic embedded but brings out his scientific moments. Thus we will discover and uncover his scientific dialectic.

Chapter Three
Marx's Onto-Epistemological Assumptions

In this chapter I investigate the ontological and epistemological assumptions that inform Marx's selection of empirical domains and his approach to studying them.[1]

Internal Relations and the Force of Abstraction

A philosophy of internal relations animates Marx's outlook (Ollman 1976, 1979a, 1993, 2003, Sayer 1987, Gould 1978). At a broad level, this philosophy assumes that everything is connected with everything else. A more refined assumption that social phenomena are the products of both historical processes and social structures also underpins this philosophy. This view has several implications. First, ontologically, history and structure are interior to each other; historical change becomes an inherent property in structural relations, just as structural relations are a central factor in historical development. Second, epistemologically, it follows that knowledge of history is requisite for knowledge of social structure (and vice versa). And therefore, third, abstractions carved at the level

---

[1] Several arguments in this chapter can be found elsewhere (see Paolucci 2001a, 2005).

of "society" can obscure such temporal-structural considerations and distort our knowledge of them.

These assumptions guide Marx's research practices. Modern society is understood as a conglomeration of changing phenomena accompanying a structure that defines it. Its periods of transformation are understood as products of real institutional processes – often in antagonistic relationships with one another – contained in that structure. Parts of the social system are thus interiorized into the whole in their mutual relations. However, though history and structure are ontologically related, epistemologically, abstraction must pull them apart and, albeit temporarily, treat them as distinct. Empirical research begins here. In historical analysis, both necessary structural relations and contingent historical events are important; in structural analysis, regular social phenomena are examined in terms of the institutional relations interior to various social systems along with the limitations and determinations making up their regular practices. Here, contingent phenomena are abstracted out of view, only to be brought back in later (more below and in chapters Four through Six).

These three issues – an ontology of society, a vision of epistemology, and how to relate one to the other – overlap in Marx's abstractive practices. Beginning with an "organic whole," the broadest units of analysis are assumed to be large enough to require abstracting out of it in order to grasp any part or parts of it, or any more general properties it may possess. We look at, conceive, and examine parts of totalities, questioning the extent to which this or that part is an essential element of the system under examination. Towards this end, each part is initially abstracted in relative isolation from each of the others. As research continues, parts are re-abstracted in their relations to other parts – as they are related to, function with, or against, each other. Sensitivity to emergent qualities of a formation is necessary to avoid reifying them into static frameworks and/or to avoid confusing transitory facts with essential characteristics.

Abstraction is a way focusing on one set of qualities, while ignoring others, much like relationships between vision, the retina, and the brain. Our eyes pick up more information than we can absorb, so the brain sorts out those things of greater importance. In producing knowledge, we often do this intentionally with agency with our minds. For example, music, poetry, cookbooks, gardening manuals, and astrology are all methods of abstracting. Their variable utility does not negate the fact that each is a product of abstraction. What makes an abstractive method scientific is its empirical, systematic, and methodological character. What identifies social-scientific abstraction is the way ontological assumptions (e.g., the existence of supra-individual level realities) translate into epistemological practices. What makes Marx's social-scientific abstraction unique is that the philosophy of internal relations extends itself into his epistemology, making his ontology and epistemology intermesh, or what I call an "onto-epistemology." His first step in moving from ontology to epistemology was grasping social systems as "totalities" and then abstracting from them in empirical research.

All science abstracts parts out from a "totality" for attention while ignoring other parts of the same. Marx used totality as a composite of interrelationships where everything takes part in the meaning of everything else, though the whole is too complex to grasp conceptually at once. Several interpretations of this usage have followed. Louis Althusser (1969, 1971), attributing a one-sided structuralist position to Marx, depicted totalities as existing over and above individual agency. Bertell Ollman (2003: 72) tells us that, in the internal relations approach, totality "is a logical construct that refers to the way the whole is present through internal relations in each of its parts." Also in this tradition, Carol Gould (1978: xx–xxi) explains that "Marx begins from what he calls a 'concrete whole,' that is, a given and complex subject matter, of which we only have an amorphous conception.... He then proceeds to analyze the concrete whole to discover fundamental principles or conceptual abstractions from which one can derive a comprehension of its workings and the interrelations within it." Evidence supports the approach of Ollman and Gould. For example, here are Marx and Engels (1976: 53–54) breaking down a totality:

> This conception of history thus relies on expounding the real process of production – starting from the material production of life itself – and comprehending the form of intercourse connected with and created by this mode of production, i.e., civil society in various stages, as the basis of all history; describing it in its action as the state, and also explaining how all the different theoretical products and forms of consciousness, religion, philosophy, morality, etc., etc., arise from it, and tracing the process of their formation from that basis; thus the whole thing can, of course, be depicted in its

totality (and therefore, too, the reciprocal action of these various sides on one another). It has not, like the idealist view of history, to look for a category in every period, but remains constantly on the real ground of history; it does not explain practice from the idea but explains the formation of ideas from material practice.

This approach allows Marx to begin with a construct large enough so that when he examines anything within the totality it is unnecessary for him to appeal to explanatory phenomena outside the abstractions the totality allows him to carve and the material available there for observation. His task is not to explain all of this totality nor does recourse to totality explain anything. Marx's more modest goal is an examination of a totality's parts and their interrelations. These interrelations change the qualities of that same totality over time. The assumption of totality thus accepts that no final conclusion for all time, place, and circumstance is possible or necessary. This does not, however, render a science of society impossible.

Naturalism

Marx's approach to uniting his ontological and epistemological assumptions is often understood as a form of "naturalism." In his dissertation's discussion of philosophical approaches to nature, Marx (1975c: 68) argues that "Our life does not need speculation and empty hypotheses" and that "the study of nature cannot be pursued in accordance with empty axioms and laws." The assumption that natural science must be grounded in observations of the real is the same assumption to be held for studying social relations:

> If abstract-individual self-consciousness is posited as an absolute principle, then, all true and real science is done away with in as much as individuality does not rule within the nature of things themselves. But then, too, everything collapses that is transcendentally related to human consciousness and therefore belongs to the imaginative mind. On the other hand, if that self-consciousness which knows itself only in the form of abstract universality is raised to an absolute principle, then the door is opened wide to superstitions and unfree mysticism.... Abstract-universal self-consciousness has, indeed, the intrinsic urge to affirm itself in the things themselves in which it can only affirm itself by negating them (Marx 1975c: 72–73).

Understanding human life through a projection of an abstract universal individual apart from nature obscures our fundamental connection to nature as animals and a species. "*Man* is the immediate object of natural science.... The *social* reality of nature, and *human* natural science, or the *natural science about man*, are identical terms" (Marx 1988a: 111; emphasis in the original). In this onto-epistemology, "The first premise of all human history is, of course, the existence of living human beings. Thus the first fact to be established is the physical organization of these individuals and their consequent relation to the rest of nature" (Marx and Engels 1976: 31). Social relations – not human nature – are the grounds for a naturalistic social science.[2]

Social production relies on appropriating nature in a reciprocal relationship with human's biological constitution as a species in their "particular historico-social contexts" (Soper 1979: 78). In such a view, a social scientist "treats the social movement as a process of natural history" (Marx 1992d*: 27). Because of their character as a species, wholly separate onto-epistemologies between natural and social science are unnecessary, even distorting. Distinctions between conditions of possibility and biological limitation make the institutional arrangements in a form of social organization, and how they compel humans to relate to nature and each other, the facts we must grasp. As an epistemological consequence of this ontology, human labor, its products, and struggles over the conditions of both play a central role in Marx's research.

This way of thinking about society and how to study it is a form of naturalism that contrasts with two types of naturalism that should be avoided. First, at the level of historical materialism in general, the "naturalistic mystification" refers to the idea where the limits imposed on individuals and society by ruling class relations are mistaken for limits imposed by human nature

---

[2] "To describe man's social activity, his spiritual, cultural, and intellectual achievements in naturalistic terms, that is, as a natural phenomenon, we have to conceive of the human individual as a social being, always living and acting in co-operation with others. The manifestations of the individual's life are in fact the manifestations of social life; the individual becomes what he is by association and interaction with other individuals. Man's spiritual, cultural, and intellectual advance is a natural process of self-creation accomplished in the course of social evolution. By acting in common with others upon his environment, man not only changes the environment but himself and his social relations with others as well. Thus he makes his own history, although he does not make it as he pleases. History is an objective, natural development, in which nothing is derived from extra-social and extra-historical factors, from an autonomous development of ideological forms, human mind, or objective spirit of which is 'real', 'empirical men', are but the bearers" (Jordan 1967: 300).

(see Mills 1985–86). Second, as applied to political economy, Marx (1847: 120–121) critiqued the standard view of modern economists, in whose eyes "There are only two kinds of institutions...artificial and natural. The institutions of feudalism are artificial institutions, those of the bourgeoisie are natural institutions." Bourgeois ideology assumes that its institutional framework overcomes past human alienation, bringing true freedom into being. History, for all intents and purposes, is *over*. By contrast, Marx sees bourgeois society as the height of alienation. When this alienation is overcome real history can *begin*. By using scientific principles within his historical materialist moments to inform to political economy, Marx believed he could expose the artificialness of capitalist institutions and, at the same time, explain their historical development. To do this, Marx had to abstract out the relationships between the human individual, the species as a whole, and the historical systems that have contained them.

The Relation of Human Individuals to their Species

Three broad ontological assumptions informed Marx's research as he went from developing his historical materialist principles to in-depth political-economic study: one on the status of the human individual, another on the essence of humans as a species, and a final one about social universals.

Eschewing a theory of a universal nature, Marx (1978b: 145) did not find the source of "the human essence" to be an "abstraction inherent in each single individual," but rather it was to be found in "the ensemble of social relations." This ensemble consists of our biological genotype, the form of society in which we live at a particular moment in its development, the family structure dominant in that society, the gender roles to which we are expected to conform, the place in the class structure into which we are born, socialized, passed up, down, and/or across. Our biological capacities are therefore dependent on social relationships for their development. If there is an abstract nature of the human individual it is as a *social* creature and all that this implies. These assumptions see the human animal as a collective *species-being*:

> Man is a species being, not only because in practice and in theory he adopts the species as his object (his own as well as those of other things), but – and this is only another way of expressing it – but also because he treats himself as the actual, living species, because he treats himself as a universal and

therefore a free being.... The universality of man is in practice manifested precisely in the universality which makes all nature his inorganic body – both inasmuch as nature is (1) his direct means of life; and (2) the material, the object, and the instrument of his life-activity.... Conscious life-activity directly distinguishes man from animal life-activity. It is just because of this that he is a species-being (Marx 1988a: 75–76).

Aware of their social relationships, humans construct formal and informal educational practices to pass on their society's knowledge and skill sets to new generations. Thus, our dependence on social relations for the acquisition of social and intellectual skills such as language, sex/gender norms, and laboring roles collectively express our species being, where "*Human nature* is the *true community* of men" (Marx 1975j: 204; emphases in the original).

Species being traits are neither immutable nor infinitely malleable, but have both fixed and relative features. Individual character traits within the species are analogous to the relation of an alphabet to specific organization of words and ideas, i.e., the general structure of alphabets and genes are relatively constant, whereas the actual outcomes in language, norms, and/or actions as squeezed through social relations are variable. The distinction is important and conflating them distorts our knowledge of each, i.e., humans as individuals and as social creatures share general forms while containing different contents. Thus, some desires "exist under all relations, and only change their form and direction under different social relations," whereas other desires originate "solely in a particular society, under particular conditions" (Marx and Engels 1976: 255). Some human wants function to "maintain...life and reproduce it," but with human "development the realm of natural necessity expands, because his wants increase; but at the same time the forces of production increase, by which these wants are satisfied" (Marx 1909: 954). What we call "human nature" is thus hooked into the development of the species as a whole, making it transformative rather than static (see Sayers 1998). The epistemological consequence is that Marx founded his research on few assumptions about human nature.

Marx objected to conventional uses of universal theory – social or individual – as a mode of explanation for two primary reasons. First, for him, "an explanation which does not provide the *differentia specifica* is *no* explanation" (Marx 1976a: 12). Because "generality and explanatory force tend to run in opposite directions" (Ruben 1979: 76), Marx rejects Hegel's tendency toward "atomism" because "A 'view' cannot be concrete when its *subject-matter* is abstract" (Marx 1976a: 79; emphases in the original). Second, "human nature" is a species-wide historical accomplishment, only knowable at the end of social development, not something to be conceptualized beforehand (as is often done in economics, religion, and philosophy). Thus, Marx (1975i: 175) held that all criticism begins with a religious critique because religion "is the *fantastic realisation* of the human essence because the *human essence* has no true reality" (emphases in the original). Humans placed in different modes of production have some powers and capacities that go unrealized because they are either repressed or because the social relations necessary for their expression have not yet come into being. Other species traits emerge over time through social development. Founding social theories on a universal human nature (i.e., an abstraction), therefore, "fails to see how the dominance of certain kinds of behaviour in our society is required by and produced by the prevalent kind of social relations" (Norman 1980: 43). If any social relation played that role for Marx, it was that humans must labor to produce a world in which to live. Because of this "emphasis on constitutive labor... Marx makes use of any universal category in only the most restricted sense" (Thomas 1976: 19–20). Before examining the role "constitutive labor" plays in Marx's onto-epistemology, we must grasp this "restricted sense" in which he uses universals.

Social Universals

Marx (1992j: 80) objected to the tendency for societal knowledge to posit its social relations as "natural, self-understood forms of social life." Thus, he does not generalize the contents of his society as universal but constructs a set of conceptual *forms* applicable across societies in general, i.e., all historical societies have had a mode of production (form), but no specific mode of production (content) has been common to them all. In rejecting the "Natural Individual" as an "illusion... common to each new epoch to this day" (Marx 1973: 83), "Marx not only identifies a set of structural categories, such as production, he also identifies the broad dynamics of social life" (Horvath and Gibson 1984: 14). In this *base-superstructure* model, the *means of production* are abstracted as raw materials and tools and technology (nature transformed by human labor). The *forces of production* include these elements as well as direct producers (peasants, slaves, proletarians). The *relations of production* encompass the direct producers plus nonproducers (chieftains, lords, owners, capitalists) as well as juridical and legal relations (armies, law, police, courts). The entire *mode of production* is the sum of the means, forces, and relations of production. This is *the base*. *Politics* and *the state* center on nonproducers' creation and use of *juridical and legal* relations in pursuit of their interests in the class struggle, broadly construed. *Ideological* discourses are articulated in *religion, art, education,* and *science* (e.g., cosmologies, standards of beauty and abstract meaning in literature, lore). Thus, the *superstructure* is composed of all that is contained in politics, the state, and ideological discourses. Not simply reflections of the base, these are nevertheless shaped, influenced, and, in part, determined by it. "The totality of human society is thus represented by the preceding elements and relations. These comprise, for Marxist scientists, tools with which we can begin to isolate a part of the social whole for analysis without having to sever it from its defining place within the complete system" (Horvath and Gibson 1984: 14–15).

These assumptions assist historical materialism in locating the essential parts in any/all social formation(s) and their mutual interactions and provide for distinguishing between society in general, primitive communalism, class systems in general, and their different types. In this model, the struggles between the classes come in "a different form in a nomadic nation than in a nation of hunters, and in an agricultural nation a form different from that in an industrial nation" (Liebknecht 1901: 49). In the relationship of superstructure to base, "there is a complex...interplay of various processes that are not mechanically and one-sidedly pre-determined" (Mandel 1980: 14), an assumption which "required the introduction of additional assumptions" (Jordan 1967: 315). In interiorizing class dynamics within base-superstructure relations, Marx's interpretation of history thus "proceeds though the strife of opposites that are interdependent and yet in conflict with one another," uses a model of an "organic totality" that includes a "hierarchical structure," "interaction," "growth," "development," and sees a "differentiated and dynamic structure rather than a static unity, or at the opposite extreme, a mere heap or collection" (Rader 1979: xvii–xxi).

For example, social knowledge can be expressed in material practices and humans ideas about them, without a rigid correspondence between them. Marx (1973: 109) asserts in one instance that there is an "uneven development of material production relative to e.g. artistic development." "In the case of the arts, it is well known that certain periods of their flowering are out of all proportion to the general development of society, hence also to the material foundation, the skeletal structure as it were, of its organization" (1973: 110). Although standards of art are superstructural, they do not develop apart from the practices of artists, a laboring activity and thus something that can be abstracted as part of the base (Rose 1984). Further, because of time lags and mediation by other social relations, Marx's assumptions do not see the ideal world as simply a reflection of the material base.[3] Some knowledge is "pure" or unentangled with material conditions. It is in such a context that we should understand Marx's (1985m: 399) speech to the General Council of the International, where he argued that...

> Nothing could be introduced either in primary or higher schools that admitted of party and class interpretation. Only subjects such as the physical sciences, grammar, etc., were fit matter for schools. The rules of grammar, for instance, could not differ, whether explained by a religious Tory or a free thinker. Subjects that admitted different conclusions must be excluded.

Because superstructural forms are not a simple reflection of material conditions, Marx's method "analyzes [base and superstructure] relations and then reintegrates them into the total movement" (Lefebvre 1968: 85, also see Hellman 1979: 145). In sorting out this movement, Marx stressed eight themes: (1) the primacy of material relations, (2) relations of necessity and contingency, (3) the power of ideological discourse, (4) reciprocal, (5) functionally interdependent, (6) contradictory, (7) asymmetrical, and (8) inverting relationships.

The Primacy of Material Conditions. In the dialectical view, different social systems have identifiable configurations of relations, "structures [that] may

---

[3] "Since the superstructure has a certain autonomy of movement, it does not necessarily correspond completely to the base at every historical moment. The parallels between the two reveal themselves over a long historical period. In some cases there may be tension between them. Political relations may change more slowly than the relations of production; law may only adjust slowly to economic changes. On the other hand, scientific developments can precede economic development. When this tension becomes too great in the long term it is relieved, and when divergences are very marked it is relieved by revolution. The solution to the contradictions of material life (i.e. revolution) does not necessarily achieve adequate expression in the social consciousness of the period; but the ideas with which men conceive of the conflicts of their times form a constituent part of the revolutionary upheaval itself" (Jakubowski 1976: 58).

be hierarchically ranked in terms of their explanatory importance" (Bhaskar 1989: 3). According to Marx and Engels (1976: 35), "The production of ideas, of conceptions, of consciousness, is at first directly interwoven with the material activity and the material intercourse of men – the language of real life." For them, "Ideas cannot *carry anything out* at all. In order to carry out ideas men are needed who dispose of a certain practical force" (Marx and Engels 1956: 160; emphasis in the original). Because humans are always born into material conditions, science must "bring out empirically, and without any mystification and speculation, the connection of social and political structure with production" (Marx and Engels 1976: 35). This view is based on the assumption that productive forces, ongoing before individuals, shape social practices individually and structurally:

> Needless to say, man is not free to choose *his productive forces* – upon which his whole history is based – for every productive force is an acquired force, the product of previous activity. Thus the productive forces are the result of man's practical energy, but that energy is in turn circumscribed by the conditions in which man is placed by the productive forces already acquired, by the form of society which exists before him, which he does not create, which is the product of the preceding generation. The simple fact that every succeeding generation finds productive forces acquired by the previous generation and which serve it as the raw material for further production, engenders a history of mankind, which is all the more a history of mankind as man's productive forces, and hence his social relations, have expanded. From this it can only be concluded that the social history of man is never anything else than the history of his individual development, whether he is conscious of this or not. His material relations form the basis of all his relations. These material relations are but the necessary forms in which his material and individual activity is realised (Marx 1982b: 96; emphasis in the original).

Material life (i.e., real practices ongoing in the real world) is where study begins and, among other qualifications, the rule of its primacy remains contingent upon empirical support in the cases examined. The difference in historical systems captured under the concept "mode of production" implies that the number of variables used in analysis, although wide, will be limited. This does not necessitate any sort of reductionism. Positing and locating a limited number of variables – and their interrelations – allows for studying complex social-structural interactions, just as a limited number of letters allows for an almost unlimited number of words.

Necessary and Contingent Relationships. Material conditions and conceptual discourse are "interwoven," marked by necessary and contingent relationships. In the capitalist mode of production, the capitalist class and working class are necessarily related, as are, by extension, capital in general and human labor-power. The degree to which materials, tools, the market, or class dynamics determine specific ideological outcomes is contingent. And, given that it necessarily true that "The tradition of all the dead generations weighs like a mountain on the mind of the living" (Marx 1978d: 595), cultural discourse is crucial in the make-up of any society, though its forms, its modes of transmission, and its ways of change are contingent.

The Power of Ideological Discourse. The layers of reality accounting for the concrete necessitates that "Marx entertains factors other than the economic" (Martin 1998–1999: 515). Although the primacy of material conditions is the general rule, Engels (1936c: 475) asserted, "if somebody twists this into the statement that the economic element is the *only* determining one, he transforms it into a meaningless, abstract and absurd phrase" (emphasis in the original). Marx several times observed instances where ideas lead or dominate social change. In 1856, he wrote to Engels about how "With the Reformation came the translation of the Bible into all the popular Slav dialectics. And thereby of course awakening national consciousness" (Marx 1983c: 21). In the *New York Daily Tribune* (February 14, 1857), he argued that conflicts between Afghans and Persians were based on "political antagonism," "diversity of race," and "religious antagonisms" (Marx 1986a: 178) – not the mode of production. On the influence of religion in the rise of capitalism, Marx (1992j: 262, note 2) presaged Weber: "Protestantism, by changing almost all the traditional holidays into workdays, plays an important part in the genesis of capital." Thus, Marx (1975i: 182) accepted that ideological conditions retain their own causal power, though always within the context of social relations: "The weapon of criticism cannot, of course, replace criticism by weapons, material force must be overthrown by material force; but theory also becomes material force as soon as it has gripped the masses."

Reciprocal Relationships. The powers of determination between material conditions and ideological discourse are often reciprocal. Human agency shapes many forms of ideological discourse and these can react back on material conditions. Engels (1936d: 482) noted that the "low economic development of the prehistoric period is supplemented and also partially conditioned and even caused by the false conceptions of nature. And even though economic necessity was the main driving force of the progressive knowledge of nature and becomes ever more so, it would surely be pedantic to try and find economic causes for all this primitive nonsense." Similarly, capitalist history has had corresponding discourses that have shaped its development, e.g., laissez-faire assumptions, development theory, and/or Keynesian economics. These ideological discourses shaped the policies put in place and regulated the growth of international capital in return. Reciprocity also may occur between material conditions or ideological conditions in themselves. For instance, on the one hand, the invention of mass production allowed for an increase in access to automobiles and this provided for the need for more road building, refineries, and all of this increased the concentration of wealth in industrialists' hands, whereas on the other hand, increased critical discourse on race and ethnic relations has influenced the rise of a wider range of public debate on these issues and others, such as gender roles and sexual mores.

Functional Interdependence. Marx often examined a part, or a special configuration of parts, their interrelations, and their functioning in the whole and/or their role in changing other relations. As a system reproduces itself, structures come to "correspond" to one another as phenomena are "determined" or "conditioned" by other phenomena (Cohen 1986: 221). For example, "The particular commodity, with whose bodily form the equivalent form is thus socially identified, now becomes the money commodity, or serves as money. It becomes the special social function of that commodity, and consequently its social monopoly, to play within the world of commodities the part of the universal equivalent" (Marx 1992j: 74). A system as a whole strives for a certain reproductive consistency through the integration of its structures, whereas other features within their internal relations drive it to change. What emerges, on the one hand, is that superstructural relations tend to correspond to the base by the way "(1) when political or ideological elements appear that are inconsistent with the continuing survival of the economic structure, they will usually be weeded out, and (2) when competing superstructural elements appear that are each compatible with but differentially suitable to the economic structure, the more suitable will generally survive rather than the less suitable" (Little 1986: 56). For example, as living labor, individuals must be fitted into the productive apparatus. Over time, educational standards were increased in order to produce a workforce better adapted to the needs of capital as it became more technologically complex. An educated workforce increased the productivity of capital, the complexity of technology grew again in turn, creating additional needs for intellectual workers. A university system grew and is now relatively institutionalized.

Several notable differences exist between Marx's approach and conventional functional analysis. First, Marx usually analyzes the functioning of a specific mode of production and usually only abstracts at the level of "society" as a point of comparison. Second, although dialectical models present the roles parts play in a system, they do so by viewing them as internally related structures rather than as externally related things. As a result, Marx's (1992j: 201–202) "explanation of the different parts played by the various factors of the labour-process in the formation of the product's value...disclosed the characters of the different functions allotted to the different elements of capital in the process of expanding its own value." Third, dialectical models are therefore not simply aggregated structural relations. The structural parts Marx examines have definite relationships with one another, e.g., "In order that variable capital may perform its function, constant capital must be advanced in proper proportion, a proportion given by the special technical conditions of each labour-process" (1992j: 207). Fourth, Marx's functionalism makes no assumptions about long-term stability. For Marx (1992j: 570), "with a given magnitude of functioning capital, the labour-power, the science, and the land...embodied in it, form elastic powers of capital, allowing it, within certain limits, a field of action independent of its own magnitude." Thus, systems change through the functioning of their parts.

Contradictory Relationships. Contradiction typically refers to (at least) two elements in the same relation whose mutual long-term development are incompatible with each other. Marx's use of contradiction stresses that "things are not indifferent to one another, but rather in interaction and conflict with each other" and that "opposition is not external and accidental...[but] internal to things and part of their nature" (Sayers 1980: 8). For example, if a class system is to reproduce itself, then its conditions of possibility must be satisfied, even to the exclusion, when crucial conflicts arise, of general social needs. This is especially so in the case of capitalism, where the process of capital accumulation undermines the system's democratic potential. So, when Marx (1976a: 31) argues that in politics, "all forms of state have democracy *for* their truth and that they are therefore untrue insofar as they are not democracy" (emphasis in the original), he is noting that the evolution of states has experienced a contradictory existence so long as the history of capitalism, changes in its material base, and corresponding changes in its ideological superstructure head in opposing directions (also see Marx 1978a: 31–32). The necessary oligarchical nature of states amid increasing concentration of wealth and demands for democratization are contradictory tendencies; each cannot survive the full development of the other. Structural relations containing significant contradictions cannot help but change each other as contradictions work themselves out, possibly destroying one or the other side of the relation (or both) in the process. The study of contradiction extends from the dialectical method to historical materialism to political economy in this way.

Asymmetrical Relationships. Many relationships are not equally balanced. Structures often organize and calibrate themselves with respect to each other, especially toward those that are more powerful, influential, or determinant. Modes of production – from the perspective of all *class* societies leading up to the present – have increasingly dominated other social relations. In capitalism's development, the commodification of more things in social life obtains the power of determination with increasing force. For example, capitalism's production and educational systems have mutual connections but the direction of determination tends toward education being shaped more by the market system than the reverse. University education has been increasingly formed toward satisfying the needs of employers and the conditions of the market via granting degrees in such areas as business, advertising, management, economics, engineering, and occupational therapy. Although these disciplines will in turn influence the system of capitalist production – indeed, this is their intended goal – the direction of causality weighs heavier in the direction of the market shaping the educational structure than vice versa.

Inverting Relationships. The concept of inversion appears throughout Marx's writings. In his critique of Hegel, Marx (1975i: 175) tells us that "This state, this society, produce religion, an *inverted world-consciousness*, because they are in an *inverted world*" (emphases in the original). In *The German Ideology*, Marx and Engels (1976: 36) claim that history, class relations, and material conditions tend to shape social knowledge in a way tantamount to a *camera obscura*, i.e., ideological knowledge becomes inverted the same way an eye or a camera lens inverts images. Here, the result is a mystified and thereby misleading form of knowledge whereby determined things are viewed as determinants themselves. For example, an idealist ontology assumes that beliefs, an ethos, and/or motivations primarily account for the outlines of social life. However, as previously noted, Marx (1911: 12) believed that one cannot "judge of such a period of transformation by its own consciousness; on the contrary, this consciousness must rather be explained from the contradictions of material life." In accounting for social action, productive forces require special consideration. What is the relationship between tools, machinery, and the displacement of human labor-power as experienced in the industrial revolution? "To pure mathematician these questions are indifferent, but they become very important when it is a case of proving the connection between the social relations of human beings and the development of these material methods of production" (Marx 1936d: 142). Other forms of inquiry might not find questions about modes of production and their dynamics worthy of consideration, though for Marx, where the inner-connections between history and system are central, these are crucial.

The Centrality of Labor and its Historical Variability

By focusing on the role of human labor in social relations and the struggle over its conditions and products, Marx traced the development of modes of production and the implications these have for the human species, its history, and the individuals its social forms contain. As a knowing subject, the human being always exists within the context of social relationships. Modes of knowing are not separated from modes of being. For Marx (1975f: 392), rather than abstract idealism or a metaphysical theory of nature or history, "correct theory must be made clear and developed within the concrete conditions and on the basis of the existing state of things." After working through his problems about philosophical approaches to nature, materialism, and dialectics, Marx (1978b) sketched out his new approach in his Theses on Feuerbach. Believing that Feuerbach had reversed the relationship between material conditions and the human subject, Marx outlines his take on the status of the individual, the materialist outlook in general, the uses of history in social analysis, and an understanding of praxis.

In Thesis I, Marx (1978b: 143) criticizes Feuerbach's view that "wants sensuous objects, really distinct from the thought-objects" because "he does not conceive human activity itself as *objective* activity" but rather "regards the theoretical attitude as the only genuinely human attitude." Knowledge and truth do not exist solely in the realm of the ideal and abstract; neither superstructural realities nor individuals possess independent existences apart from their material relations. Although Marx (1978b: 144) accepts the premise that "men are products of circumstances and upbringing" and change with those circumstances, he adds that it is "men that change circumstances" and that these realities can and should "be conceived and relationally understood only as revolutionising practice" (Thesis III). For Marx (1978b: 145), history must be brought to materialism, as well as to structural examination, given that "'religious sentiment' is itself a social product, and that the abstract individual whom [Feuerbach] analyses belongs in reality to a particular form of society" (Thesis VII). Because "All mysteries which mislead theory into mysticism find their rational solution in human practice and in the comprehension of this practice" (Marx 1978b: 145, Thesis VIII), social research requires starting not with "an abstract – *isolated* – human individual" (Thesis VI) but rather with what people do in their material and historical contexts.

Because practice is central to our mode of being and both (our practice and our mode of being) are in a necessary relationship to our laboring capacity, the question of labor interiorizes materialism into Marx's dialectic and his conception of history (Arthur 1979, Gould 1978). On what sort of justification does this assumption about the centrality of labor stand? For Marx (1988a: 76), an "animal is immediately identical with its life activity. It does not distinguish itself from it. It is *its life-activity*" (emphasis in the original). The production of material-social life *is what labor is* and its relationship to the human individual is reciprocal: no people, no labor; no labor, no society; no society, no people. Marx (1988a: 79) concludes that the "*alien* being, to whom labor and the produce of labor belongs, in whose service labor is done and for whose benefit the produce of labor is provided, can only be *man* himself" (emphases in the original). He translated such assumptions into onto-epistemological rules. Because humans must produce themselves by their labor in, through, and against nature...

> ...the first premise of all human existence and, therefore, of all history, the premise, namely that men must be in a position to live in order to be able to 'make history.' But life involves before everything else eating and drinking, housing, clothing and various other things. The first historical act is thus the

production of the means to satisfy these needs, the production of material life itself (Marx and Engels 1976: 41–42).

This ontological assumption of labor's centrality justifies the epistemological focus on the material basis of social relations. Although materialism does not require a focus on labor – military apparatuses or technology have been claimed to be leading factors in history – the reverse is true: The assumption that labor is a central social relation immediately invokes the problems of materialism, i.e., How is labor performed? What are its terms, means, and products? What are the legal and social relations between owning classes and laboring ones? How is wealth appropriated through the class structure? What does this mean for political relations? What does this mean for the types of knowledge produced and institutionalized? Are labor's social relations a constant across various modes of production? How does this or that set of class relations help us understand historical development? What are the implications of all these for the individual under capitalism as a class relation? Questions forged under historical materialism inform political economy in this manner.

For Marx (1975i: 182), "To be radical is to grasp the root of the matter. But for man the root is man himself." Because all societies are made by laboring activity, labor is at the root of many other social relations. The ability and need for labor is a mutually transformative mediation between human beings and nature. Because of this internal relation between humans, labor, and history, there is no need for a theory of a timeless abstract "Man" in order to understand the role of labor for individuals, the human species in general, its history, and for the history of capitalism. Rather than a theory of human nature, following the course of labor and its struggles in this way is a more informative and fruitful inquiry, one that treats history as a variable.

What does it mean for history to be a variable? Marx and Engels (1956: 125) argue that "*History* does *nothing*, it 'possesses *no* immense wealth,' it 'wages *no* battles.' It is *man*, real living man, that does all that, that possesses and fights; 'history' is not a person apart, using man as a means for *its own* particular aims; history is *nothing but* the activity of man pursuing his aims" (emphases in the original). Thus, given that "history" as an abstraction cannot be a causal agent (a metaphysic), what is the proper subject of a *historical* materialist study? According to Marx and Engels (1956: 205), "it is the *real human activity* of individuals who are active members of society and who suffer, feel, think, and act as human beings" (emphasis in the original). It is historically material social relations that are the primary forces in this drama, and these always involve the exigencies of human labor.

When the human individual and human labor are abstracted from observations made more narrowly in time and space, what can be observed is more restricted and therefore their concrete characteristics allow for permanence and stability to appear to be the normal state. But Marx's approach opts for bringing wider temporal and spatial relations into view. Thus, in his critique of Proudhon, Marx (1982b: 98) complains, "In discussing the division of labour, he feels no need to refer to the world *market*" (emphasis in the original). A sufficiently broad historical vantage point makes social change the normal state of affairs and brings into view how individuals and labor have changed over time. This tension between apparent stability and change based on the abstractive framework allows an analyst to treat history as a variable, whereas tracing the social relationships around the role of labor allows for the discovery of temporally dynamic causal agents.

How does Marx treat labor as a historical variable within his materialist dialectic? Postone (1993: 5) argues that "Marxian theory should be understood not as a universally applicable theory but as a critical theory specific to capitalist society." Though there is some truth here, this is not entirely correct. Marx holds that labor is universally important and is therefore valid as a sociological category, but with two stipulations. First, it is only in capitalism that "abstract labor" emerges as an ordering principle in social life and therefore labor attains its highest level of development here. Second, it is because of this development that labor arrives to our consciousness in its most congealed expression and thus makes an adequate scientific conception of it possible, i.e., as "Labor" (Marx 1973: 102–105). Although Postone endorses these two points, he does so by jettisoning Marx's assumption that labor exists as a vital cross-cultural agent, e.g., "Far from considering labor to be the principle of social constitution and the source of wealth in *all* societies, Marx's theory proposes that what uniquely characterizes capitalism is precisely that its basic social relations are constituted by labor and, hence, ultimately are of a fundamentally different sort than those that characterize noncapitalist societies" (Postone 1993: 6). Although labor relations might be different in various modes of production, it nevertheless remains true that labor is the source of social value under tribal, slave, feudal, and capitalist modes of production, though it may not have crystallized in the former three to the point of being able to be discovered and conceptualized in the same way capitalism makes possible. Postone, however, repeatedly claims that Marx's mature approach to labor was "historically specific," meaning applicable to only capitalism, which overlooks how Marx points out that the mode of appropriation of the wealth extracted through surplus-labor is the signature feature that differentiates *class* societies from each other (see Chapter Four). This position is inconsistent with the idea that labor holds meaning for Marx only in relation to capitalism.

Labor does not step forward as the most important social variable only under capitalism. Rather, as capitalism emerged from feudalism, its structure transformed the division of labor into a more crystallized set of calculative exploitative relations. Labor was the source value previously but both value and labor *also* emerge fully mature in capitalism (as wealth – a continuation of class dynamics – and as profit and capital). This development allows Marx to read history retrospectively, partially, and selectively as a process of the evolution of labor and value. These are not teleologically operative but, from point of view of the present, the question is how they emerged more fully formed in our present. Other issues can be treated as background and secondary details in the history of class struggles. From the point of view of capitalism, the social-structural importance and powers of determination of labor and value have grown over time. Marx stresses their historical importance not because they are historically universal without qualification but because capitalism brings them forward with unique and significant powers as wages and profit in the concrete and labor-power and capital in the abstract. Thus, it is necessary that a history of the present examine the past in terms of labor and value's common evolution – during which they acquired new powers – toward their climax as modern forms of wealth.

For Marx, the nature and quality of social variables are neither universal nor constant. This means, as Gould (1978: 71) explains, if labor finds concrete expression in different modes of production, then "causality itself [undergoes] a development through the various stages of history." This history can be divided into epochs by using structures unique to their periods of origin, development, maturation, and decline. This helps reveal how the magnitude of labor's centrality has been variable *between and within* different modes of production. The "bookending strata" of historical modes of production, the earliest and the latest – i.e., primitive communalism and capitalism – have been dominated by the central power of labor. Other relations in other systems may mediate the causal powers of labor, though labor's powers of determination are never completely overridden. The "middle level strata" – i.e., tribalism, the Asiatic mode, slavery, and feudalism – although containing salient labor relations, are marked by religious and cultural norms that influence politics and class dynamics. Thus, the centrality of labor finds its highest powers of determination, on the one side, when humans struggled with nature as a species-being, and on the other, when, as a species divided, labor struggles with capital.

Focusing on the division of labor to grasp this history, Marx views labor as central for understanding three key dimensions of human life: (1) labor is an essential trait of the human species; (2) labor has played a significant role in the unfolding of human history; and, (3) labor is important for the realization of humans as individuals (Gould 1978). Marx addressed these dimensions at levels of abstraction carved at the level of society in general and how each is related to capitalism in particular.

The Importance of Labor for the Species

Labor is *the* trait that provides for the species' survival as it interweaves nature, objects, individuals, and humans as a species in an internal relation of sociality. As Gould (1978: 41, 75) explains, "Marx characterizes labor broadly as the distinctive activity of...their species.... [I]t is labor that brings things into being and connects one thing with another." The "elementary factors" of the "labor-process" included in this conception are: "1, the personal activity of man, i.e., work itself, 2, the subject of that work, and 3, its instruments" (Marx 1992j: 174). Over their history, societies increasingly developed a division of labor where various life activities were broken up into constituent parts and divided up among various social strata. Understanding these parts and strata helps us understand a variety of social relationships and processes:

> And every child knows, too, that the amounts of needs demand differing and quantitatively determined amounts of society's aggregate labour. It is self-evident that this *necessity* of the *distribution* of social labour in specific proportions is certainly not abolished by the *specific form* of social production; it

can only change its *form of manifestation*. Natural laws cannot be abolished at all. The only thing that can change, under historically differing conditions, is the *form* in which those laws assert themselves (Marx 1988b: 68; emphases in the original).

One general historical law is that forms of knowledge – e.g., values, norms, even gods – became separated from their origins in productive activity as modes of production changed. These superstructural elements sometimes survive over time, especially if they are compatible with a new structure. If they survive, they are likely to take on the normative and metaphysical appearance of tradition, especially as knowledge of their origins recedes into the past (e.g., religion). This is why tracing human relations historically and materialistically tends to demystify traditional knowledge and poses a threat to received wisdom.

The Species and the Division of Labor Under Capitalism. Living in alien material conditions, humans are ruled by abstractions of their own making. "For Marx," Gould (1978: 48) tells us, "objectification is the intrinsic character of every productive activity and is alienated only when the relation between the subject and the object becomes an external one... [i.e., when the object of labor belongs] to another." The possession of labor's products by others is an inherent quality of class relations, creating a divide in human sociality, setting the species against itself. In these conditions, labor and the legal ownership of its products become abstracted through objective and alien powers. Class societies and their objectification of labor produce both great magnitudes of socially recognized value *as well as* alienation. This divide between value production and human alienation reaches its highpoint in capitalist society, where the "community from which the worker is isolated by *his own labor* is *life* itself, physical and mental life, human morality, human activity, human enjoyment, *human* nature" (Marx 1975j: 204; emphases in the original).

In capitalism, alienation is expressed in class antagonisms via calculations of profitability in the wage-system. Labor under capitalism "is sold... not with a view of satisfying... the personal needs of the buyer. His aim is augmentation of his capital, production of commodities containing more labour than he pays for" (Marx 1992j: 580). Capital is a caretaker of labor only to the extent labor can augment capital; otherwise it has no use-value for it: "Production of surplus-value is the absolute law of this mode of production. Labour-power is only saleable so far as it preserves the means of production in their capacity of capital, reproduces its own value as capital, and yields in unpaid labour a source of additional capital" (1992j: 580). The species-wide centrality of labor is transformed into the tool of one class in pursuit of anti-social ends over and against another class. Labor in capitalism becomes a weapon used against its possessors as individuals and as a class rather than something that frees them from the dictates of nature. The resulting system leaves human beings estranged from their work, the products of their work, each other, and the species as a whole (Marx 1988a). It is here Marx's ontological concerns about labor unite with his epistemological questions of how to study it and where all of them inform and combine in his central moments of inquiry (i.e., historical materialism, political economy, and the communist project).

The Importance of Labor for Human History

Given that "objects in turn constitute the objective conditions for subsequent laboring activity" (Gould 1978: 75), Marx's perspective "views history neither as an accumulation of accidents, of the deeds of great men, nor as a process of constantly recurring ebbs and tides, an eternally self-repeating pattern – nor as the work of mysterious forces, predestined by some other-worldly plan of redemption or damnation or by the destiny of the World Spirit, but as the development of the human race determined by the nature of *labor*" (Fischer 1996: 90; emphasis in the original). Without labor, there would be no society, different cultures, or social change. Gould (1978: 41) goes so far to assert that, for Marx, "labor is the origin of time – both of human time-consciousness and of the objective measure of time." That is, she continues, "in describing activity as the synthesis or connection of" activity guided by some purpose in which a later stage of production preserves an earlier stage...

> ...Marx is presenting this activity as the origin of the three constituents of time – namely, past, present and future – and sees it as providing the ground for their interconnection as the process of time.... This synthesis...is not an event, but a process.... [T]he product or the made object is the result of the process of production which therefore precedes it. Thus the relation of before and after is established as an asymmetrical one in this laboring activity (1978: 58–59).

By erecting abstract models of objects in the mind to be forged into reality through labor, the experience of marking the passing of time comes into conscious reflection. As a result, the possibility of systematic historical knowledge arose. Labor thus accounts for what Kant thought was an *a priori* universal, i.e., the exigencies of time.

As the division of labor and the development of material forces became more complex, and regulated, more precise notions of time emerged and ways of conceptualizing and recording passing eras developed, e.g., lore, migration of herds, stars, the moon and sun, history books, calendars, and clocks. Labor is, therefore, the source of all manner of human social knowledge, including abstract time (for discussion, see Postone 1993: 186–225). Thus, one onto-epistemological conclusion is that "time [is] a social creation and that the historian [sh]ould not allow himself to be imprisoned by the utilization of only one variety of time" (Wallerstein 1991: 136). By examining what labor has made possible, and what other possibilities these conditions entail, a reading of history is no longer condemned to the chaos of all that has occurred in the past, and thus "a relatedness in the history of man" (Marx 1982b: 96) can be uncovered.

The Importance of Labor for the History of Capitalism. Because social structures have comparatively uneven spatial and developmental qualities, their causal powers are similarly uneven. The quality of social development influences forms of knowledge and action. Capitalism's growth, temporally and spatially uneven, has become increasingly regularized in the form of state laws and regulations, precise measurements in the production process, and financial calculations in the pursuit of shares of value. In Marx's time, the processes of commodification and proletarianization were less advanced than today, whereas the sociocultural influence of ideological factors was probably greater. However, with continual capitalist development on a world-scale, commodification and proletarianization are greater today and thus the influence of material forces continues to grow in scope.

In Marx's (1992j: 301) analysis, "The real value of a commodity is, however, not its individual value, but its social-value; that is to say, the real value is not measured by the labour-time that the article in each individual case costs the producer, but by the labour-time socially required for its production." When measured at the class level, labor is the source and measure of value within capitalist relationships, where...

> ...labor becomes qualitatively undifferentiated; any part of it is like any other part. It becomes abstract labor.... This homogenous abstract labor now

can be measured by a universal measure and divided into standard units. The measure of this abstract labor is time, but time now conceived as itself a universal and homogenous quantity, any part of which may be substituted for any other part in equivalent measure. Thus units of time can be mapped onto units of abstract labor (Gould 1978: 64–65).

If labor creates time but can be alienated by the measure of it, then capitalism's drive for efficiency plus profit requires the reduction of necessary labor-time and a corresponding increase in surplus labor-time devoted to the creation of surplus-value. This change produces a radical alteration in human relations, where "the history they constitute is beyond their control" (Postone 1993: 295). Exploited and alienated, treated as a standardized measure of work, human beings are forced to devote increasing hours per day to surplus-labor, losing, as a consequence, time available for cultivating their creative capacities. Thus, not only is history beyond their control in alienated society, but so is the form of labor the individual is forced to perform and the manner in which this shapes them.

The Importance of Labor for the Individual

The ability to create abstract ideas in the mind and forge them into reality is an inherent trait of the human individual but one that is always materially grounded: "Since the reasoning process itself arises from the existing conditions and its itself a *natural process*, really comprehending thinking can always only be the same, and can vary only gradually, in accordance with the maturity of development, hence also the maturity of the organ that does the thinking" (Marx 1988b: 69; emphasis in the original). As new forces, means, and/or modes of production develop, new needs and capacities develop. Expressing reflexive creativity through the division of labor, individuals cultivate nature, objects, and themselves. This is the path to self-realization. Without creativity only that which has already been accomplished can be realized anew; without the ability to act on creative inquiry there is no freedom. An increased division of labor and creative reflexivity therefore make freedom and individuality possible.

Movements of change, uneven development, and structural dynamics across time/space shape the types of individuals created. In human history, there has been a tendency toward increasing technological complexity, diversity and allotment of tasks, and types of production systems. A double movement springs from this. First, property has moved away from communal ownership and toward private property. This is related, second, to the movement of humans away from (what Marx viewed as) the "herd-like" aggregates in communal society toward greater individuality. Over time, as modes of production were transformed, the coherence of private property increased, the division of labor became more complex, forms of individuality grew, and the means of production became possessed by an increasing number of persons making up a class with separate and identifiable interests. However, class societies, because of their systems of property and methods of appropriation, undermine labor as the basis of universal human creativity and freedom through dominating the terms of labor, e.g., what sorts of work, goods, value, wealth, and leisure are made available and on what terms. Thus, historical class dynamics have a profound impact on the prospects for individuals: "When society is in a state of decline, the worker suffers most severely. The specific severity of his burden he owes to his position as a worker, but the burden as such to the position of society" (Marx 1988a: 26).

The Individual and Their Labor Under Capitalism. Class history, a history of human division and estrangement from itself, and class struggle as an attempt to combat the more readily apparent aspects of this, reach their climax and denouement in capitalism. Within capitalist development, the alienation of people from their land through the private property system occurred with a movement in the opposite direction – i.e., increasing conditions of possibility for human individuality, creativity, and freedom. At the same time, because *class* antagonisms reach their climax in capitalist class relations, the internal relation between the individual and social structure means that something similar should happen to *individuals* under these relations – i.e., "Only when the life of society is fragmented into isolated acts of commodity exchange can the 'free' worker come into being; at the same time his fate becomes typical of the whole society" (Lukács 1971: 91). As a standardized measure of human work, the individual is transformed into a commodity through the buying and selling of their labor-power, a fundamentally alienating condition of life that infects all those exposed to it. If human activity *is* socially productive activity, and if alienation in capitalism "is manifested not only in the result but in the act of production, within the producing activity itself" (Marx 1988a: 73), then alienation is inherent, if not unique, to the capitalist mode of production.

Alienation, a product of class systems, takes a special form under capitalism. Viewing alienation from the viewpoint of class exploitation of labor, Marx (1988a: 76) concluded:

> In estranging from man (1) nature, and (2) himself, his own active functions, his life activity, estranged labor estranges the species from man. It turns for him the life of the species into a means of individual life. First it estranges the life of the species and individual life, and secondly it makes individual life in its abstract form the purpose of the life of the species, likewise in its abstract and estranged form.

Under class relations, the entire social being, and therefore the individual, is in a state of alienation: "The *alienation* of the worker in his product means not only that his labor becomes an object, an *external* existence, but that it exists *outside him*, independently, as something alien to him; and that it becomes a power on its own confronting him; it means that the life which he has conferred on the object confronts him as something hostile and alien" (1988a: 72; emphases in the original). In moving from this vision (under historical materialist principles) to the form of alienation experienced within capitalism, Marx (1992j: 77) concluded that the "social relation between men...assumes, in their eyes, the fantastic form of a relation between things." Under a system where labor is transformed into a commodity to be bought and sold, human social relations are transformed into relations similar to those between other commodities. In such a state, freedom is negated, creativity is undermined, and knowledge is distorted.

As a contradictory system, capitalism works in opposing directions. The rise of the capitalist market brings the destruction of the social bonds of prior systems as well as a modicum of social freedom for both worker and capitalist. The production and distribution of commodities for sale, the search for work, and the creation of spaces for buying and selling all facilitate, along with their harsh results, a cultural-social space that provides for a freer flow of people and things. Its chaotic nature shapes social life in such a way that "All fixed, fast-frozen relationships, with their train of ancient and venerable prejudices and opinions, are swept away, all new-formed ones become antiquated before they can ossify" (Marx and Engels 1978a: 476). To the extent that traditional forms survive, they are transformed by their contact with modernity (e.g., consider the relationship Christianity and Judaism each have with the practice of usury versus what one finds in their texts). With the standardization of material conditions, commodities, and work comes the standardization of the individuals. As capitalism developed into a technologically complex system, higher levels of education and knowledge were increasingly needed, first by owning classes, then managerial classes. Engineers and scientists were cultivated and workers were eventually encouraged to be literate and educated. This expresses an internal relation between social structure and the individual. These processes work in the direction of the cultivation of the free individual unconstrained by the bonds of pre-modern superstition. But strapped to the modern industrial apparatus, scientifically planned, dissected, and inspected, individuals become homogenized. Thus, "Man" is a product of modernity and individuality is a social-historical event (see Seve 1978, Foucault 1977, 1980).

Essentialism, Organicism, and Evolutionism

It is impossible to achieve a scientifically correct conception of things unless there exist essential parts to which concepts can refer. Just as chemistry uses "reagents" to extract purer forms and geology demarcates epochs of natural history, political economy and historical materialism do the same through the force of abstraction (Marx 1992j: 19, 351). With recourse to nothing extra-social or extra-historical, a dialectical – and materialist – abstraction is grounded in observations of real practices. Sets of relationships indicative of a social formation are extracted in order to explain its observed empirical regularities. These essential characteristics distinguish one set of structural relations from others in such a way that each deserves separate conceptual terms. The concept of structure refers those features of a social formation that exist across all its forms, though "Knowledge of the essential structure of a thing...is the fruit of scientific inquiry, not an *a priori* philosophical speculation" (Ruben 1979: 72–75). And, as Andrew Sayer (1984: 92) reminds, "the more a social object is internally related to other objects, the less likely is it to be invariant across time and space." Thus, an "essential structure" refers to those specifiable social relations that must be in place in order to be able to analyze a social system *as a system*.

To get at such essences, a science must critique the level of appearance. In the same way that astronomers demystified religious explanations of heavenly bodies, Marx believed his method allowed for revealing the causal mechanisms behind historical and structural phenomena. This research inspired his critique of idealism. However, not all appearances mask material relations in the same way. Rather than masking real relations, some appearances suggest how certain relations can be studied. In *Capital*'s section on commodity fetishism, for example, Marx claims that it appears that commodities "naturally" attract money. This is because use-values (as commodities) become embodiments of value and, as a manifestation of the circuit of capital, this appearance is real but also allows further investigation into the essence of capital as a structural and historical relation. This analysis assumes the full development of the money form. The form of money in capitalism reduces all forms of money into fungible currencies, which is less possible with pre-capitalist forms. Although pre-capitalist forms contain a germ of the mature form (in embryo), it is only with the equality of commodities in capitalist material relations, where the average value of a commodity is based on the socially average abstract labor-time necessary for creating it, that this appearance provides evidence as to its historical essence.

With his sifting through appearances and essences, Marx, concluding that societies possessed structural relations that created historical change, adopted organic and evolutionary models. In the historical materialist approach, "economic life offers us a phenomenon analogous to the history of evolution in other branches of biology" (Marx 1992d*: 28). Marx (1992a: 18–19) thought that, like biological creatures, capitalism is made up of internally related parts that make up "an organic whole," where "the commodity-form...is the economic cell-form...of the same order as those dealt with in microscopic anatomy." His objective is to examine this mode of production and the conditions of production and exchange corresponding to it and to extract its laws and tendencies. The dynamics of any social formation depend on its conflicting social relations at given points in time. Given the breadth of the temporal lens, a system may change within its own form or may be transformed into a new one. This view assumes change is inherent in things (conception, embryonic growth, maturity, passing away). Rooted in the centrality of labor and the nature of class struggle, capitalism's dynamic is often determined by the contradictions between its productive forces and relations, making change its natural condition and implying that it will end at a certain point.

With the assumption that the "present society is no solid crystal, but an organism capable of change, and is constantly changing" (Marx 1992a: 21), political economy's organicism is evolutionary. Darwin's work had a "deep influence" on Marx (Jordan 1971: 201).[4] He wrote to Engels, for instance, "Darwin's book on Natural Selection...developed in the crude English fashion...is the book which, in the field of natural history, provides the basis for our views" (Marx 1985c: 232). Marx (1985d: 246–247) admired Darwin's theory of evolution as dialectical and non-teleological, remarking, "Darwin's work is most important and suits my purpose in that it provides a basis in natural science for the historical class struggle.... Despite all shortcomings it is here that, for the first time, 'teleology' in natural science is not only dealt a mortal blow but its rational meaning is empirically explained." Given the theology of the time (then as today), Darwin was widely read as arguing that the human species was the endpoint toward which evolutionary processes strove, a metaphysical teleology. Today, biologists understand natural selection as a nonteleological process: "natural selection has no eye to the future. [Things] are eliminated by natural selection...because they are not adapted to their present environment" (Nagel 1979: 303). Evolution, natural and/or historical, is a process by which those forms best adaptable to changing conditions will tend to survive over those less so. Moreover, evolution, like history, does not contain forward-looking goals that drive it, i.e., "Humans are not the end result of predictable evolutionary progress" (Gould 1995: 329).

Marx employed a similar views.[5] In his anthropological studies, he examined history – including the earliest communal societies – as a part of natural

---

[4] "*Charles Darwin*, in the introduction to his *On the Origin of Species by Means of Natural Selection, or the Preservation of Favoured Races in the Struggle for Life* (5th thousand), London, 1860, says the following: 'In the next chapter the *Struggle for Existence* amongst all organic beings throughout the world, which inevitably follows from the high geometrical ratio of their increase, will be treated of. This is the doctrine of *Malthus* applied to the whole animal and vegetable kingdoms' (pp. 4–5). In his splendid work, *Darwin* did not realise that by discovering the 'geometrical' progression in the animal and plant kingdom, he *overthrew* Malthus's theory. Malthus's theory is based on the fact that he set Wallace's geometrical progression of man against the chimerical '*arithmetical*' progression of animals and plants. In Darwin's work, for instance on the extinction of species, we also find (quite apart from his fundamental principle) the detailed refutation, based on natural history, of the Malthusian theory. But in so far as Malthus's theory rests upon Anderson's theory of rent, it was refuted by *Anderson himself* " (Marx 1968b: 121; emphases in the original).

[5] "Owing to the fact, therefore, that Hegel makes the elements of the state the predicate, whereas in historical reality the reverse is the case, the state idea being instead the predicate of those forms of existence, he expresses only the general character of the period, its *political teleology*. It is the same thing as with his philosophical-religious pantheism. By means of it all forms of unreason become forms of reason. But essen-

history, using Darwinian principles in the analysis of social forms – e.g., "*Communism in living* seems to have originated *in the necessities of the consanguine family*" (Marx 1972: 115; emphases in the original). The development of familial norms – including the "gradual exclusion of own brothers and sisters from the marriage relation" – begins in "isolated cases, introduced partially at first, then becoming general, and finally universal among the advancing tribes... [which] illustrates the operation of the principle of natural selection" (Marx 1972: 109). This approach was not new to him. Marx used evolutionary thinking in trying to establish the relationship between internal social change and change from one social form to another while accepting no *a priori* universal theory about the results of historical development. Each mode of production has its own tendencies. Thus, when Marx (1992a: 19) states that the societies "more industrially developed only shows, to the less developed, the image of its own future," he is comparing concrete forms of the capitalist mode of production, not positing a transhistorical, teleological theory. Lower modes of production do not necessarily evolve into predetermined higher ones. They evolve internally but future conditions or ends do not produce causal interactions in the present.

Marx's approach to teleology and evolutionary models (and interpretations of that approach) brings us back to the issue of appearance and essence and the traditional reading of Marx's so-called theory of history (see Chapter One). While Marx's model was not universally evolutionary, he does see something peculiarly evolutionary about capitalism and its internal relation to other social forms. For him, societies adjacent to those with a capitalist mode of production *appear* to evolve into capitalism because once it matured, the "categories which express its relations, the comprehension of its structure, thereby also allow insights into the structure and relations of production of all the vanished social formations out of which whose ruins and elements it built itself up, whose partly still unconquered remnants are carried along within it, whose mere nuances have developed explicit significance within it" (Marx 1973: 105). That is, capitalism makes world history appear linear and evolutionary because it absorbs all others into its dynamic. Because of

---

tially here in religion reason is made the determining factor, while in the state the idea of the state is made the determining factor. This metaphysics is the metaphysical expression of reaction, of the old world as the truth of the new world outlook" (Marx 1975h: 130; emphasis in the original).

this dynamic, coupled with its associated uneven development, the historical process of the maturation of value into the commodity form is a process that "is not a unilinear progression of forms, of course, but of a more complex process in which forms overlap, interpenetrate and develop each other" (Meikle 1979: 17–18). Moreover, because historical materialism accepts that modifications of social structures may come at different speeds for different reasons, a "Marxian approach should encourage questions about both gradual evolution and rapid revolution" (Sherman 1995: 226). Marx's evolutionism is therefore employed in a restricted sense, i.e., societies do not contain a teleological metaphysical inner-drive that eventually brings them to capitalism but capitalism has an inner-logic that drives it forward into new forms and it changes other systems along with this drive. However, ultimately, because of its contradictions, capitalism reaches the point of its own dissolution.

The Necessity of a Dialectical Science

Naturalism stayed with Marx throughout his entire intellectual career and extends across his various moments of inquiry. It was his view that "all science would be superfluous, if the appearance, the form, and the nature of things were wholly identical" (Marx 1909: 951). Much of science is an exercise in tautology to the extent it remains on the level of appearances (Marx 1973: 87). The form of tautology Marx (1982b: 102) warns against is related to his concern about making appropriate abstractions and not mistaking these for causal agents themselves:

> To M. Proudhon...the prime cause consists in abstractions and categories. According to him it is these and not men which make history. *The abstraction, the category regarded as such,* i.e., as distinct from man and his material activity, is of course immortal, immutable, impassive. It is nothing but an entity of pure reason, which is only another way of saying that an abstraction, regarded as such, is abstract. An admirable *tautology*! (emphases in the original).

Marx does not negate the tautological nature of social knowledge but reflexively acknowledges it by creating abstractions that match up with the social processes he investigates. Any social scientist's job is to construct a series of interrelated concepts that reflect concrete practices. Such concepts must be both logically congruent with the way social phenomena function and develop and logically related to each other. If social reality is best grasped as a product of internally related historical and structure relations, then an external relations philosophy of science is less able to crack open the tautological nature of popular knowledge and move beyond it. It is thus the task of science to not accept surface appearances – i.e., what a society says about itself – but rather to grasp things at the root. When done well, all science is, by Marx's definition, radical. However, the social relations that prevail in capitalism produce appearances that belie what lies behind them. We shall see in the following chapters how Marx translated this assumption about the need for a dialectical science into his research practices.

This chapter has argued for several things. First, a set of ontological and epistemological assumptions operates in Marx's work, assumptions that are internally related – i.e., his internal relations philosophy of science extends from his assumptions about social ontology and these inform his epistemological practices, a view best understood as an "onto-epistemology." Second, Marx's model of social relations incorporates a complex interplay of material and ideal relations. Third, a central assumption in this onto-epistemology is the centrality of labor for history, the human species, and the individuals that make it up. Thus, fourth, these assumptions inform Marx's analysis of what happens in capitalism. Fifth, Marx believes that only by uniting historical, dialectical, materialist, and scientific methods can an adequate grasp of these collective realities be achieved. Sixth, although material conditions are primary but not sole causal factors, it is nevertheless true that we live in a capitalist society and to treat this category as unimportant is to treat modernity as a timeless constant and thus as something where a special set of tools is not needed for its analysis – a mystified form of knowledge. When understood in this manner, Marx is telling us how to begin an inquiry into our present, an inquiry that demands that capitalism to be front-and-center. The remainder of this book tries to show how this form of inquiry unfolded in Marx's studies. It does so by, first, examining how these assumptions are put to work in Marx's analytical procedures. Thus, second, the results of this investigation will put us in a better position to examine Marx's central terminological strategies and explanatory models and how these were informed by his onto-epistemological assumptions and analytical procedures.

Chapter Four
Marx's Analytical Procedures

When Marx (1992c: 26) noted "That the method employed in *Das Kapital* has been little understood, is shown by the various conceptions, contradictory to one another, that have been formed of it," it was a testimony to the multiple interpretations of his approach. He had previously referenced his views on the study of political economy: "For an estimation of [Proudhon's] work, which is in two fat volumes, I must refer you to the refutation I wrote. There I have shown, among other things, how little he had penetrated into the secret of scientific dialectics and how, on the contrary, he shares the illusions of speculative philosophy" (Marx 1985l: 29). The term "scientific dialectics" suggests Marx believed several conventions of science could and should be united with dialectical method. To help us understand how he accomplished this, this chapter introduces Marx's central analytical procedures. These procedures include the logic and order of investigation, methods of abstraction within structural and historical analysis, the use of certain positivist methods, and the presentation of results.

The Order and Logic of Investigation: Critique, Empirical Inquiry, and Analysis

"Reason has always existed, but not always in a reasonable form," Marx (1975k: 143) tells us. As an antidote to this problem, Marx employed "critique...[as] a specific form of analysis" (Sayer 1979: 105). In his moments of critique, "Marx rarely collected data himself; he almost always worked as a critic of others' theoretical positions or reinterpreted their findings within a different theoretical framework" (Beamish 1992: 17). In this work, he distinguished between "vulgar" criticism, which is dogmatic, and "true" criticism, which examines the "inner" and "specific" logic of theoretical views (Marx 1976a: 91). "True" criticism has both negative and positive moments. In negative critique, Marx provided "minatory effusions on what views everyone should *avoid*" (Carver 1983: 84; emphasis in the original). Among those things Marx advises us to avoid are the following:

- Avoid approaches that "invent mystical causes" and "high-flown expressions, such as universal reason, God, etc." (Marx 1982b: 96).
- Do not be one who "confuses ideas and things" (1982b: 96).
- Avoid positing "eternal laws" (1982b: 100).
- Do not be "dogmatic" (1982b: 100) or "doctrinaire" (1982b: 103).
- Avoid "sentimentalizing" and "banalities" (1982b: 104).
- Avoid "vague reasoning" and "magniloquent phrases" (Marx 1975e: 394).[1]
- Estimate theories on their "substance" and not by "casting suspicions on [an author's] civil character" (Marx 1982d: 402).
- One should "distinguish between what a particular author actually says and what he believes he says" (Marx 1979d: 324).
- A "question...badly formulated...cannot be answered correctly" (Marx 1985l: 27).[2]

---

[1] "'Phrase-monger' was in his mouth the sharpest censure – and whomever he once had recognized as a 'phrase-monger' he ignored forever. To think logically and to express your thoughts clearly – this he impressed on us 'young fellows' on every occasion" (Liebknecht 1901: 84–85).

[2] Elsewhere Marx (1992e: 66) wrote: "We cannot solve an equation that does not comprise within its terms the elements of its solution."

The goal of critique is self-clarification and, one hopes, demystification (Appelbaum 1988: 69–86). As such, these principles inform our understanding of Marx's critique of other traditions and, by extension, the manner in which he developed his own approach to dialectical method, historical materialist principles, and political-economic investigation.

After reading Lassalle's *Heraclitus*, Marx (1936b: 105) wrote to Engels that Lassalle "will learn to his cost that to bring a science by criticism to the point where it can be dialectically presented is an altogether different thing from applying an abstract, ready-made system of logic to mere inklings of such a system." To Lassalle, Marx (1983f: 316) noted, "I should, moreover, have liked to find in the text proper some *critical* indications as to your attitude to Hegelian dialectic. This dialectic is, to be sure, the ultimate word in philosophy and hence there is all the more need to divest it of the mystical aura given it by Hegel" (emphasis in the original). Later, in a note in an early edition of Volume Two of *Capital* (omitted in later editions), Marx explained that over the course of his work "I have taken the liberty of adopting towards my master a critical attitude, disencumbering his dialectic of its mysticism and thus putting it through a profound change, etc." (cited in Anderson 1992: 68). Marx's dialectic was not to be a mystical-metaphysical system simply applied to or imposed upon the facts.

Marx had returned to Hegel after his critique of Kant (see Chapter One). He found in both an internal contradiction that was expressed in their tautological arguments. For example:

> The proofs of the existence of God are either mere hollow tautologies. Take for instance the ontological proof. This only means: 'that which I conceive for myself in a real way, is a real concept for me', something that works on me. In this sense all gods, the pagan as well as the Christian ones, have possessed a real existence (Marx 1975c, Appendix).

This concern was again advanced in Marx's (1976a: 11) *Critique of Hegel's Philosophy of Law*, where he argues that Hegel's position "That the various aspects of an organism stand to another in a necessary connection arising out of the nature of the organism is sheer tautology. That if the political constitution is defined as an organism, the various aspects of the constitution, the various authorities, behave as organic features and stand to one another in a rational relationship, is likewise a tautology." Hegel's argument is a self-referential, illogical, form of knowledge, "an illusion of practical consciousness" (Marx 1976a: 62).

Rejecting tautological reasoning is but one moment in Marx's critique of the Hegelian dialectic (as well as of Kant's speculative philosophy). Marx (1976a: 9) criticized Hegel's belief that the material world is "produced by the actual idea." This is indicative of his tendency toward the "inversion of subject and predicate" (1976a: 12), a view which fails because of its *"uncritical, mystical way of interpreting* an *old world-view* in terms of a new one which turns it into nothing better than an unfortunate hybrid.... This *uncritical approach,* this *mysticism,* is...the mystery of the Hegelian philosophy" (1976a: 83; emphases in the original). In depicting modern social relations as expressing an abstract universal, Hegel's dialectic, "standing on its head" (Marx 1992c: 29), inverts subjects and predicates, forwards tautological arguments, offers false dualisms, mystifies his subjects through atomistic, idealist, and *a priori* assertions, falsely identifies some things and fails to specifically identify others, and bases his philosophy on abstract generalities. "Turned right side up again" (1992c: 29), a non-metaphysical, non-mystical dialectic should not be an *a priori* theory in general application but rather a framework for reasoning mobilized during research as necessary.

Marx's critiques continued with his engagement of Feuerbach. For him, Feuerbach's approach "gives too much importance to nature and too little to politics" (Letter to Ruge, cited in McLellan 1973: 68) and "never arrives at the actually existing, active men, but stops at the abstraction 'man' and gets no further.... He gives not criticism of the present conditions of life" (Marx and Engels 1976: 41). Moreover, Feuerbach's "discoveries about the nature of philosophy required still, for their *proof* at least, a critical settling of accounts with philosophical dialectic" (Marx 1988a: 17–18; emphasis in the original). Feuerbach ultimately failed because his materialism and historicism did not inform one another. Thus, "As far as Feuerbach is a materialist he does not deal with history, and as far as he considers history he is not a materialist" (Marx and Engels 1976: 40–41). Feuerbach's approach was not just incomplete; it was distorting.

The Poverty of Philosophy, one of Marx's early forays into political economy, helped him hone his "dialectical critique of theoretical concepts and the use of concrete historical material to refute Proudhon's position" (Beamish 1992: 47). Marx (1982b: 97) complained that Proudhon "no longer feels any need to speak of the seventeenth, eighteenth or nineteenth centuries, for his history takes place in the nebulous realm of the imagination and soars high above time and place." It was Proudhon's failure where, "instead of regarding *economic categories as theoretical expression of historical relations of production, corresponding to a particular stage of development in material production,* he garbles them into pre-existing *eternal ideas,* and how in this roundabout way he arrives once more at the standpoint of bourgeois economy" (Marx 1985l: 29; emphases in the original). For Marx (1982b: 102), many "categories are no more eternal than the relations they express. They are historical and transitory products."

This critical stance toward Proudhon was extended to conventional political economy. Marx (1973: 83) criticized Adam Smith and David Ricardo because they "still stand with both feet on the shoulders of the eighteenth-century prophets, in whose imaginations this eighteenth-century individual – the product on one side of the dissolution of the feudal forms of society, on the other side of the new forces of production developed since the sixteenth century – appears as an ideal, whose existence they project into the past. Not as historic result but as history's point of departure." Smith and Ricardo foisted their economics upon an ahistorical theory of abstract universal human nature, a form of individualistic reductionism (Paolucci 2001b).

In sum, Marx rejected Kant's *a priori* speculative philosophy, Hegel's mystical idealist dialectic, Feuerbach's ahistorical materialism, Proudhon's metaphysical political economy, and Smith and Ricardo's ahistorical atomism. Hegel and Kant failed because their philosophies were not empirically rooted, and Hegel's particularly because its mysticism fundamentally obscured real relations. Feuerbach failed because his ahistorical materialism disallowed a proper conceptualization of change. Both speculative philosophy and traditional political economy reversed the most important relationships in their models (inversions), held that mainly ideas structure the world (idealism), and posited unchanged transcendent forces across time and space (metaphysics). This critique helped Marx see what to avoid in his own work.

Marx also advocated several principles for *positive* critique:

- "Ruthlessness – the first condition for all criticism" (Marx 1936m: 346).
- Do not be "afraid of the results" or "conflict with the powers that be" (Marx 1975k: 142).

- Criticism is against "raising a dogmatic banner [but rather should] help the dogmatists clarify their propositions for themselves" (1975k: 142).
- The starting point is "criticism of politics, participation in politics, and therefore *real* struggles" (1975k: 144; emphasis in the original).
- Criticism should also target "the theoretical existence of man...religion, science, etc." (1975k: 143).
- Understand "present social conditions in their [intermeshing]" and do not ignore "the historical development of mankind" but rather study "the real course of history" (Marx 1982b: 95–97).
- It is important to "see our social institutions as historical products and to understand...their origin and development" (1982b: 100).
- Thus, make sure to "grasp the bond linking all forms of *bourgeois* production" and "understand the *historical* and *transitory* nature of the forms of production in any one epoch" (1982b: 100; emphasis in the original).
- Provide more "definiteness" and "attention to the actual state of affairs" (Marx 1975e: 394).
- True science requires "more expert knowledge" than is available in commonsense (1975e: 394).
- By extension, one should respect the inherent "fairness of making yourself at least sufficiently acquainted with the subject of your criticism" (Marx 1992i: 162).

In positive critique, even when objections were made about another's work, if there were elements to fruitfully incorporate, Marx would do so. Though he objected to idealism, he held that the "outstanding thing in Hegel's *Phenomenology*" is viewing "the dialectic of negativity as the moving and generating principle" (Marx 1988a: 149), a move which "sensitized him theoretically to contradictions and changes through opposition, to grasp the division of labor more fully as he incorporated more historical material into his knowledge interests" (Beamish 1992: 44). Marx (1988a: 15–16) saw Feuerbach's *Preliminary Theses for the Reform of Philosophy* as the starting point for "*positive*, humanistic and natural criticism" and "a real theoretical revolution," given his premise that "*Sense-perception*...must be the basis of all science" (1988a: 111; emphases in the original). He wrote to Feuerbach to praise him for allowing a new "concept of the human species brought down from the heaven of abstraction to the real earth, what is this but the concept of *society*!" (Marx 1975l: 354; emphasis in the original). In political economy, although guilty of atomism, he adopts from Smith and Mill the labor theory of value and a study of capitalism's law-like features. Finally, in relating political economy to his communist project, Marx (1982b: 104) again referred to Proudhon: "The *only point* upon which I am in complete agreement with [him] is the disgust he feels for socialist sentimentalising" (emphasis in the original).

In order to facilitate its improvement, analysts have the duty to critique on their own work. Marx (1975b: 12) criticized his early work in law for its "metaphysical propositions," an approach "characteristic of idealism" and "a serious defect...the source of hopelessly incorrect division of the subject-matter." He faulted his "definitions of concepts, divorced from all actual law" and his use of an "unscientific form of mathematical dogmatism," which when combined with the problems listed above, resulted in a "shallow classification" (1975b: 12, 15). In his self-critique of his 1844 *Manuscripts*, Marx (1988a: 13–14) explained that the "intermingling of criticism directed only against speculation with various subjects themselves proved utterly unsuitable, hampering the development of the argument and rendering comprehension difficult. Moreover the wealth and diversity of subjects to be treated, could have been compressed into *one* work in a purely aphoristic style; while an aphoristic presentation of this kind, for its part, would have given the *impression* of arbitrary systematizing" (emphases in the original). His Afterword to the Second German Edition of *Capital* critiques his first edition by holding that the new edition has a "clearer arrangement" with previous topics "only alluded to...now expressly emphasized," not to mention addressing "occasional slips" (Marx 1992c: 22–23). Reflexive critique pushes analysts to new insights and provides clues as to how to shore up their work's weaknesses, to make needed improvements, and to head off avoidable criticisms. Thus, positive and negative critiques animate Marx's engagement of several schools of thought, including his own, and helped him establish and pursue his research agenda. (See Tables 4.1 and 4.2.)

During and after his critique of intellectual traditions, Marx's journalism and political-economic research studied relationships in the real world, things that were not esoteric, e.g., money, factories, wages, the working-day, the Paris Commune, the Beer Tax, bureaucracy, etc. These things indicated relations of class interests and political power. In this work, he appealed to an array of scholars, books, periodicals and newspapers, the findings of government commissions, the Blue books printed by the government in the British

Table 4.1
Marx's Relationship to Selected Contemporaries

Theorist	Marx's Critique	Marx's Alternative/ Correction	What Marx Kept
Kant	Conceptualization and empirical analysis are reversed; Speculation; Tautology; Metaphysics	Induction; Rejection of metaphysics	Categories such as time, space, and causality
Hegel	Idealism; Metaphysics; Speculation; Tautology	Reversal of Hegel's idealism and materialism; Rejection of metaphysics	Dialectical reason for a study of change
Feuerbach	Ahistorical	Historical Analysis	Materialism
Proudhon	Metaphysics; Formulaic dialectic (i.e., thesis, antithesis, synthesis)	Analysis of the historically concrete; Scientific dialectic	Political-economic focus
Smith & Ricardo	Capitalist behavior mistaken for human behavior in general; Ahistoricism; Atomism	Analysis of capitalism as a unique form of society; Class analysis; Historical analysis	The labor theory of value; Search for laws

Table 4.2
The Phases of Maturation of Marx's Research

Negative Critique	Negative and Positive Critique	Self-Clarification
Letter to his father	*Economic and Philosophic Manuscripts of 1844*	*A Contribution to the Critique of Political Economy*
Contribution to the Critique of Hegel's Philosophy of Law	*The German Ideology*	*Grundrisse*
The Holy Family	Theses on Feuerbach	*Mathematical Manuscripts*
The Poverty of Philosophy	*Theories of Surplus-Value*	*Ethnological Notebooks*

Museum, journals from the sciences, and even insider information: "Engels, indefatigable student of statistics, kept [Marx] supplied with all available information about the state of markets" (Sprigge 1962: 81). Marx used these sources as empirical evidence of structurally rooted patterns and constructed categories to analyze them.

Within analysis one must pick over previous theories, relevant data, and make sense of the relations between patterns and those things that seem chaotic. The initial inquiry has two primary goals: self-clarification and systematic category construction (more on the latter below and in Chapter Five). In this process, Marx looks for forms of discourse that can be evaluated against what was learned through the moment of criticism. This is why he did not publish several of his works: Self-clarification – about concepts, forms of logical reasoning, and methods of analysis and presentation – was often his goal. After self-clarification and additional investigation, he would move toward writing up his results. Thus, *Capital* offers us "a systematic presentation, an apodictic arrangement of the concepts in the form of that type of demonstrational discourse that Marx calls *analysis*" (Althusser 1970: 50; emphasis in the original). Marx's order of research, then, is inquiry first (critique, observation, analysis of data), and then presentation of findings (description and explanation of movement).

Abstracting Levels of Generality, the "Backward" Study of History, and the Synthesis of Historical and Structural Analysis

In his discussion of the dialectical method, Bertell Ollman (2003) stresses Marx's methods of abstraction. He explains how Marx abstracted data into seven "levels of generality" (2003: 86–99), used abstractions of "extension" to help him analyze their interrelationships (2003: 73–86), and employed a "backward" study of history (2003: 115–126). It is crucial to understand these methods if Marx's overall analytical tack is to be grasped. Later, when we examine Marx's use of conventional scientific methods, it will be important to note how he used what Ollman (2002: 99–111) calls abstractions of "vantage point" in this analytical work.

Two of the most important moments of analysis are the investigation of the relations between history and structure (Schmidt 1981). Marx's research on such questions does not begin without a sense of where to start. During and after critique, Marx looks for those things in the past that made the present possible. The hidden qualities of realities in the present require an act of scientific abstraction to get at them, practices we see him struggle with (and progressively master) in his *Critique of Political Economy* and the *Grundrisse*. In *Capital*, Marx isolated the economic structure out of the total social formation in which it is found and broke it down into productive (Volume One), circulative (Volume Two), and distributive (Volume Three) spheres in order "to systematically isolate from the complex whole particular relations for examination" (Horvath and Gibson 1984: 18). The force of abstraction helped him to do this.

The force of abstraction allows inquiry to focus on specific features of an object while providing a method for removing extraneous material from view. This assists in estimating the features of a social structure. However, because there have been various types of social structures in history, the institutional structure that makes a social formation unique is first grasped as a particular level of structural-historical experience and is then broken down through abstraction into its constituent parts, after which their internal relations can be pieced together.

The abstraction of "extension" conceptualizes spatial relations while specifying a level of generality in time "for treating not only the part but the whole system to which it belongs" (Ollman 2003: 74). In abstracting the extended interrelationships of parts and systems, although the "movement is from the most specific, or that which sets it apart from everything else, to its most general characteristics, or what makes it similar to other entities," depending on the level of generality, some things will be brought into or out of view (2003: 74–75). The force of abstraction allows refocusing investigation so that which was previously abstracted out of view can be reincorporated back into a broader view where "each abstraction can be said to achieve a certain extension in the part abstracted, and this applies both spatially and temporally" (2003: 74). In this process, one establishes the context from which data are extracted and thus to which level (or levels) of historical generality subsequent abstractions refer or extend. Next, once abstractions are identified with a level of historical generality, one can then manipulate them in a way that corresponds to the simplicity or complexity of the given phenomena (Horvath and Gibson 1984: 12). There are six general rules:

1. Abstractions should be commensurable with the time/space location of data.

2. Analysts should not conflate what is specific with what is general.
3. Abstractions should capture the essential structures of the system under examination.
4. Do not assume that abstractions that capture time and space variables possess the same qualities across the range of their historical and structural application.
5. Be sensitive to what data abstractive choices shut off from inspection.
6. Be skeptical of the manner in which systems depict their own institutional forms.

These rules help Marx organize a wide array of human social relations. They do this, first, by helping him compare humans as a species to other forms of nature. Next, these rules ask which social forms – broadly construed without regard to content – are common to all societies. If deemed central to life in society in general, a concept is developed for each part. Marx then examines how these essential parts are uniquely configured across different time/space locations and creates subcategories for them. This method of abstracting interrelationships between levels of historical and structural generality allows "the study of anything [to] involve one immediately with the study of its history and encompassing system" (Ollman 2003: 13). The broadest levels (Seven and Six) are those things humans share with matter (e.g., weight, motion) and with animals (e.g., the need for food, sex, etc.). Marx's naturalism uses tools from the natural sciences to study these realities and applies them, by extension, to what is unique to "humans as humans" (Level Five), including the centrality of labor, material forces, social institutions, and "society in general." Based upon the division of labor, Marx breaks down society in general into the history of class systems (Level Four). What we now call "historical materialism" is an inquiry that examines the general sociological laws found across these systems – e.g., the role of class struggle and base-superstructure relations. Of class systems in general, Marx is interested in capitalism as a system with unique properties. This inquiry into "political economy" is broken up into abstract models of capitalism in general (Level Three) and the investigation into capitalism's recent developments (Level Two). Finally, Marx (at Level One) directs attention to unique individuals and events, e.g., specific people, places, dates, and names (Ollman 2003: 88–91).

In abstracting levels of generality, abstractions of extension help Marx trace the interrelations of history and structure, examine their internal dynamics, discover the ways of social change, and mobilize his activity to facilitate the changes he desires (the communist project). The relationships between these levels of generality and Marx's conceptual strategies associated with each are depicted in Figure 4.1.

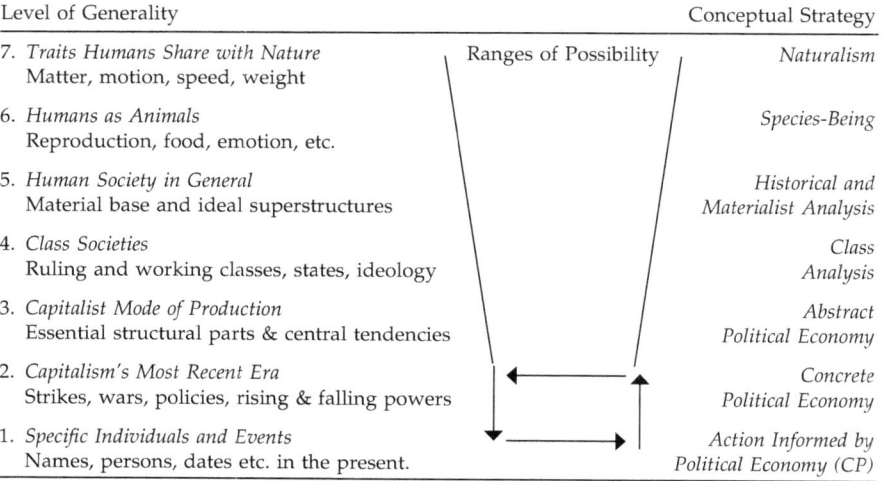

Figure 4.1
Marx's Levels of Generality and their Associated Conceptual Strategies

Level of Generality	Conceptual Strategy
7. *Traits Humans Share with Nature* Matter, motion, speed, weight	*Naturalism*
6. *Humans as Animals* Reproduction, food, emotion, etc.	*Species-Being*
5. *Human Society in General* Material base and ideal superstructures	*Historical and Materialist Analysis*
4. *Class Societies* Ruling and working classes, states, ideology	*Class Analysis*
3. *Capitalist Mode of Production* Essential structural parts & central tendencies	*Abstract Political Economy*
2. *Capitalism's Most Recent Era* Strikes, wars, policies, rising & falling powers	*Concrete Political Economy*
1. *Specific Individuals and Events* Names, persons, dates etc. in the present.	*Action Informed by Political Economy (CP)*

Marx's concern about the inversion of knowledge has its corollary in his "backward" study of history (Ollman 2003: 115–126). This is a sort of historical-materialist synthesis of induction and deduction in which one starts from the present and looks backward. Here, the structure of the present is first grasped in order to study the past as a way of prioritizing what to describe in the present and how to explain it. Marx (1973: 460–461) writes:

> [O]ur method indicates the points where historical investigation must enter in, or where bourgeois economy as a merely historical form of the production process points beyond itself to earlier historical modes of production. In order to develop the laws of bourgeois economy, therefore, it is not necessary to write the *real history of the relations of production*. But the correct observation and deduction of these laws, as having themselves become in history, always leads to primary questions – like the empirical numbers e.g. in natural science – which points toward a past lying behind this system. These indications, together with a correct grasp of the present,

then also offer the key to the understanding of the past – a work in its own right which, it is to be hoped, we shall be able to undertake as well. This correct view likewise leads at the same time to the points at which the suspension of the present form of production relations gives signs of its becoming – foreshadowings of the future. Just as, on one side the pre-bourgeois phrases appear as engaged in *suspending themselves* and hence in positing the *historic pre-suppositions* of a new state of society (emphases in the original).

Marx is telling us that historical and structural analyses have specific analytical relationships that must be recognized – i.e., before a historical account of a system can take place one must grasp its essential structural features, and this understanding helps in projecting its potential futures.

A structural analysis of capitalism was not possible until it was mature to a discernable degree. As Marx expressed this idea, "the method of rising from the abstract to the concrete is only the way in which thought appropriates the concrete, reproduces it as the concrete in the mind. But this is by no means the process by which the concrete itself comes into being." He continues:

> ...the simple categories are the expressions of relations within which the less developed concrete may have already realized itself before having posited the more many-sided connection or relation which is mentally expressed in the more concrete category; while the more developed concrete preserves the same category as a subordinate relation. Money may exist, and did exist historically, before capital existed, before banks existed, before wage labour existed, etc. Thus in this respect it may be said that the simpler category can express the dominant relations of a less developed whole, or else those subordinate relations of a more developed whole which has already had a historic existence before this whole developed in the direction expressed by a more concrete category. To that extent the path of abstract thought, rising from the simple to the combined, would correspond to the real historical process (1973: 101–102).

Marx is asking: How do studies of change and social structure relate to and inform each other? For example, similar systems (e.g., class systems) can have qualitatively different features over time/space (e.g., state forms), features that develop in ways where some elements change (e.g., the specific classes) and others stay intact (e.g., exploitation of labor). Moreover, systems change and develop in ways that are often "uneven," making the spatial location of a phenomenon as important as its temporal moment (see Mandel 1970). Methodological strategies are needed to handle such complexity.

For Marx's ontology, all objects of social science have historical features internal to their essence; they are an integral part of each other and must be conceptualized as such. This means that the history included in creating a social phenomenon is as much part of its definition as its structure. Marx does not begin in the past and trace concrete examples of abstract concepts forward through time, as might Durkheim, Simmel or general historical sociology. Rather, he pieces together the relation between human activity and the structure of modern society by looking in the documentary record for evidence of the historical evolution of the essential constituent parts of the social structure of the present. Just as the "Human anatomy contains a key to the anatomy of the ape," Marx (1973: 105) holds that "subordinate...species...can be understood only after the higher development is already known" – i.e., the capitalist mode of production contains relations within it that aid in studying its evolution from prior forms.

Although history usually does not announce the change of one epoch of a social structure to another, changes occur and establishing them is not purely an act of abstraction. Marx needs to locate both what *had* to occur (the realm of necessity) and what *did* occur (the realm of contingency) in order for the phenomenon in question to emerge. Here, abstraction starts with the most important structural elements in the historical period in question (capitalism, in Marx's case). This provides the criteria for examining the historical record and sets priorities for weighing the relative importance of evidence. Empirical-analytical concerns here include:

- What are the ongoing empirical regularities within the system in question?
- What are the most essential structural relations in this system?
- What structural relations account for specific empirical regularities?
- What historical events account for the rise of this or that set of relations?
- How have these empirical regularities and structural relations changed over time?
- What are the primary causal forces in this change?

"Preconditions" are those things *that* occurred and thus made other things possible. Whether a precondition originated in the structure of a previous social formation or whether it occurred historically "by chance" is contingent. "Presuppositions" refer to those things that logically *had to* occur for the new structure to take form at all. That is, structures in a social formation are presupposed by other social phenomena that have determined them and made them possible. Uncovering such inner-connections helps one discover the causal factors of historical development and systemic change. How is this discovery accomplished if the situation is seen as in dynamic flux? Ollman (2003: 117) explains:

> Viewing precondition and result as movements, in the process of their becoming, and both of these movements as aspects of a single movement, requires in the first instance an abstraction of their extension (of what all is included) that is large enough to encompass the interaction of the elements referred to over time.... [Then integrate] the separate movements – in which [one] serves as precondition for the [other] and simultaneously becomes a result of [it] – within a single combined movement without losing the distinctive character of each...by changing vantage points for viewing them in mid-analysis.... [After viewing an element as a precondition of another, switch analytical vantage point and view it as a result of that element, because] we only know that one thing is a pre-condition for another when that latter has emerged in some recognizable form.... [I]nvestigating how something that exists came to be proceeds from its present form, the result, backward through its necessary preconditions.

By grasping the result as the outcome of presuppositions and preconditions, history can be examined with an eye toward what is explanatorily important, where the particulars of a prior system (and its dissolution) contain the keys to understanding the beginning of a new one. For example, the necessary relations that make a new system possible are often presupposed by prior structural transformations – e.g., the "capitalist system pre-supposes the complete separation of the labourers from all property in the means by which they can realise their labour" (Marx 1992j: 668); "The transformation of *state activities* into *official posts* presupposes the separation of the state from society" (Marx 1976a: 52; emphases in the original). Among the historical preconditions allowing capitalism's growth was a specific event – e.g., "the rounding of the Cape" (Marx and Engels 1978a: 474) – an outcome not necessarily predicated on the structural dynamics of feudalism. And, among the technological preconditions of modern manufacture are gunpowder, the compass, the printing press, the clock, and the mill (Marx 1936d).

We can now better see how Marx synthesizes and changes abstractions within and between historical and structural moments of analysis. Because methodology must be able to shift focus if/when data is collected at various points of historical/structural development, Marx "frequently changes both the perspective from which he sets out and the breadth of units (together with the meaning of their covering concepts) that come into his analysis" (Ollman 1979a: 111). Shifting abstractions helps clarify the range of phenomena to which concepts and generalizations extend (time/space, level of historical generality), and, moreover, helps build a model of the inner-connections between systems and their parts (more below and in Chapter Six). This synthesis of historical and structural analysis has its priorities. "As a general rule, the interactions that constitute any problem in its present state are examined before studying their progress over time. The order of inquiry, in other words, is *system before history*, so that history is never the development of one or two isolated elements with its suggestion, explicit or implicit, that change results from causes located inside that particular sphere" (Ollman 2003: 15, emphasis added). Abstracting a social form as a system of interrelated parts requires knowledge of the formal properties of the system, and this therefore requires a careful comparison between models constructed and concrete realities observed before the researcher may do any scientific writing of the system's history or exposition of its contemporary unfolding.

Dialectical Method and the Scientific Method[3]

Marx's view of science is not always clear across the range of his work – he positioned himself toward "science" and "positivism" in multiple ways, e.g., "In the course of a single letter...we find the word science being used with reference to his own conception of science and to erroneous, dogmatic conceptions, such as pseudo-positivism and positivism itself" (Thomas 1976: 3). So optimistic was a young Marx (1988a: 111) that he thought "Natural science

---

[3] An earlier version of portions of the following discussion has been published elsewhere (see Paolucci 2003b).

will in time subsume itself under the science of man, just as the science of man will subsume under itself natural science; there will be *one* science" (emphasis in the original). This interest in science in general continued throughout his life. His colleague Liebknecht (1901: 91) noted: "Especially on the field of natural science – including physics and chemistry – and of history Marx closely followed every new appearance, verified every progress."[4] Exposure to these traditions provided Marx with a view of scientific developments as he drew on calculus, algebra, Darwin, and even anatomy for his own work (see Gerratana 1973, Ball 1979, Gerdes 1985, Struik 1948). Eldred and Roth (1978: 75) argue that when "science" is first used in *Capital* – save its introductions – it is "implicitly taken to mean natural sciences and mathematics." It is therefore not without reason that McLellan (1975: 38) notes that Marx "was certainly not immune to the general enthusiasm for a supposedly value-free natural science that reached its high point in late Victorian society." But we should not confine our understanding of Marx's view of science to simply natural science, mathematics, technology, or even positivism.

In the application of natural science methods to a study of social relations, Marx, I believe, viewed "science" as a method that helped one arrive at logically sound and empirically driven knowledge. For him, science was a form of inquiry that assisted in avoiding the pitfalls awaiting speculation, illogic, cultural prejudice, and misleading impressions taken from surface appearances (Zeleny 1980). It could be said that these views do not differentiate Marx's view from traditional positivist thinking. However, Marx did not succeed in explaining how scientific methods fit into his conception of dialectical method or how his method of inquiry differed from traditional positivism in important ways. These issues can be better understood by first grasping his critique of the application of positivist methods to social questions.

---

[4] "[In July-Dec. 1963,] Marx engages in mathematical studies (differential and integral calculus).... During his illness [in Sept.–Dec. of 1964] Marx studies physiology and anatomy. [In May–Aug. 1865, Marx is] tormented by illness...but his reading is not neglected: this time it is astronomy. [In early-mid1878] he will study the problems of money, agronomy, and geology [while by the end of the year] Marx reads some studies of Leibniz, and also Descartes' posthumously published writings on physics and mathematics. [In 1882] Marx pays close attention to the Deprez experiments on the long-distance transmission of electricity" (Rubel 1980: 65, 68, 73, 111, 113, 122).

Marx's Critique of and Alternative to Positivist Sociology

Marx held a positive view of science (in principle) throughout his life and had no doubts that he could bring scientific methods to the study of social organization. Concerned with uncovering their experiences as scientifically as possible, a questionnaire he once handed out stated its goal as an "exact and positive knowledge of conditions in which the working class...lives and works" (see Wiess 1973, "Enquete Ouvriere," cited in Bottomore 1973: 178; also see Marx 1989f). During his later years he "had good, though distant relations at the time with some members of an English positivist society" (Kovalevsky 1909, cited in McLellan 1981: 128). However, as we have seen, Marx objected to ahistorical universalist theories – hallmarks of positivist thinking – for social science. Acton (1967: 30), for instance, holds that "Marx was undoubtedly a positivist, although he would not have called himself by that name, as he disliked many of the social views of Comte, the leading positivist of the nineteenth century." Of August Comte, Marx (1936h: 210) remarked:

> I am also studying Comte now, as a sideline, because the English and French make such a fuss about the fellow. What takes their fancy is the encyclopædic touch, the synthesis. But this is miserable compared to Hegel (Although Comte, as a professional mathematician and physicist, was superior to him, *i.e.*, superior in matters of detail, even here Hegel is infinitely greater as a whole). And this Positivist rot appeared in 1832!

Bottomore (1975: 10) thus suggests that Marx "considered that his own social theory was closer to being a positive science than was Comte's positivism." Given these observations, how should we understand Marx's views on a positivist philosophy of *social* science?

Kant's speculative *a priori* approach influenced positivist philosophies of natural and social science; Comte (1974) influenced the type of sociological inquiry later thinkers would adopt or reject. Positivist sociology assumes that social facts can be explained via "predictive-theory" coverall laws about "society" in general (Alexander 1982). In their form of "predictive-theory naturalism," social sciences are deductive enterprises that describe unobservable mechanisms that predict and explain observable events that can be empirically corroborated, all of which allows theory to be reduced to a set of principles systematized as general laws, though "Even a cursory survey of *Capital* shows that Marx's system does not possess a unified deductive structure of this sort " (Little 1986: 16). Marx's theories are not set up to be tested in the same way as those in natural science, because predictions require "situations of closure" (Isaac 1987: 65), a characteristic which capitalism has never had. Marx's historical materialism stipulates relationships between one structure (the mode of production) and others (social, political, and intellectual life), not specific predictions about what will or will not occur. Marx's political economy uses capitalism's essential structure to account for a range of regular phenomena, such as the concentration of wealth, economic cycles, and crises. By extension, historical materialist principles stipulate that the capitalist mode of production shapes social norms, political relations, and intellectual life. These conditioning relationships are the targets of Marx's inquiry and analysis. Within these he locates variables that help him provide descriptions and explanations of causal forces.

All scientific inquiries use the concept of "variables." Predictive theories of positivism use operational definitions of variables inserted into testable hypotheses. This approach tends to assume that "*we can measure anything that exists. There are no exceptions*" (Babbie 1995: 110; emphasis in the original). Here, if researchers "are not exactly clear about which dimensions of a variable they are really interested in," they are instructed to "be pragmatic" and to operationalize variables according to the "expected distribution of attributes among subjects," defining them in a way that is "exhaustive" and "mutually exclusive" (Babbie 1995: 138–140). In this approach, conceptualization comes first, definition second, and measurement third. But instructing researchers to exhaustively define variables when they are not sure of their dimensions is inherently contradictory. One cannot define variables exhaustively if one does not know their dimensions (speculation?); if one defines them as mutually exclusive, then one assumes their relationships are discrete and external prior to data collection; and, if one operationalizes them, they lock variables into static *a priori* frameworks. In Marx's view, these methods unnecessarily restrict and distort inquiry and analysis.

For Marx's internal relations ontology, mutually exclusive definitions obscure the reality in question. In the internal relations approach, "Marx is arguing that one cannot form an adequate principle of explanation until one knows what it is that has to be explained; and that one cannot know what is to be explained until one has the actual circumstances and the experience of them" (Gould 1978: 29). If one rigidly defines variables before empirical inquiry, this mode of abstraction may obscure pertinent detail or confuse concepts with causal agents. In his critique of Hegel, for example, Marx (1976a: 7–8) holds that "The so-called 'actual idea' (mind as infinite and actual) is presented as if it acted on a specific principle and with specific intent...whereas the form of existence of the actual idea is not an actuality evolved from itself, but ordinary empirical fact." That is, if concepts are mistaken for "the determining principle" they risk becoming "hypostatised abstractions" (1976a: 15). This concern approximates "the accidental error of mistaking the abstract for the concrete" or, the Fallacy of Misplaced Concreteness (Whitehead 1960: 52). Abstractions are not causal agents and operational definitions that lock dynamic realities into static frameworks are likely to distort them.

Operational definitions and predictive models mirror the speculative philosophies of Hegel and Kant, approaches Marx found wanting. When Marx (1992c: 28) warned – as we have already seen – not to interpret *Capital* as forwarding "a mere a priori construction," he was informing readers that his political-economic models were not formulated prior to empirical investigations. He explained to a critic:

> In the first place, I do not start out from 'concepts,' hence I do not start out from 'the concept of value,' and I do not have 'to divide' these in anyway. What I start out from is the simplest social form in which the labour-product is presented in contemporary society, and this is the 'commodity.' I analyse it, and right from the beginning, in the form in which it appears (Marx 1975m: 198).

Rather than positing abstractions at the level of society in general (as deductive predictions), for Marx (1982b: 100), "*categories* are but *abstractions* of those real relations, that they are truths only in so far as those relations continue to exist" (emphases in the original). Marx's method of grounding concepts in historically concrete observations proceeds in reverse course of (*a priori*) operational definitions and relieves him of the demand for definitional permanence. In the internal relations philosophy of science, "the empirical referents of Marx's concepts may neither be mutually exclusive, nor consistent across space and time. An empirical particular – a form of division of labor, for example – might figure as a production relation under one description and a productive force under another" (Sayer 1987: 22). As Gould (1978: 27–28) explains, "it is only in retrospect that one can reconstruct this logic as a contingent one based on what has in fact happened." Thus, dialectical methodologists "reject cookbook prescriptions of method which allow one to imagine that one can do research by simply applying them without having a scholarly knowledge of the object of study in question" (Sayer 2000: 19). Be all this as it may, Marx believed that the conceptual precision requisite for a social science could be reached.

Typifying Objects of Study

Appropriate questions help locate appropriate data; both facilitate appropriate answers.[5] In this process, abstractions reflect light on realities that make knowledge possible and critique examines the logical and empirical soundness of abstractions and concepts. Although concepts should unite necessary relations and institutional forms that define a specific historical context into thought, structural relationships cannot be directly observed, nor are they articulated apart from one another. Where does the mode of production end and the mode of reproduction begin? In one instance, Marx (1973: 93) goes so far as to equate them: "Production is consumption, consumption is production. Consumptive production." Given that a whole's constituent parts are inner-connected and mutually changing, Marx (1973: 94) criticizes abstractions pitched at the level of "humanity in the abstract." Concepts at this broad abstract level are less able to make contact with the real world. In *Capital*, Marx conceptualizes the core characteristics that define capitalism – e.g., private property, commodity exchange, circulation, the organic composition of capital, the wage-system, and all that these entail. "Nothing is made up of whole cloth, but at the same time Marx only finds what his abstractions have placed in his way. These abstractions do not substitute for the facts, but give them a form, an order, and a relative value; just as frequently changing... abstractions does not take the place of empirical research, but does determine, albeit in a weak sense, what he will look for, even see, and of course emphasize" (Ollman 2003: 74). The "mode of production" – an assumption about the base of wider social relations – is a dynamic concept that helps Marx observe and organize the relationships within and between

---

[5] "Just as the solution of an algebraic equation is found the moment the problem is put in its purest and sharpest form, any question is answered the moment it has become an actual question" (Marx 1967: 106–107).

historical systems. As a configuration of various ideal and material relations, some extending from the past and others newly developed, he abstracted modernity with capitalist productive activity as its fundamental base. Tracing modernity's interrelations required abstracting modes of production across history and comparing them (and their material-ideal relations) with what is obtained under capitalism.

Now we can better answer the question: If operational definitions are eschewed under Marx's internal relations approach, then what is the character and role of his concepts? Engels (1909: 24) warned of one critique that "starts out from the mistaken assumption that Marx wishes to define where he is only analyzing, or that one may look into Marx's work at all for fixed and universally applicable definitions. It is a matter of course that when things and their mutual interrelations are conceived, not as fixed, but as changing, that their mental images, the ideas concerning them, are likewise subject to change and transformation; that they cannot be sealed up in rigid definitions, but must be developed in the historical and logical process of their formation." Because social phenomena are composed of multiple material and ideal realities across time/space, "some events or some aspects can more profitably be described in one language, whereas others are more fruitfully described in another" (Israel 1979: 14). It is not a matter of reality matching our categories, but of our categories approximating reality.

Though their construction is not *the goal* of his research, Marx uses concepts to mediate the relations between his onto-epistemological assumptions and the observation, analysis, description, and explanation of socially regular practices. Thus, his concepts change as his empirical referents change and are more or less justified by their descriptive adequacy, logical structure, and explanatory power – e.g., *Capital* (Volume One) is about the modern mode of production and we should not mistake what Marx writes there for either an analysis of "society," all that is associated with modernity as a totality, or even all things associated with capitalism. As goals and foci change, new data must be typified and placed into relational categories. Some elements in a research program will be retained over time, whereas others will be set aside.

In historical work such as *Class Struggles in France*, Marx strives to explain a particular set of events. In his political-economic work, his goal was the presentation of the causal relationships indicative of general systemic phenomena. Both goals required that he set boundaries for his concepts. The conventional operational definition "spells out precisely how [a] concept will be measured" (Babbie 1995: 116). In the internal relations approach, the "starting-point of the analysis lies not in abstraction ('simple' or otherwise) but in a concrete social form...as...it presents itself phenomenally" (Sayer 1979: 112). Patterns uncovered during research are seen as expressions of a general type and categories are created from these empirical realities so "that actual conditions are represented only to the extent that they are typical of their own case" (Marx 1909: 169). In settling on typifying conditions that make a category meaningful, one compares cases that share common characteristics. If those characteristics represent an essential part (or parts), then a range of phenomena they regularly determine should be present (unless there are overriding factors) and concepts are constructed to register this. However, because structural relationships can change, Marx changes his abstractions to fit the quality of data he examines.[6] For example, money, wages, labor, circulation, and the state are typified by Marx and reconceptualized at different moments in exposition depending on the level of generality and/or the time/space location he is analyzing. In relating conceptualization in political economy to the principles of historical materialism, the relevant concerns include:

- What material-ideal relations typify class systems?
- What are the typical relations indicative of capitalism?
- What characteristics do capitalist relations share with other class systems?
- How do capitalist relations differ from other class systems?
- What presupposition and preconditions in prior class systems account for capitalism?

---

[6] "Marx had the two qualities of a genius: he had an incomparable talent for dissecting a thing into its constituent parts, and he was past master at reconstituting the dissected objects out of its parts, with all its different forms of development, and discovering their mutual inner relations. His demonstrations were not abstractions – which was the reproach made to him by economists who were themselves incapable of thinking; his method was not that of the geometrician who takes his definitions from the world around him but completely disregards reality in drawing his conclusions. *Capital* does not give isolated definitions or isolated formulas; it gives a series of most searching analyses which bring out the most evasive shades and the most elusive gradations" (Lafargue 1890; cited in McLellan 1981: 73).

- How are these extended historical-structural relations to be captured conceptually?

In Marx's approach, categories are less right or wrong than they are more or less useful, more or less uncovering or concealing, more or less explanatorily powerful and/or descriptively adequate. As the degree of usefulness increases (or decreases), the more (or less) a concept approximates what might be counted as wrong or right (correct or incorrect) in the traditional sense. Two initial rules for concept construction are suggested. First, abstractions that unite multiple observations into shared conceptual categories "should distinguish incidental from essential characteristics." For example, wage-labor, or an otherwise in-kind contract labor, is an essential component of capitalism as a system, whereas slavery is not. Second, abstractions "should neither divide the indivisible nor lump together the divisible and the heterogenous" (Sayer 1984: 82). For instance, labor exploitation under capitalism and under slavery are not of the same nature and quality.[7] In discovering causal properties between such essential parts, Marx used a form of multivariate analysis and the controlled comparative method. (Marx's conceptual practices will be addressed in greater detail in chapters Five and Six.)

Multivariate Analysis, Controlled Comparison, and the Experimental Model

It is standard in science to isolate and examine the effects of variables on one another. A materialist social science must handle the fact that a historical event "may itself be viewed as the product of a power which, taken as a whole, works *unconsciously* and without volition" (Engels 1936c: 476; emphasis in the original). Some causal variables, then, will be less apparent than others. In moving from this assumption to research practice, Marx accepted "the general view (obviously derived from Hegel) that no element in the total

---

[7] "In the corvee, the labour of the worker for himself, and his compulsory labour for his lord, differ in space and time in the clearest possible way. In slave-labour, even that part of the working-day in which the slave is only replacing the value of his own means of existence, in which, therefore, in fact, he works for himself alone, appears as labour for his master. All the slave's labour appears as unpaid labour. In wage-labour, on the contrary, even surplus-labour, or unpaid labour, appears as paid. There the property-relation conceals the labour of the slave for himself; here the money-relation conceals the unrequited labour of the wage-labourer" (Marx 1992j: 505).

process of history could be isolated and given a significance unaffected by the other elements" (McLellan 1975: 38). In such a conception – i.e., a multivariate analysis – Marx needed to distinguish and compare different historical systems, isolate variables of interest, and then estimate their causal influence, order, and mutual interaction. For example, in the article "Population, Crime, and Pauperism," Marx (1980: 488–489) lists statistics for these variables over time and concludes:

> By comparing these three tables of population, crime and pauperism, it will be found that from 1844 to 1854 crime grew faster than population, while pauperism from 1849 to 1858 remained almost stationary, despite the enormous changes worked during that interval in the state of British society.... There must be something rotten in the very core of a social system which increases its wealth without diminishing its misery, and increases in crimes even more rapidly than in numbers.

Although using controlled comparison allowed him to make such claims, it was his manipulation of the levels of generality and various analytical vantage points that helped him focus and refocus his inquiries, much like adjusting a microscope.

As noted in Chapter Two, in Marx's (1992a: 19) "analysis of economic forms...neither microscopes nor chemical reagents are of use. The force of abstraction must replace both." Like microscopes, abstractions help expose features of an object that were hitherto unobservable. Reflexive abstraction, recognition of various levels of generality, and awareness of a measurement's limits are important when focusing the abstractive lens. "Vantage point" abstractions – shifting the analysis from one relationship to another – allow Marx to piece interrelationships between parts and wholes together. The flexibility in altering one or more vantage points when collecting and analyzing data helps him respect the historical and/or structural location on which data falls.[8] Carved from both inner-connected parts and their historical development, vantage point abstractions thus assist in clarifying the range

---

[8] "What were treated in [Book] I as *movements*, whether of capital in a given branch of production or of social capital – movements changing the composition, etc. of capital – are now conceived as *differences* between the *sums of capital invested in the various branches of production*" (Marx 1936I: 242; emphases in the original).

of historical phenomena to which concepts and generalizations extend and help the researcher build a model of the connections between structural parts (see Ollman 2003: 99–111).

Marx uses a controlled comparison in this process. Because realities exist at different spatial-temporal moments, Marx (1979c: 321–322) tells us, "events of a striking analogy, because they took place in a different historical milieu, led to entirely different results." Thus, "one would never thereby attain a universal key to a general historical-philosophical theory, whose greatest advantage lies in its being beyond history." Variable levels of development, real historical change, multiple structural interrelationships, countervailing and/or undiscovered forces, and interactions between material forces and forms of social knowledge all combine to make analysis of any social reality incredibly complex. Variables may exhibit different properties at different times.[9] Variables' properties may change over time, and change their relations to other variables and their overall causal effects. Similar events may have different causes; different outcomes may have similar causes. How, then, is controlled comparison possible, especially without operational variables? The answer lies in changing the vantage point from the analysis of historical social structures at one point to examining a particular social structure at another. In historical analysis, controlled comparison is used to identify and sift out different modes of production and crucial historical events. In structural analysis, variables are isolated and their action upon one another is examined through an approximation of the experimental model (more below). Thus, controlled comparison has a central place both *within* and *between* the levels of generality that Marx studies (for discussion, see Van Den Braembussche 1990, McMichael 1990, Haydu 1998).

Controlled Comparison from the Vantage Point of Historical Analysis. History does not announce wholesale systemic changes as they occur; a dynamic of change and stability exists, though the distinctions can be difficult to disentangle. Marx (1992j: 351) claimed that successive "epochs in the history

---

[9] "I remark *en passant* that during the past year – so bad for all other business – the *railways* have been flourishing, but this was only due to extraordinary circumstances, like the Paris exhibition, etc. In truth, the railways keep up an appearance of prosperity, by accumulating debts, increasing from day to day their *capital account*.... However the course of this crisis may develop itself – although most important to observe in its details for the student of capitalistic production and the professional *theoricien* – it will pass over, like its predecessors, and initiate a new 'industrial cycle' with all its diversified phases of prosperity, etc." (Marx 1991c: 355; emphases in the original).

of society are no more separated from each other by hard and fast lines of demarcation, than are geological epochs." Although one era ends and another comes into being, when exactly does this happen? As historical epochs end, the next ones, paradoxically, never really "begin" – i.e., they are always partial and in potential until they are fully developed. Historical stages can best be abstracted and identified conceptually when they have matured and are generalized, though some parts may already be dying or dissolving, if only to a minor degree. The transition to capitalism has been understood most clearly, even if only partially, in the present, after feudalism's full dissolution. Because some things are found in all types of societies (e.g., families, sex norms, language), some in many types of societies (e.g., religion, class struggle), and others in just a few (e.g., no formal class structure), and because change happens within and between types of societies, analysts must be sensitive to the variability and stability of relationships at different levels of generality. Not all class societies can be analyzed in the same terms as capitalism, and we cannot be sure how far into the future capitalism will extend; it is easier to discover how far back in the past it existed. "That is, we cannot predict the future concretely, but we can predict the past" (Wallerstein 1974: 389).

Zimmerman (1976: 70) reminds us that Marx attempted to create "a method able to grasp the socio-historical forms of changing categories in related but different modes of production." This method has a dual movement. First, it attempts to capture the essential structures of a formation while remaining attentive to the length of time relations originating in one period extend (or fail to extend) to others. Second, in the comparison of historical realities, "attention is paid not only to the contrasts as such, but more emphatically to the uniqueness of the studied phenomenon" (Van Den Braembussche 1990: 192). Although all societies share certain structures (e.g., production in general), what sets them apart are their unique configurations (e.g., class relations). If we understand that the "distinction between a society where [the commodity] form is dominant, permeating every expression of life, and a society where it makes an episodic appearance is essentially one of quality" (Lukács 1971: 83), then it becomes clear why Marx (1992j: 209) argues that "The essential difference between various economic forms of society, between, for instance, a society based on slave-labour, and one based on wage-labour, lies only in the mode in which this surplus-labour is in each case extracted from the actual producer, the labourer." The mode of appropriation is thus the criteria Marx used to distinguish class systems from one another. Ancient/tribal society, slavery, the Asiatic mode, and feudalism are among the class systems that include capitalism – all have modes of appropriation – while their different modes of appropriation set them apart, i.e., "distinctions within a unity" (Marx 1973: 99).

The abstractions of essentialism, historical generality, and vantage point combine with cross-cultural comparison to help the researcher recognize the historical-structural relationships out of which data is extracted. As Van Den Braembussche (1990: 192) explains:

> Marx...stressed the unique characteristics of modern capitalism and made a thorough use of the comparative method to point to the distinguishing features of Western developments *vis-á-vis* non-Western ones.... Here, capitalism is already conceived, even at a purely heuristic level, as a more or less irreducible whole. So, though social formations...are already assumed from the beginning, the comparative method is used here exclusively to demonstrate the *specificity* of different but, at least in principle, comparable developments on a world scale (emphasis in the original).

Marx (1992j: 300), in urging that the "general and necessary tendencies of capital...be distinguished from their forms of manifestation," uses the force of abstraction and the comparative method to specify the generalizations he makes (and thus to periodize history). Because abstractions of extension are about ontological assumptions as much as epistemological practices, Marx compares the extended interrelationships between Levels Two and Three (recent capitalism and capitalism in general) with those of Four (class society) and Five (society in general) to construct categories, make comparisons, and tentatively extend conclusions to other social phenomena in other concrete expressions of the same general type.[10] In weighing variable relationships, Marx sifted out their connections by a controlled comparison of independent and dependent variables within an experimental model constructed in the abstract.

The Experimental Model from the Vantage Point of Structural Analysis. Once the essential elements of a system are grasped, analysis can shift abstractions from

---

[10] "While division of labour in society at large, whether such division of labor be brought about or not by exchange of commodities, is common to economic formations of society the most diverse, division of labour in the workshop, as practised by manufacture, is a special creation of the capitalist mode of production alone" (Marx 1992j: 339).

one vantage point to another, "placing them within a more comprehensive framework" (Beamish 1992: 167). As temporal variables change, the quality of empirical events, with concepts already constructed for them, may change in a manner whereby the analytical construct previously in use no longer matches the empirical concrete to which it was once attached. This method provides for more detail and depth. For example, although all economies have production and distribution of social value, capitalism has capital and labor as the central social/class relation. Production from the vantage point of capital can be abstracted as profit and growth. Production from the vantage point of labor means work. The objective for both is survival in the market. But for labor the identity with capital is offset by a difference – the lack of survival in the market for labor is truly a matter of life and death. From the vantage point of capital, distribution of value and wealth entails the search for markets, buying labor cheap, and selling commodities dear. From the vantage point of labor, distribution means finding affordable housing, food, daycare, etc. Abstracted in its similar identities for both labor and capital, the subject is competition for scarcity, a literal life and death struggle for the former, a figurative life and death struggle for the latter. Vantage point abstractions allow for the incorporation of variables with multiple identities in this way. Because of their multiple identities and multiple relations, such variables must be placed within an analytical framework that allows their causal forces to be isolated for particular moments in their development.

The experiment is perhaps the key aspect that makes science qualitatively different from other modes of knowing. In an experiment, "variables are manipulated and their effects upon other variables observed" (Campbell and Stanley 1966: 1). The central analytical maneuver is that "when estimating the effect of X_i on X_j, control all prior and intervening variables; that is, control all variables not consequent to X_j," where "the central notion is 'control' and the central problem is when and when not to introduce variables as controls. The issue is 'whether', and not 'how'" (Davis 1985: 37, 9). Holding variables constant allows for their effects to be observed independently of one another and this allows the researcher to estimate their hierarchy of importance. Understanding how Marx used the experimental model requires grasping the way he introduced variables and how he isolated their causal effects.

In real life, structural parts emerge in different historical periods or develop at different places and rates. Because real experiments are often unavailable in social science, they must be approximated *in abstractio*. In *Capital*, Marx

(1992a: 19) claims his goal is to observe "phenomena where they occur in their most typical form and most free from disturbing influence," where the powers of abstraction will allow him to make "experiments under conditions that assure the occurrence of the phenomenon in its normality." He introduces variables in a way so they can be examined within the proper level of historical and structural development, where other variables – those previously introduced and those not yet introduced – are held constant. For example, "if commodity production, or one of its associated processes, is to be judged according to its own economic laws, we must consider each act of exchange by itself, apart from any connexion with the act of exchange preceding it and following it" (Marx 1992j: 550). This point of entry allows for introducing variables in a controlled order. Marx (1975m: 199) explains: "If we have to analyze the 'commodity' – the simplest economic concretum – we have to withhold all relationships which have nothing to do with the present object of analysis." In such an approach, "inquiry will confine itself to the confrontation and the comparison of a fact, not with ideas, but with another fact" (Marx 1992d*: 27). Marx's experiments are performed by either 1) examining concrete cases with several similar variables at different levels of development; 2) examining concrete cases that share the same variables except one (or more); or, 3) by constructing model relationships based on empirical observations and deducing how outcomes are likely to change if variable relationships change.

Models of causal systems distinguish relationships between independent (causal) and dependent (caused) variables. In a multivariate causal analysis, variables must be fitted together so that each relates to a relevant temporal-spatial plane, i.e., a level of generality, a geo-historical location, plus specific structural configuration. Marx's texts contain numerous instances where he pauses to present a mini-cluster of relations in a causal system and examines the consequences for the causal cluster when variables are allowed to vary. For instance, Marx (1992j: 296) reviews his analysis of absolute and relative surplus-value:

> That portion of the working-day which merely produces an equivalent for the value paid by the capitalist for his labour-power, has, up to this point, been treated by us as a constant magnitude, and such in fact it is, under given conditions of production and at a given stage in the economic development of society. Beyond this, his necessary labour-time, the labourer, we saw,

could continue to work for 2, 3, 4, 6, &c., hours. The rate of surplus-value and the length of the working-day depended on the magnitude of this prolongation. Though the necessary labour-time was constant, we saw, on the other hand, that the total working day was variable.

After presenting this multivariate causal system in mid-analysis, Marx holds the length of the working-day constant, allows the length of necessary labor-time to vary, and then examines the outcome for workers, technology, the cost of living, etc. These variables are not carved at the level of the individual, the capitalist, the factory, industry, nor even the nation-state, but at the level of class relations as a whole where capitalist relations obtain concrete expression. This examination is not about abstract concepts, marginal or unique events, nor patterns of regularities that are next to impossible to observe. Marx strives to explain real life under capitalism in a way people have not had explained to them before.

Within structural analysis, clusters of variables function as explanatory models, continuously amended as data is collected and as the vantage point is changed. Therefore, some concepts are not constructed as constants across their range of application. How, when, and in what way to introduce variables into analysis is thus very important if we are not to misconstrue explanatory and interpretive frameworks. The simplest come first and others cannot be introduced unless they have reached a certain level of maturity. If re-introduced later in analysis, they may have changed their function. For example, money was developed in a variety of societies to facilitate economic exchange. Under capitalism, an increasing number of social institutions and practices became subordinated to the expansion of money as profit, and production of surplus-value and the ceaseless accumulation of capital became the central purpose of the system. However, a limited amount of money does not yet function as capital and the pursuit of financial gain in market exchange does not yet express capitalistic behavior. Money must be of a certain magnitude in order to purchase the means of production. Capitalist behavior is defined by the exploitation of labor-power resulting from this purchase. Referring to simply "buying cheap and selling dear" as capitalist activity does not fit with what makes capitalism a *differentia specifica*: What characterizes capitalism is the private ownership of the means of production and production for the sake of surplus-value and the expansion of capital, with these latter realities based upon the exploitation of labor-power through the wage-system.

If simple trading, buying, and selling (and even hoarding) are equated with capitalism, as is often done in conventional wisdom and social science, then historically general practices are conflated with capitalism as a unique system, making it appear both ahistorical and more a product of human nature than is the case.

It is instructive to re-read *Capital* with the experimental model and controlled comparison in mind. Marx's examination of the commodity is presented first. Variables – causal, caused, and controlled – are added and relaxed as various chapters and analyses unfold. By Volume Three, the Falling Rate of Profit is examined with a range of variables held constant in Volume One. We know this is an instance of controlled comparison rather than a *post hoc* addressing of an error because Volume Three was written first. Marx sets up his arguments in Volume One knowing that the rules and laws found there will be set in motion and change in other volumes as new variables are introduced. Volume One presents an analysis of surplus-value, showing its origins in production, Volume Two reintroduces exchange (placed in the background in Volume One), although retaining Volume One's conclusion that commodities exchange at their values, and Volume Three examines prices, interest, credit, and rent. Marx's method is to bring everyday knowledge into play via a careful sequencing, testing everyday knowledge by way of taking one of its elements and temporarily excluding others in his mode of presentation and analysis and then later relaxing the previous assumptions and readmitting what was previously excluded (Roth and Eldred 1978: 11). Thus, *Capital* approximates an experimental model through the way each volume increasingly adds variables to those models previously established, often while relegating previous variables to the status of constants. Examining these volumes and their chapters with the knowledge that they represent an experimental model would go a long way in settling many debates over the intended meaning(s) of the laws Marx sets out in these texts.

Controlled Comparison in the Synthesis of Historical and Structural Analysis. If structure and history are internally related, then failing to incorporate their inner-connections into abstractions is to misappropriate empirical detail. Marx's levels of generality attempt to handle this problem. However, though internally related in time and space, levels of generality cannot be measured with the same tools. They must be reconstructed in thought through categorical analysis, entered into controlled comparison, and the variables they contain must be put together in an order and quality that acknowledges their dynamic character (see Ollman 2003: 117–118).

In researching the internal relations of history and structure, Marx (1973: 817) recognizes that the "fixed pre-suppositions themselves become fluid in the further course of development. But only by holding them fast at the beginning is their development possible without confounding everything." Marx examines clusters of relations that account for the variability in observed events and focuses on the action of variables on one another. "If one studies each of these developments by itself and then compares them with each other, one will easily find the key to each phenomenon" (Marx 1979c: 322). There are no first causes within the historical-structural relation, so Marx's structuralism is not an iron-clad determinism and, in fact, "one misreads this history if the structures are hypostatized or related to a principle or metastructure supposedly determining all that history may produce" (Zimmerman 1976: 75). The variables that make up an explanatory framework might have different structural-historical origins and might have variable quality of effect. Dialectical variables, therefore, are not conceptualized as constants across their range of application and their causal relations are not strictly one-way.

In establishing causal relations, Ollman again provides us several principles to keep in mind. First, after specifying generality and modeling of innerstructure, a vantage point abstraction "sets up a...place within the relationship from which to view, think about, piece together the other components in the relationship. Meanwhile, the sum of their ties – as determined by the abstraction of extension – also becomes a vantage point for comprehending the larger system to which it belongs, providing both a beginning for research and analysis and a perspective in which to carry it out" (Ollman 2003: 100). Second, abstractions made at one level of generality can bring into focus "qualities that can now serve individually or collectively (depending on the abstraction of extension) as vantage points, just as other possible vantage points, organized around qualities from other levels of generality, are excluded." Although extension, level of generality, and vantage point are abstracted simultaneously, "on any given occasion one or another of them tends to dominate" (Ollman 2003: 100–101). Finally, in piecing together historical-structural relations, cause and effect are linguistically replaced: cause and/or reciprocal effect; determine and are determined by; precondition and result; and, presupposes and is presupposed by (see Ollman 2003: 116–125).

Statistical Analysis, Deductive Reasoning, Formulating Laws, and Model Building

Marx's political economy can and must be read in light of his study of mathematics (see Marx 1983i). In discussing his political-economic studies, Marx noted, "I occupy myself with statistics" (reported by Duff 1879, cited in McLellan 1981: 142). During the 1858–1883 period, the extent to which Marx worked on algebra, analytical geometry and differential, integral, and infinitesimal calculus "is shown by a large amount of manuscript material which was found among his papers" (Struik 1948: 181). Marx (1985h: 484) once wrote to Engels that his "spare time is now devoted to differential and integral calculus." Lafargue (1890) reported that Marx "held the view that a science is not really developed until it has learned to make use of mathematics.... Algebra even brought him moral consolation and he took refuge in it in the most distressing moments of his eventful life" (cited in McLellan 1981: 71). For example, in November 1860, as he was nursing Jenny Marx to back health, Marx (1985b: 216) confessed, "The only occupation helps me maintain the necessary quietness of mind is mathematics." Not simply a passing hobby, his work on "infinitesimal calculus [was], according to the opinion of experts...of great scientific value" (Lafargue 1890, cited in McLellan 1981: 70), as it measured favorably against the standards in the field of the time (Gerdes 1985).

Capitalism, in many ways, works by a ledger sheet of wages, prices, and profits. This material relation must be accounted for epistemologically. Quantitatively measured relationships reveal information that other methods of data collection might miss, e.g., the distribution of wealth. Although Marx's worker questionnaire was not intended to test ideal concepts, it is likely he would have used statistics on the data it revealed. Moreover, capitalists fund agencies to generate aggregate data about markets and it is important to be able to grasp such information. It is therefore incorrect to assume that Marx's approach is antithetical to mathematical and quantitative methods. *Capital*, a deeply dialectical work, regularly refers to statistical concepts and allows for sophisticated statistical modeling (e.g., see Maarek 1979, Wolff 1984). As such, it is as much a mathematical analysis as it is a dialectical one (see Chapter Six).

A correlation, a "measure for describing the strength of the association between the two variables" (Agresti and Finlay 1986: 263), is a statistical concept with which Marx was familiar.[11] For example, the rate of accumulation of capital changes when there is a change in the "mass of exploitable labour-power.... To put it mathematically: the rate of accumulation is the independent, on the dependent variable; the rate of wages, the dependent, not the independent, variable" (Marx 1992j: 581). In another example, Marx (1992j: 574) notes that between the relationship between technological complexity and the productivity of labor "there is a strict correlation" that has a meaning for capital and labor "in so far as it is determined by its technical composition and mirrors the changes of the latter." Sometimes Marx (1992j: 581–582) presents correlations without using the exact terminology, where changes in the magnitude, mass, quantity, size, and/or amount of one variable account for "corresponding" changes (positive correlations) and/or "inverse ratios" (negative correlations) elsewhere (1992j: 602–603). These abstractions isolate variables and establish their relationships with others. Marx estimates these with both quantitative phenomena (e.g., wages) and qualitative social facts (e.g., value, means of production, and labor-power).

Marx looked at multiple variables and understood their relationships as always being partial and examined them in terms of their mutual correlations, e.g., "The total circulation of commodities in a given country during a given period is made up on the one hand of numerous isolated and simultaneous partial metamorphoses" (Marx 1992j: 121). Because the composition of forces may change each individual causal power when taken alone, there is a need in social science to use a "multiple regression model...[where the 'partial correlation coefficient' is] a set of measures that describe the partial association of Y with each of the dependent variables, controlling for others" (Agresti and Finlay 1986: 329). For example, Marx (1992j: 487) pauses to explain the relationship between several variables: "Very different combinations are clearly possible, according as one of the three factors is constant and two variable, or two constant and one variable, or lastly, all three simultaneously variable. And the number of these combinations is augmented by the fact that, when these factors simultaneously vary, the amount and direction of their

---

[11] "I recently had an opportunity of looking at a very important scientific work, Grove's *Correlation of Physical Forces*. He demonstrates that mechanical motive force, heat, light, electricity, magnetism, and chemical affinity are all in effect simply modifications of the same force, and mutually generate, replace, merge into each other, etc." (Marx 1985j: 551).

respective variations may differ." Many of the relationships Marx presents in *Capital* approximate multiple regression in this way. After discovering such relationships, Marx used deductive reasoning to construct explanatory models for his empirical observations.

Though Z.A. Jordan (1967: 311) believes Marx was "aware that a mere assertion of logically related general propositions is not sufficient to establish a theory on a sound scientific basis," Marx *did* use deduction in his work. Not only did critique help him deduce the validity of other theories, deduction helped him work through provisional questions and hypotheses in his analysis and its presentation. Requisite to Marx's deductions are knowledge of an empirical reality, grasp of a concept's limits in relation to this empirical reality, and whether an empirical reality has matured to a level where an adequate conceptualization is possible (Therborn 1976: 47). Marx's (1992j: 75) form of dialectical deduction can be seen in the following:

> The difficulty in forming a concept of the money-form, consists in clearly comprehending the universal equivalent form, and as a necessary corollary, the general form of value, form C. The latter is deducible from form B, the expanded form of value, the essential component element of which, we saw, is form A, 20 yards of line = 1 coat or x commodity A = y commodity B. The simple commodity-form is therefore the germ of the money form.

Marx (1992j: 282) also uses the phrase "it follows" to present deductive conclusions: "Without, however, anticipating the subsequent development of our inquiry, from the mere connexion of the historic facts before us, it follows... *First*..." (emphasis in the original). Such deductions helped Marx construct structural models of capitalist production and estimate its laws.

After Engels urged him to brush up on political economy, Marx took the idea of an economic system regulated by laws from the literature (Brewer 1984: 4). His approach to laws was forged through his critique of writers such as John Stuart Mill, David Ricardo, Pierre Proudhon, and Adam Smith. He concluded that capitalism is governed through internal relations in contradictory relationships and tendencies that determine a range of concrete outcomes. These are the ultimate targets of his study. Thus, Marx (1992a: 20) used the details of capitalist development of his time period as data from which "to lay bare the economic law of motion of modern society." To discover these laws, Marx had to answer the following questions:

- What are the essential parts of the capitalist mode of production as a whole?
- What are their internal relationships?
- How are these relationships related to empirical observations?

The more rigidly rule-based a social structure is, the more likely it manifests a law-like tendency. Capitalism's commodity markets, wage-system, and search for profits reduce many social relationships to material calculations. Marx formulated his political-economic laws by examining the empirical relationships between two or more variables in this system and deduced the implications of their interactions and mutual changes. A law for Marx does not start with predictive syllogisms (e.g., "if A, then B"). "Rather, it is composed of a complex of...causal mechanisms, which operate in an unpredictable, but not undetermined, manner" (Isaac 1987: 47–48). Marx's laws are not therefore carved with exact mathematical precision with strict statements of causation and predictability, nor are they carved at the level of "society" in general. Rather, Marx (1992j: 591–592) carved many of his laws as "peculiar to the capitalist mode of production; and in fact every special historic mode of production has its own special laws...historically valid within its limits alone." Even with this separation of spheres, Marx's approach (shared with Darwin) provides general hypotheses that define objects of investigation and suggests the corroboration one is likely to find in particular cases, though any such hypothesis "will not strictly reflect actual circumstances in every case to which we attempt to apply it and to which it is in principle applicable" (Carver 1982: 36–37). Generalized to the level of a system, the processes and the relationships expressed by laws have tendencies that must be approximated – i.e., "With the varying degree of development of productive power, social conditions and the laws governing them vary too" (Marx 1992d*: 28). Thus, Marx's belief that capitalism's laws can be uncovered in a way similar to natural science's laws is a qualified position.

Each part of a relation does not have equal effect on the parts with which it shares an essential inner-connection. Some parts tend to determine the range of outcomes for other parts more than the other way around. Although it is true that in *Capital* Marx often refers to "iron laws," these were formulated with many variables held constant and are not meant to apply with strict and invariable application. Because the array of variables Marx holds constant in Volume One are later relaxed, the laws he targets take on the character of structural tendencies. This is the general logic of Marx's laws. What is their specific logic?

Marx (1992j: 104) sees capitalism as a "mode of production whose inherent laws impose themselves only as the means of apparently lawless irregularities that compensate one another." That is, laws are bookended in the real concrete by cases that often appear to be different but, if averaged out, we find that they are ends of a continuum: The "general law of value enforces itself merely as the prevailing tendency, in a very complicated and approximate manner, as a never ascertainable average of ceaseless fluctuations" (Marx 1909: 190). In discovering such laws, one must "take long periods into consideration" (Marx 1992j: 123). Central tendencies are therefore the average around which cases fluctuate (see Figure 4.2). However, because there is a "fixed minimum [which] deviates from the average" (1992j: 307), overriding conditions make it possible that, in a given case, the essential part and/or its function may be obscured from view. As a result, "laws can be true even if the tendencies of which they speak do not become manifest" (Ruben 1979: 71). In this way Marx "always leaves room for countervailing tendencies that may impede or reverse the working of the law" (Little 1986: 29). A countervailing factor, Marx (1909: 275) explains, does "not suspend the general law. But it causes that law to become more of a tendency, that is, a law whose absolute enforcement is checked, retarded, weakened, by counteracting influences." Marx's laws – even those he refers to as "iron laws" – are therefore tempered by determinable negations, suspensions, and even reversals. The reality of such transformations is weighed per case. These stipulations allowed Marx to make provisional generalizations on their basis.

A generalization is a characterization of how far in time/space statements about facts, functions, causal relations, and laws can and/or should extend. Generalizations provide for extending the estimation of the relative influence of one set of relations over others. Although Marx seeks to discover generalizable truths derived from the inspection of empirical reality, he is critical of approaches that abstract from more narrow samples while making generalizations pitched at the level of the human condition in general. These risk dehistoricizing phenomena that are structurally specific, mistaking contingent historical facts for necessary ones. However, although Marx uses generalizations, their construction is not his method's reason for being. A dialectical science's ultimate goal is not pronouncement of abstract correlational relationships between two or more variables (though Marx uses

Figure 4.2
Marx's Laws as Scatter Plots*

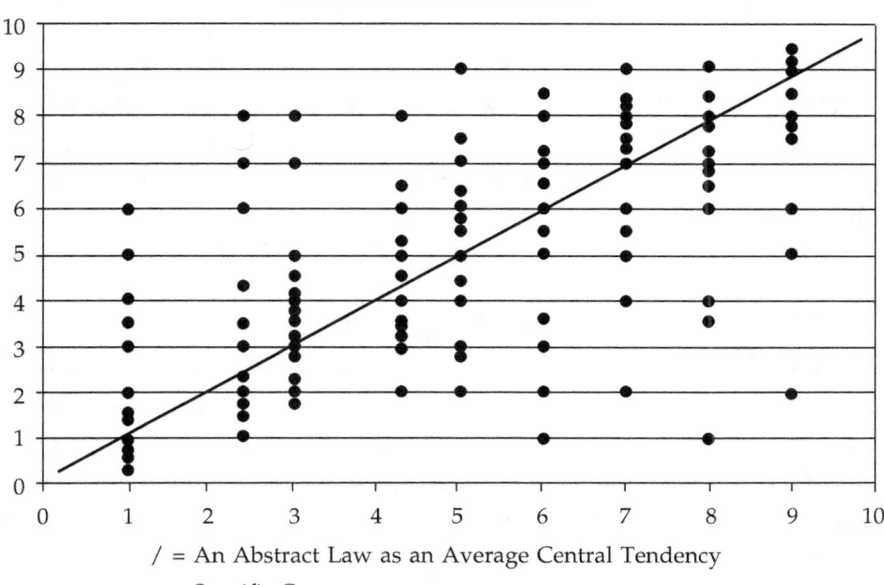

/ = An Abstract Law as an Average Central Tendency
• = Specific Cases

* This graph was developed with the help of Maya Becker.

these), but the description and explanation of historico-structural relations and processes via pertinent detail. Once processes and relationships are grasped at a level of maturation that allows for a study of them as sets of internally related parts, then their patterns of interaction suggest the extension of generalizations to other similarly composed phenomena. It is in these statements where correlations, laws, and generalizations do their work. Their collective interrelations function analytically as a form of model building.

Model building, a method of locating a complex series of variables and putting them together with their attendant laws, is a key feature of scientific method. When Marx incorporated multiple variables into working models, he altered these as his research proceeded (see Marx 1985h). These served as the bases for grasping the general social structure of capitalism, its essential parts, their interrelations, and their contemporary institutional configurations. In the application of a model, it is possible that a structural relation's functions, or tendencies, are present in essential form, but given the nature of internal relations, its configuration may render what normally would be easily observed into an unrecognizable form. Thus, one cannot assume that similar institutional frameworks function in the same way in different cases. Explanatory models only approximate concrete reality and a perfect fit between a model and concrete action is not always possible. However, the model itself should contain the essential parts of a structure, delineate their interrelations, and estimate the laws that follow from these interrelations, all inductively generated first and deductively pieced together afterward. (The approach Marx used in his model building is developed in more detail in Chapter Six.)

Presentation of Data and Findings

There are better and worse ways to present one's research. Marx's presentation was not always clear and he often remained unsatisfied by it. Demands from publishers sometimes limited the space afforded him to present his findings (see Marx 1987b). Hindsight teaches us that he would have rewritten some things had he known which ones would take on the level of interpretive importance they would later acquire. This tension seems insuperable. For example, Marx (1936f: 204) asserted that "Whatever shortcomings they may have, the merit of my writings is that they are an artistic whole, and that can only be attained by my method of never having them printed until they lie before me as a *whole*" (emphasis in the original). This was untrue. Marx often had a picture in mind of what he wished to convey but could not do so in a manner he felt appropriate, e.g., "But for all that...the thing is rapidly approaching completion. There comes a time when one simply has forcibly to break off" (Marx 1982c: 377). Rather than investigating this observation-*cum*-grievance, it is more useful to ask what can be learned about what to do and not do by looking at Marx's method of presentation.

Marx's presentation in political economy is a systematic attempt to reveal the social complexity of the sway of capital. Though beginning his inquiry from the whole backward, Marx started his presentation with relatively simple conceptions that displayed the logic of the system's parts in the abstract (e.g., the commodity) and, in working out their interconnections, he progressed to increasingly complex models that brought in variables in a particular articulation and order, introducing complicating and independent factors that assisted with an empirical account of the whole, extending the structural moment of analysis, and/or bringing historical developments into play (he had planned to write with increasing historical and structural complexity to reach the level of the world market and world history; see Marx 1973: 108). This is in line with his claim that the order of his investigation and presentation were opposite of one another. That is, when "investigation is ready for presentation it is presented not as beginning with the concrete but with the abstract. The method of inquiry moves from concrete to abstract, but the method of presentation moves from abstract to concrete" (Fraser 1997: 98–99).

Conclusions about systems are presented first in the form of a series of abstract concepts that set up a model that puts the data in context, i.e., system before history. Once a rudimentary model is developed, covering concepts are introduced, beginning with simple, abstract detail gleaned from the system's concrete moments. After the structure is mapped out, its historical origins can be presented. Observed regularities are explained by piecing together the mechanisms of the system's development. In Volume One, Marx's inquiry begins with abstract historical regularities (typifications of real activity in a structural context) and moves to concrete representations of that system (model building). His presentation begins with abstract systemic parts and moves to concrete historical description in an "attempt to work through the argument, raising and resolving objections along the way" (Eldred and Roth 1978: 10). These rules of presentation allow for a provisional and delimited view of future change and developments (Sekine 1998). The evolution of these moments in Marx's inquiries is presented in Figure 4.3.

Figure 4.3
The Phases of Inquiry Leading to Marx's Presentation of Political Economy

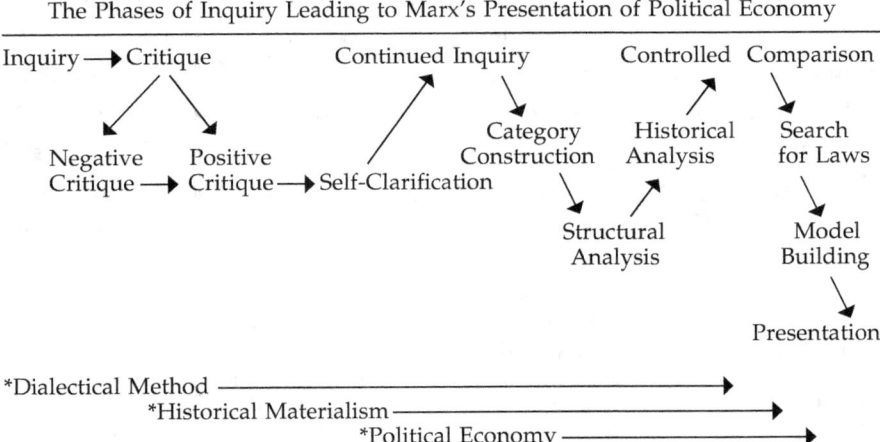

* Asterisk indicates the fact that Marx's presentation of this moment of inquiry remained partial.

Marx's presentation runs into several difficulties, some he could deal with only in minimal ways, some he could not foresee, and others he could not change. For example, he confessed to a friend:

> In the first part, the method of presentation [in *Critique of Political Economy*] was certainly far from popular. This was due partly to the abstract nature of the subject, the limited space at my disposal, and the aim of the work. The present part [*Capital*] is easier to understand because it deals with more concrete conditions. *Scientific* attempts to revolutionise a science can never be really popular. But, once the scientific foundations are laid, popularization is easy (Marx 1985g: 436; emphasis in the original).

Presentation of analytical work must grapple with issues related to the prior knowledge of the audience, complexity of the subject matter, and the illusion of the text as a finished form of thought. An author cannot be expected to be general yet specific, detailed yet abstract, and eclectic yet ecumenical, so that readers of all levels and capability can easily understand her or his narrative. Thus, ruthlessness criticism "is impossible in such company; besides which constant attention has to be paid to making things easily comprehensible, *i.e.*, exposition for the ignorant. [It would be hard to i]magine a journal of chemistry where the readers' ignorance of chemistry is constantly assumed as the fundamental presupposition" (Marx 1936m: 346). Given the desire to reach both a popular and a scientific audience, Marx was destined to confront audiences with various levels of prior knowledge, a situation about which he could do nothing to amend but write, publish, speak out, and organize.[12]

The difficulty of a subject matter creates problems for the development of arguments, data, and the presentation of each. If a system is multidimensional with changing and contingent variables, then no one model can perfectly capture the same configuration of variables with the same explanatory significance; there always remains some openness about what to present and when. Two difficulties arose in Marx's presentation: Where to make the

---

[12] Upon the request that he give a rebuttal to another at a meeting of the International, Marx (1936e: 203) commented to Engels: "Inane though this is, only attaching itself to the most superficial external appearance, it is nevertheless not easy to explain to ignorant people all the economic questions which compete with one another here. *You can't compress a course of political economy into one hour. But we shall do our best*" (emphasis in the original).

breaks? In what order should the narrative be presented? Marx changed his mind several times on these issues. As late as September-December 1866, "Marx thought of publishing Books I and II [of *Capital*] in a single volume" (Rubel 1980: 77). He was torn between preparing the entire multi-volume set for publication, or following Engels's advice to publish his findings more quickly. After reviewing the manuscript, Engels (1936a: 220) noted its deficiencies – such as the "rather strong marks of the carbuncles" – and suggested that the dialectical transitions needed special headings in order to be "more easily comprehensible to a very large class of readers." Marx (1936i: 222) responded to these concerns:

> As to the development of the *form of value* I have and have not followed your advice, in order to behave dialectically in this respect as well. That is to say I have (1) written an *appendix* in which I describe *the same thing* as simply and as much like a schoolmaster as possible, and (2) followed your advice and divided each step in the development into paragraphs, etc., *with separate headings* (emphases in the original).

Still, Volume One might not be in Marx's ideal form, e.g., early editions contained "a long chapter entitled 'Results of the Immediate Process of Production'...[that] Marx seems to have intended to put at the end...but left out at the last minute.... [T]his chapter – some 200 pages long – was only published in 1933" (McLellan 1973: 350). It appears that *Capital* could be presented in multiple ways and, in its final form, certain parts seem out of place, out of order, or in a peculiar order (Eldred and Roth 1978: 28–32). We cannot be confident that Volume One is a "textbook" case of Marx's ideal mode of presentation without qualification. It does not contain Marx's entire analysis of capital and the entire work as a whole is incomplete, though once published his texts took on the appearance of an organic whole. Marx's complaint rings true: The need for him to publish as soon as possible and the limited presentation of the whole spoiled his argument's dialectical method of development. We are left only a partial picture of Marx's vision of the whole and the procedure that should be used in its presentation.

Considering his use of scientific method in conjunction with dialectic, it was perhaps predictable that readers would find Marx's work confused or confusing. Though Hegel believed dialectical logic surpassed traditional mathematics, Marx embraced both schools of thought. He certainly would not have jettisoned dialectical methods in the face of modern statistical methods, though he would have objected to the undue focus conventional attitudinal surveys train on individual mental states and their frequent failure to grasp the capitalist structural context. Marx would also have objected to using quantitative models as a master-key to unlock all research questions. Still, the techniques of abstraction and category construction, the use of tools to focus and refocus his investigative lens, the controlled comparison of variables and constants, statistical analysis, the discovery of laws, and model building were all crucial features of his approach. If we are to understand what Marx is trying to get us to understand, then we must grasp these features of his scientific dialectic. But this is not enough. We must also grasp the logic of the models and language Marx used to present his views and the special way he went about constructing them. This is the goal of the next two chapters.

Chapter Five
Marx's Conceptual Doublets

Science endeavors to advance our understanding of reality by breaking it down into its elementary parts and explaining their interrelationships. This requires that an author depict the complexity of the real, while using, at least to some degree, terms with which her or his audience is familiar. However, a subject matter's complexity may limit a science's accessibility. Moreover, any discourse only a handful of specialists can grasp risks foundering in obscurity. In striving for a wide audience, Marx used terms specialized to dialectics, the natural sciences, and political economy, as well as terms more familiar to the public. This strategy was shaped by his internal relations philosophy of science (see Israel 1979, and chapters Three and Four) and principles from scientific realism (Sayer 1984, Isaac 1987). These traditions help reveal how a series of "conceptual doublets" operate in Marx's work. These doublets advance our understanding of the interrelationships between Marx's overall epistemology, his methods of abstraction, and forms of conceptualization, where his goal is to re-invert our thinking to counter the inverted knowledge he finds in popular and scientific discourses.

Conceptualization through Realist and Internal Relations Philosophies of Science

Concepts do not exist independently of us, nor do we develop them in a vacuum. Marx (1992c: 29) held that "the ideal is nothing else than the material world reflected by the human mind, and translated into forms of thought." In other words, interrelationships between concrete realities form the conditions of possibility for concepts. But concrete social realities are changing realities. So, for Marx (1992j: 80), traditional political-economic "categories... are forms of thought expressing with social validity the conditions and relations of a definite, historically determined mode of production." In his view, "men who establish their social relations in conformity with their material productivity, produce also principles, ideas and categories, in conformity with their social relations.... Thus these ideas, these categories, are as little eternal as the relations they express. They *are historical and transitory products*" (Marx 1847: 109–110; emphasis in the original). In the broad view, the only constant is change.

If we accept that "common-sense language, through which people interact in their daily activities, is the basis of epistemological analysis" (Israel 1979: 3), then, by extension, a social science's language must intersect with everyday language if it is to be intelligible to the average reader. For his part, Marx (1992j: 43) tells us that one way to "discover the various uses of things" is the work of both "history" and "the establishment of socially-recognised standards of measure." In this approach, social scientific thinking "involves the transformation of preexisting materials" and descriptions and explanations must be "parasitic upon the lay identification of social practices. On the basis of these [lay] identifications... social scientists develop theoretical abstractions... [and] a dialogue with and a critique of other theories as well" (Isaac 1987: 66).

Foucault (1984) reminded us that an author is always engaged in a discursive horizon comprised of other people, ideas, and work. Marx also believed that the "biography of a single individual can in no way be separated from the biographies of previous and contemporary individuals: indeed, it is determined by them" (cited in McLellan 1981: xi). Not working in isolation from either society or other scientists, how far and wide the horizon a scientist's discourse can extend is partly determined by the discursive network within which they exist and do their work. Having concluded that the concerted influence capitalism had on knowledge – popular and scientific – was a mystical and metaphysical inversion, Marx's goal was a scientific critique of standard political economy and a revolutionary critique of capital (i.e., an internal relation between his science and his politics). Intending to broaden his audience while accomplishing both tasks, he needed to engage both popular and scientific discourses in terms recognizable to the average readers in these camps but at the same time push them beyond prior conditioning.

In this work, Engels (1992: 14) tells us that Marx used "certain terms in a sense different from what they have, not only in common life, but in ordinary Political Economy." Some of these terms would have been familiar to the everyday reader, e.g., wages, working-day, money, and machinery. Others were used in a new way, e.g., forces, relations, modes, and processes. These must be juxtaposed with Marx's dialectical terms, e.g., contradiction, metamorphosis, expressions, and essence. In trying to re-invert our thinking, "dialectic has successfully reached its conclusion where everyday knowledge has recognized its imaginary character and its origins in essential relations" (Eldred and Roth 1978: 12).[1] Thus, Marx's scientific realism worked *with* and *against* bourgeois ideology, popular knowledge, and the dominant scientific paradigms of his day. This is one reason his work is essentially critical – i.e., if successful, it provides new insight into everyday life and challenges prevailing assumptions.

---

[1] "Dialectical thinking thinks about...everyday language in a particular way and the starting point and base for *Capital* is to be found in everyday language itself. But not every articulation of everyday language is relevant to dialectical thinking; dialectical thinking focuses on those articulations which express knowledge about bourgeois society in general, the character of these articulations as knowledge being based on their adequacy to practical living in our society. Dialectical thinking, therefore, has a positive attitude to everyday knowledge even though in the course of the dialectic the character of everyday knowledge is shown to be limited and mystifying.... The presentation develops certain categories of analysis in the course of the dialectic, and in order for everyday knowledge to continue its dialectic with the presentation, it must transform its articulation into terms that can be understood by the presentation. A special language is used by the presentation to articulate its categories and the progress of the dialectic leads simultaneously to an extension of the categories of the presentation and a delimitation of the way everyday knowledge can express itself to the presentation" (Eldred and Roth 1978: 10, 12).

But Marx not only had to satisfy the standards of science in general while engaging the work of political economists, he was also critiquing and providing *an alternative* to standard scientific knowledge. Moreover, with no measure to perfectly compare them, Marx's science would not be uniformly comparable with scientific conventions, i.e., they are "incommensurable" (Kuhn 1970). At the same time, with his overall concern to make a case on behalf of the working-class, he needed to satisfy basic scientific standards. However, with material conditions working against a clear view of capital, we would only be able to "strip off its mystical veil" when production becomes based on "freely associated men, and is consciously regulated by them in accordance with a settled plan," the attainment of which is a "product of a long and painful process of development" (Marx 1992j: 84). Without a scientific dialectic animated by communist activity, the alienating conditions of capitalism undermine accurate knowledge of it. Marx would need help from history in order to be better understood.

For Marx, conceptualization does not start from individual people as personalities nor with individuals as abstract representatives of ready-made ahistorical categories. According to Ollman (1979a: 109), Marx approaches category construction by "conceptually interiorizing [the] interdependence [of objects] within each thing, so that the conditions of [their] existence are taken to be part of what" they are in an ontological sense (Chapter Three). In internal relationships between phenomena, tensions and mutual interaction result in a certain dynamism and "the particular ways in which things cohere become [part of their] essential attributes" (Ollman 2003: 72). Marx's goal was thus to decipher the relationships among internally tied parts, how they change the whole, and how they are changed with it.

Traditional realism confounds an internal relations philosophy to the extent that it requires precisely conceptualizing objects as *clearly defined separate things*, whereas in an internal relations approach the *relationships between things* take priority. For example, Marx (1992c: 28) tells us his focus is on the "inner connexion" between phenomena. And Engels (1980: 476) noted that Marx's "economics is not concerned with things but with relations between persons, and in the final analysis between classes; these relations, however, are always *bound to things* and *appear as things*" (emphases in the original). What appear as things should not be the place to end one's abstractions but their multifaceted relationships should be the focus. In respect to this view,

Marx's approach to capturing relations with precision "intended to supply a series of flexible structural concepts" (McLellan 1975: 40).

Abstracting Flexibly and Conceptualizing Precisely

Grasping social relations dialectically requires a corresponding act of conceptualization. This means that language mediate and be mediated by ontological assumptions of internal relations and the aforementioned epistemological concerns (Israel 1979: 27–29). In response to Lange's comment that he "move with rare freedom in empirical matter," Marx (1988c: 528) claimed that "He has not the slightest idea that this 'free movement in matter' is nothing but a paraphrase for the *method* of dealing with the material – that is, the *dialectical method*" (emphasis in the original). Rather than "a complete and closed system" or "a picture of the universe (an ontology) or dogma or set of laws of any sort," Marx's dialectic is used as "a flexible tool of analysis" (Sherman 1995: 235). Therefore, what needs to be shown is how Marx achieved his free movement in matter though flexibility in abstraction and precision in conceptualization. These epistemological concerns are as internally related as the objects of study to which Marx's ontological assumptions direct him (see Chapter Three).

Flexibility in abstraction does not mean that conceptualization involves an "anything goes" attitude (Feyerabend 1975). Marxist-realism puts forth the principle that "while it is true that we cannot get outside *some* interpretive framework, this does not license any particular framework" (Isaac 1987: 67; emphasis in the original). When relations are abstracted out of a whole, their most essential elements are conceptually reconfigured in the mind according to their important historical and structural interrelationships. This is how concepts attain their truth in relationship to the conditions from which they are abstracted:

> In short, the world consists only in its interrelations. Any thing that is torn out of its relations with the world ceases to exist. A thing is anything 'in itself' only because it is something for other things, by acting or appearing in connection with something else (Dietzgen 1906: 75).

Sharing this view, Marx "did not see a thing singly, in itself and for itself, separate from its surroundings" (Lafargue 1890, cited in McLellan 1981: 73).

Marx offers a different form of truth, a *relational* rather than an *ultimate* truth. In translating this view into research practice, "there are different levels of analysis and abstraction which must be combined in order to understand the concrete social world" and the free movement in material must proceed in a way where "these levels of analysis are not reducible to one another" (Isaac 1987: 61). Rigid analytical tools, inflexible abstractions, and/or imprecise concepts increase the odds of a mystifying construct, especially should it be abstracted outside of its material-historical relations.

Scientific abstraction requires logical justifications and criteria. Analytical philosophy's dictum of The Excluded Middle is meant to achieve precision by drawing discreet boundaries around concepts. However, if concepts strive to capture both the fluid and stable qualities of phenomena, then they "cannot be understood except in terms of [their] interrelations" (Israel 1979: 27). Here, according to Derek Sayer (1987: 20)...

> ...drawing boundaries to concepts – particularly to general concepts – is evidently going to be a problem. The problem is compounded when...the relations at issue are viewed as being in the process of constant formation and transformation.... Words must be 'like bats' [– at once birds, at once rodents –] if they are to be able to grasp this complexity. From the standpoint of this philosophy [of internal relations], one which differs in fundamentals from the whole analytic tradition, to use concepts otherwise would be singularly unrigorous, since it would entail systematically distorting reality.

Because "Marx perceives the elasticity and alterability of concepts" (Zeleny 1980: 19), he "extends his abstractions...to include how things happen as part of what they are.... [In this way], each of the elements that come into Marx's analysis includes as aspects of what it is all those other elements with which it interacts and without which it could neither appear nor function as it does" (Ollman 2003: 82, 116). In this philosophy of science, concepts are not invented as free constructions of the mind tested deductively (Popper 1983), but are constructed inductively. If precision is what Marx's flexible methods of abstraction and concept formation reflect, then it is this elasticity and alterability we must understand.

If a system is characterized by inner-connections, mutual dependence, contradiction, relations of negativity, and cyclical change, flexibility in abstraction is required in order to achieve conceptual precision. Recognizing this helps us understand seemingly contradictory claims from Marx's colleagues. From Liebknecht (1901: 77) we hear, "In regard to purity and precision of language, he was of painstaking conscientiousness." At the same time, Engels (1992: 16) warned readers that it "is...self-evident...a theory which views modern capitalist production as a mere passing stage in the economic history of mankind, must make use of terms different from those habitual to writers who look upon that form of production as imperishable and final." To reconcile these views, it is necessary to grasp how "dialectic develops in a to-and-fro between presentation and everyday knowledge through which the content of everyday knowledge is systematically taken into account and categories developed" (Eldred and Roth 1978: 12). Starting with the observable world, Marx's method entails "abstraction from concrete social circumstances which allows a common element amongst phenomena to be focused on" (Fraser 1997: 81–107). This occurs through "a double movement: concrete => abstract, abstract => concrete" (Sayer 1984: 80–81). After the inquiry uncovers data to use in concept construction, these concepts then lead continued inquiries and are changed if and when investigation shifts to different levels of generality. We will examine this process in some detail below.

Given that "scientific encounters with reality are necessarily mediated by language and interpretation" (Isaac 1987: 44), Marx's flexible precision (or precise flexibility) can be understood as "a deliberate attempt to find the philosophically appropriate language for expressing the ontological structure of the social world" (Wolff 1988: 20). This demands that the most important and defining systemic and historical relations are captured within concepts. Though language should reflect the most important characteristics of social life, material conditions of a constantly changing quality make absolute precision difficult, if not impossible (even undesirable), to achieve in terms of finality. Thus Marx's precision counter-intuitively "consists of demolishing every preconceived and crystallized concept, to stop any atrophying of concepts that would impede the capacity of seizing human wholes or ensembles *en marche* and grasping simultaneously these totalities and their constituent parts" (Bosserman 1995: 50). Any social science that fails to accomplish these tasks would be less precise than is possible.

Dualities and Identity/Difference

Given ontological dualities (Sayers 1980), incorporating flexibility into concepts is necessary. Marx (1976a: 32) tells us that our experience, "modern times, one of *abstract* dualism" (emphasis in the original), tends to bifurcate knowledge. How do Marx's concepts grasp such dual properties? What sort of duality is Marx talking about? The "identity/difference" abstraction, which brings contrasting features into a single conception, helps us make sense of these questions. Marx (1976a: 88–89) explains:

> north pole and south pole are both *pole*; their *essence* is identical; similarly, *female and male* sex are both one *species*, one *essence*, human essence. North and south are opposed aspects of *one* essence – the differentiation of one *essence* at the *height of its development*. They are *differentiated* essence. They are what they are only as a *distinct* attribute, and as *this* distinct attribute of the essence. *True actual* extremes would be pole and non-pole, human and *non*-human species. The difference in one case [i.e., between north and south poles, women and men] is a *difference of existence*; in the other [between pole and non-pole, human and non-human] a difference of *essences* – between *two* essences (emphases in the original).

Thus, not simply any feature can be brought into an identity/difference abstraction. Some identities abstracted out of a dual-essence are "very superficial" (1976a: 48), "false," or "fragmentary, patchy" (1976a: 105). For example, Marx (1976a: 6) accuses Hegel of an "unsolved *antimony*" when he equates as "identical" the "unity of *ultimate general purpose* of the state with the *particular interests of individuals*" (emphases in the original). In such an abstraction, it is crucial to establish differences:

> The identity Hegel is asserting was at its most complete, as he himself admits, in the *Middle Ages*. Here the *estates of civil society* as such and the *estates in the political sense* were identical.... This separation does indeed *really* exist in the *modern* state.... Or rather, only the *separation* of the civil and political estates expresses the *true* relationship of *modern* civil and political society (1976a: 72; emphases in the original).

If a concept that "where it succeeds seeing *differences*, it does not see *unity*, and that where it sees *unity*, it does not see *differences*" (Marx 1976b: 320; emphases in the original), then its "identity is illusory" (Marx 1976a: 82). Observing identity/difference thus requires abstractive and investigative practices appropriate to the reality in question.

Dualities and identity/difference – where one set of properties is united under a given set of criteria (identity), whereas another set offers a different but related set of criteria (difference) – resonate throughout Marx's moments of inquiry. For instance, at the level of production in general (HM), "every social process of production is, at the same time, a process of reproduction" (Marx 1992j: 531). In capitalism's circulation of commodities (PE), the "apparently single process is in reality a double one" (1992j: 110). Marx (1987e: 514) also explains, "if the commodity has the double character of use value and exchange value, then the labour represented in the commodity must also have a double character." He thus argued that...

> The best points in my book are: 1 (this is fundamental to *all* understanding of the facts) the *two-fold character of labour*, according to whether it is expressed in use-value or exchange-value, which is brought about in the very *First* Chapter; 2. the treatment of *surplus-value* regardless of its particular forms as profit, interest, ground rent, etc. This will be made clear in the second volume especially. The treatment of the particular forms in classical political economy, where they are forever being jumbled up together with the general form, is an *olla potrida** [*hotchpotch] (Marx 1987d: 407; emphases in the original).

As an identity/difference, value in capitalism reflects dual relations that can be grasped as both quantitative (costs of machinery, wages, prices, and profit rates) and qualitative realities (nature, labor-power, surplus-labor, and capital). This abstraction allows the multiple relations within a structure's cluster of ties to be captured conceptually, which, in turn, produces better analysis and understanding by uncovering the common functions differences serve or different functions a specific identity serves.

Thus identity/difference is internally related to the abstraction of vantage point. According to whether identity/difference (one side of a relation) or duality (both) is the focus, the researcher is led to differing emphases in data collection, conceptualization, conclusions, and presentation. At one moment of analysis, shared qualities in a set of cases might be of concern (i.e., a focus on identity), such as capital and labor as classes. In this moment, two seemingly separate objects are united in thought because of the similar function each plays in the operation of a part – a dualism, such as capital and labor as productive relations. At another moment, objects of study are compared in terms of essential qualities they do not share (i.e., registering a difference), such as labor, not capital, being the source of value. Making such distinctions between internally related objects dialectically solves the problem of the excluded middle while satisfying the need for precision by helping abstractions carve out different levels of historical or structural generality, i.e., all class systems versus capitalism specifically. Duality and identity/difference abstractions are thus important when considered in juxtaposition to operational definitions, which possess only a singular identity and, it can be argued, unnecessarily close off from inspection both duality and identity/difference as united conceptual categories. As a general rule, it is only possible to capture identity under traditional operational definitions. But if the cost is separating one-side of relation from its ontological partners – duality and difference – then the identity captured is a false identity.

Having concluded that Marx's conceptual form used many terms as internally related companions, we may now ask: What is their content? What are those meanings? And how do they demonstrate flexibility? Of the strategies Marx employed to grasp the dynamic between history and system and capture such realities conceptually, three stand out. Each contains its own "conceptual doublets." The first strategy entails "epistemological moments." These doublets represent his research practice as a whole. The second strategy, "moments of abstraction," extends from the first and represents how the mental process of abstraction takes place, i.e., how wholes are broken down in thought for research and study. The third strategy, "moments of conceptualization," involves naming abstractions, i.e., specifying concepts with identifiable meanings. These concepts connect the important features of collected data with the central relations used within explanatory frameworks (also see Chapter Six).

Marx's Conceptual Doublets: Epistemological Moments

Marx's overall epistemology has four constituent moments, each themselves a doublet: (1) inquiry and presentation, (2) observation and conceptualization, (3) analysis and interpretation, and (4) description and explanation. These moments should not be understood as being in a lock-step order. Sharing each other's meanings, they are paired here with their logical partners, not necessarily the order in which they are accomplished (which is listed in List 5.1).

Research begins with inquiry and ends with presentation. However, "the method of presentation must differ in form from that of inquiry" (Marx

List 5.1

Marx's Epistemological Moments

Inquiry

Observation and Conceptualization
Analysis and Interpretation

Presentation

Description and Explanation

1992c: 28). Inquiry includes establishing onto-epistemological assumptions (Chapter Three), the method of critique, using the abstractive method, and historical and structural analysis (Chapter Four). Presentation involves presenting abstract structural relationship only to move to examining concrete examples and historical origins (Chapter Four). This relation between inquiry and presentation may make it appear as if Marx's theory of capitalism was arrived at first and the data presented was selectively chosen to prove his case, though he warns against this reading.

Observation is connected to conceptualization through, first, data collection and then, second, the act of abstraction. With the concrete as "the point of departure for observation and conception" (Marx 1973: 101), what is empirically observed, the result of many determinations, must be abstracted into thought and conceptualized in way commensurate with their essential characteristics. This helps introduce variables in a systematic way (more in the next section).

After observation and conceptualization, new data can be analyzed and interpreted. This is done through examining material in terms of the general form it represents for a type of society, its institutional configuration, and regular concrete practices. In this work, the researcher "has to appropriate the material in detail, to analyse its different forms of development, to trace out their inner connexion" (Marx 1992c: 28). In appropriating empirical detail, Marx (1973: 83) warns against an abstract individualism "detached from the natural bonds etc. which in earlier historical periods make him the accessory of a definite and limited human conglomerate." To counter this tendency, he reminds us that certain "abstract categories" are the "product of historic relations, and possess their full validity only for and within these relations"

(1973: 105). This explains why Marx (1992c: 23) thought, as we have seen, "German professors remained schoolboys" who approached political economy through a series of "dogmas, interpreted... in terms of the petty trading world around them, and therefore misinterpreted."

One cannot always control how they are interpreted, though they can provide guidelines. In the Preface to *Capital*, Marx (1992a: 18) tells us that his "analysis of commodities" will be presented first and refers readers to his previous studies in the "analysis of the substance of value and the magnitude of value." As to the question of which organism possesses an "economic cell-form" most appropriate for his analysis, he holds that "their classic ground is England. That is the reason why England is used as the chief illustration of the development of [his] theoretical ideas" (Marx 1992a: 19). His strategy is to find the highest form of empirical development relevant to his object of study to increase the odds for its essential features to be observed more readily.[2] Unhappy with *Capital*'s initial reception, Marx (1992c: 26–28) lists "metaphysics," "the deductive method," "analytic," "realistic," and "German-dialectical" as "various conceptions of it, contradictory to one another." Demonstrating his concern about how to interpret (and not interpret) his work, Marx (1992c: 26) chides the Paris *Revue Positiviste* for interpreting him as a metaphysician. Later, he allows for the assertion that, "old economists misunderstood the nature of economic laws when they likened them to the laws of physics and chemistry" (Marx 1992d*: 28). Certain natural science methods have their place, he believes, but positing universal law-like regularity in social processes is not one of them. Finally, before concluding his Afterword, Marx attempts to explain his dialectic in relation to Hegel's, which again brings up the inversion metaphor. In each instance, Marx provides tips and guidelines to readers about his methods and the difficulties they might have in interpreting them.

In preparing work for presentation, two central tasks are description and explanation. By abstracting reality into its constituent parts, Marx could locate and describe the essential causal mechanisms that explain the way reality presents itself to our consciousness. Thus, the analytical goal is "to

---

[2] Following a similar tack, Foucault (1977) chose prisons as an empirical example in his study of modernity's disciplinary techniques and the historical-structural roots of productive power.

find the law of the phenomenon...[as well as] the law of their variation, of their development.... This law once discovered, [one] investigates in detail the effects in which it manifests itself in social life...following and explaining from this point of view the economic system established by the sway of capital" (Marx 1992d*: 27–28). However, only after the data is analyzed "can the actual movement be adequately described" (Marx 1992c: 28). Regularities are established first; then the laws and variations that explain the patterns observed are determined. Discovered through inquiry, analysis, and interpretation, these laws and causal forces are offered to the reader through presentation. Done well, description and explanation should expose the "inner-physiology" of the social formation as a whole (Little 1986: 93), bringing the reader to a "revelation" (Ollman 2003: 4).

Marx's Conceptual Doublets: Moments of Abstraction

Marx examined wholes, their interrelated parts, and the systemic outcomes produced through four relationships: general relations, specific relations, abstract frameworks, and concrete facts. In conceptualizing and analyzing reality at these four dual moments – i.e., *general* versus *specific*, *abstract* versus *concrete* – Marx moved from the study of real history to a study of political economy to developing abstract models of each, i.e., from the specific and observably concrete to the general and conceptually abstract...and then back to continued research.

Physical, sensuous, observable reality is the *concrete*. The *abstract* refers to interpretive frameworks erected in our minds to think about the concrete. Empirical inquiries begin not with abstract concepts (having been targeted in the moment of critique) but with concrete detail understood as determined by "Mutual interaction...between the different moments...[in an] organic whole" (Marx 1973: 100). Although observation begins with "real individuals, their activity and their material conditions of life, both those which they find already existing and those produced by their activity" (Marx and Engels 1976: 31), concrete relations – i.e., material conditions of life – exist within various levels of generality, providing for material-conceptual levels that range from the general to the specific.

Marx (1973: 85) points out that "all epochs of production have certain common traits, common characteristics," such as "production in general."

A general category contains the essential elements found across all social formations of interest. The specific is a form of a general category but one whose unique traits mark it as a special case, i.e., identity/difference. For example, though labor exists in previous production systems, "Labor" as an abstract social category regulating social relationships becomes fully formed and mature in capitalism. Marx (1973: 104–106) tells us that such forms of knowledge are made historically and materially possible and must be grasped methodologically:

> As a rule, the most general abstractions arise only in the midst of the richest possible concrete development, where one thing appears common to many, to all. Then it ceases to be thinkable in a particular form alone.... Bourgeois society is the most developed and the most complex historic organization of production.... In the succession of economic categories, as in any other historical, social science, it must not be forgotten that their subject – here, modern bourgeois society – is always what is given, in the head as well as in reality, and that these categories therefore express forms of being, the characteristics of existence, and often only individual sides of this specific society, this subject, and that therefore this society by no means begins only at the point where one can speak of it *as such*; this holds *for science as well* (emphases in the original).

General categories are not *a priori* universals applied to observations, given that they develop over time. It is their maturation that allows us to observe, discover, and conceptualize them.

The realities captured between general and specific concepts can range from a specific form of a general category to a concrete instance that expresses the category and allows the researcher to distinguish between different subcases of the general form. Production in general is a broader category, whereas its historically constituted narrower subforms include primitive communalism, tribalism, Asiatic despotism, feudalism, slavery, and capitalism. Conflating unique traits of the specific with more general abstractions produces a false identity.[3]

---

[3] "All production is appropriation of nature on the part of an individual within and through a specific form of society. In this sense it is a tautology to say that property (appropriation) is a precondition of production. But it is altogether ridiculous to leap from that to a specific form of property, e.g., private property" (Marx 1973: 87).

The method of abstracting at successive levels of determination was Marx's way of avoiding this problem.

Abstractions of Determination: The Method of Successive Abstractions

Marx (1973: 101) assumed that the "concrete is concrete because it is the concentration of many determinations, hence unity of the diverse." These determinations need to be pulled apart through abstraction and conceptualized. For Marx (1973: 85), "the elements which are not general and common, must be separated out from the determinations valid for production as such, so that in their unity...their essential difference is not forgotten." The concepts with which such realities are captured must be inner-connected so they may depict the different "moments of one process" (1973: 94). Not a deductive maneuver, explanatory frameworks must be built-up from observations. Concepts that reflect these causal forces are *determinate abstractions*, incorporating the abstract and the concrete, the general and the specific (1973: 83–111; also see Montano 1971, 1972, Horvath and Gibson 1984, Fraser 1997). Before these are discussed in more detail, a few words about the method of successive abstractions are in order.

One method of connecting abstract categories to concrete observations has been "what modern theorists have called the method of 'successive approximation,' which consists in moving from the more abstract to the more concrete in a step-by-step fashion, removing simplifying assumptions at successive stages of the investigation so that theory may take account of and explain an ever wider range of actual phenomena" (Sweezy 1964: 11). Sweezy (1964: 11–12), however, notes that "abstraction is itself powerless to yield knowledge; the difficult questions concern the matter of its application." He continues:

> In other words, one must somehow decide what to abstract from and what not to abstract from. Here at least two issues arise. First, what problem is being investigated? And, second, what are the essential elements of the problem? If we have the answer to these questions, we shall surely know what we cannot abstract from, and, within these limits, we shall be able to frame our assumptions according to criteria of convenience and simplicity (1964: 12).

Marx (1975b) long criticized arbitrary concept construction and abstract system building outside empirical referents. For him, the essential elements of the concrete are investigated and concepts are arranged in their appropriate historical-structural context and explanatory order.

Marx's use of successive approximations, moving through a series of stages, "starts with a vast mound of concrete data (and some preconceived notions) and tries to extract, by successively broader generalizations, some abstract principles to be used in all future cases" (Sherman 1995: 229). In helping us in "individuating objects, their attributes and relationships" the process of conceptualization "must 'abstract' from particular conditions, excluding those which have no significant effect in order to focus on those which do.... What we abstract *from* are the many other aspects which together constitute *concrete* objects such as people, economies, nations, institutions, activities, and so on. In this sense an abstract concept can be precise rather than vague" (Sayer 1984: 80; emphasis in the original). Concrete objects (e.g., people, economies, nations, institutions, activities) are not *the* structures of structural analysis but their *inner-connections and/or internal relations* are. Thus, Marx's presentation in *Capital* begins with small, simple conceptions and moves toward building more abstract conceptions of the whole pieced together in their mutual relations. "This means that concepts introduced at one stage of the analysis are often altered when new elements are taken into account" (Roth and Eldred 1978: 12). At a certain point a more complete view of the whole emerges.

Although Sweezy's essay is acknowledged as a Marxist classic, it is useful to re-tool it in light of the moments of observation, abstraction, and conceptualization. It is therefore renamed here the "method of successive abstractions." This method of using determinant abstractions – abstracting from the concrete, creating interpretive-analytical frameworks, only to return again to research – was a movement from inquiry and observation to abstraction and conceptualization to analysis and interpretation, i.e., from carving parts out of the world for analysis and naming them and then reconstructing their interrelations.

With historical and structural internal relations and the issues of duality and identity/difference in mind, several general rules for conceptualization are suggested. First, concept construction should systematically describe reality in a way reflexively cognizant of the material and ideal relations it is attempting to capture. Second, because terms attain their meaning only in relation to their empirical referents, and if referents are dynamic in quality, then concept formation must be flexible to avoid undue distortion. Third, concepts should emphasize the most important relationships and downplay those that have little or no meaning for the objects of study. Fourth, conceptualization must capture both the historically changing nature of variables and their role in a structural context. Fifth, the internal relations between history and structure require an appropriate extension of concepts over time and their appropriate specification within space.[4] Sixth, the correspondence between concepts and reality is never absolute. Although a scientist wants a conceptualization "that reflects more relations of the object" and one which "contains fewer superfluous and empty relations" (Hertz and Boltzman, cited in Feyerabend 1981: 8–9), there are degrees of precision, accuracy, and effectiveness in constructing language. Seventh, always assume that data is found "at a definite stage of social development" (Marx 1973: 85). Concepts should be carved at the level of historical generality – including its inner-connections with other levels – from which the data is collected data. This aids in distinguishing between what is specific to class societies, what is representative of modern capitalist society, and what is indicative of societies in general. Thus, treat "concepts as grounded in a historically specific form of life" (Bologh 1979: 34). Marx's concepts "are necessarily historical categories.... Their content is historically specific, and their validity historically circumscribed" (Sayer 1987: 21). These rules help abstractions avoid mystifying, obscuring, reifying, and/or otherwise misrepresenting the object of study.

We can now better understand Marx's specific conceptual procedures. An initial, broad abstraction "fixes the common element and thus saves repetition" (Marx 1973: 85). Afterward, differences within identities across totalities are to be "sifted out by comparison" (1973: 85). A concrete referent often can be "segmented many times over" and thus sometimes "splits in different determinations" (1973: 85). When sifting out the segmented concrete references that split into different determinations, the analyst must keep in mind that some of these "determinants belong to all epochs, others only a

---

[4] "But none of all this is the economists' real concern in this general part. Their aim is, rather to present production – see e.g., Mill – as distinct from distribution etc., as encased in natural laws independent of history, at which opportunity *bourgeois* relations are then quietly smuggled in as the inviolable natural laws on which society in the abstract is founded" (Marx 1973: 87; emphasis in the original).

few" and others "will be shared by the most modern epoch and the most ancient" (1973: 85). Failure to differentiate these two facts, i.e., to "leave out just the specific quality which makes [an object specific to an epoch]," will create a fundamentally distorted concept (1973: 86). Adequate concepts must "single out common characteristics" and in doing so they must avoid "bringing things which are organically related into an accidental relation" (1973: 87–88). Only then can "definite relations between these different moments" be examined adequately (1973: 99).

The categories that make up determinant abstractions – the abstract and the concrete, the general and the specific – allow for four different "synthetic unities," ranging from the most specific and concrete to the most general and abstract. They can be understood as unified sets of doublets – i.e., general abstract, general concrete, specific abstract, and specific concrete (Table 5.1). Using these abstractions helped Marx incorporate flexibility into conceptual precision and precision into abstractive flexibility.

After finding general abstractions too broad and obfuscating, Marx moved to the study of real empirical observations before returning to the construction of abstract categories for analytical work. From there he moved to additional empirical research, but now more fully armed with conceptual tools. Next, Marx began using these tools for constructing interpretive, descriptive, and explanatory frameworks for general concrete forms as a whole. But, as noted previously, Marx imbued his practice with an internally flexible form. Thus, it should be noted that the categories in Table 5.1 are not simply a synthesis of these moments of abstraction but are also relational and flexible. As we shall see, categories constructed from observations that fit into one dimension can be re-abstracted and treated as another dimension. How this is done should become clearer as we move through the explanation of each synthetic unity.

Table 5.1
Synthetic Unities within Successive Abstractions

	THE ABSTRACT	THE CONCRETE
THE GENERAL	The General Abstract	The General Concrete
THE SPECIFIC	The Specific Abstract	The Specific Concrete

Synthetic Unity (1): The General Abstract. In the broadest category, *general abstractions* are initially carved at the level of the social in general. The qualities historical systems share, their mutual one-sided partialities, unite them in reality. Therefore, a conceptual term is used to unite them in thought, e.g., production in general (Marx 1973: 85). These abstractions fix a range of determinants across all formations captured by the concept. However, empirical cases studied under such abstractions often "are and remain uncomprehended, because they are not grasped in their specific character" (Marx 1976a: 12). Abstractions pitched at the level of society in general based on realities of the present threaten to mislead us as to the nature of the present. For example, what happens in the capitalist market is often reported in the mass media in terms of how well "the economy" is doing, i.e., what is properly capitalist in nature is conceptualized at the level of "society in general," transforming the present into something ahistorical and universal. As an antidote, Marx began abstracting analytical concepts from the specific concrete.

Synthetic Unity (2): The Specific Concrete. The specific concrete is sensuous, observable reality – it is here where empirical data is gathered and it is here the conceptualization of historical-structural relations begins. In the movement from empirical data to concept formation, a "specific concrete abstraction is produced by analyzing the impure, unique, and specific occurrences...in a given time and place...whereby the specific *individuality* of a mode of production and its superstructure are identified" and the vehicle by which it is observed "may take the form of a record, text, or data series pertaining to a specific concrete setting" (Horvath and Gibson 1984: 18). The specific concrete is not necessarily the most recent, but refers to actual data, events, people, and places that can be observed, counted, and measured. For example, in *Capital,* Marx's research on 18th and 19th century English capitalism included the Blue books published by the government in the British Museum, reports from inspectors, doctors, and other representatives of industry, and the laws and regulations passed in Parliament.

From observations made at the specific concrete, Marx analyzed the internal relations between the specific abstract models he developed and the general concrete cases he found.

Synthetic Unities (3 & 4): The Specific Abstract and The General Concrete. The specific abstract and the general concrete are closely related, so discussing them in tandem better reveals their inner-connection. Once specific concrete detail is examined, research moves to determine how general it is. If patterns found in the study of specific concrete detail extend to wider social relations, one can construct a specific abstract category to capture these relations and inform a study of their general concrete expressions. Marx's study of English capitalism (the specific concrete) allowed him to model the capitalist mode of production in general (one specific abstract form of class society, itself abstracted out of the general abstract category of production in general) in order to inform research on other capitalist societies existing in real time/space (the general concrete). To be a submode – class systems as a submode of production in general and/or capitalism as a submode of class systems – a concrete reality must possess all the essential qualities of a general category (identity), and possess a unique institutional configuration that sets it off from other submodes of the same category (difference). Marx was interested in how one general concrete category (i.e., capitalist society) is a submode of a specific abstract category (i.e., class history in general) of the general abstraction "production in general."

This procedure has flexibility built into it. What was once abstracted as a general concrete category of a broader specific abstract category can be re-abstracted as a specific abstract category. For example, once capitalism (general concrete) is abstracted out of class systems (specific abstract), it can be re-abstracted as a specific abstract category and then broken down again – always using observations of the specific concrete – into its own general concrete subcategories (should this be warranted). Although a specific abstract category informs a general concrete category and its developmental features, the patterns unique to a general concrete form indicate the social dynamics of a specific structure at contemporary moments in its evolution. Thus, a general concrete category (e.g., industrial capitalism) functions as a submode of a broader specific abstract category (e.g., capitalism) and is a manifestation of that category found with such regularity that it is common in other observable examples. By extension, the practice of re-abstracting can be repeated – i.e., a general concrete category (e.g., industrial capitalism) can be re-conceptualized as a specific abstract category and all empirical realities it informs can be treated as its general concrete manifestations. In this manner the flexibility built into abstractions of identity/difference helps facilitate an increased level of precision (Table 5.2).

Table 5.2

Abstract Model of the Relations between Synthetic Unities

	IDENTITY	CRITERIA FOR CATEGORICAL DIFFERENCE
GENERAL ABSTRACT	Mode of Production	NA
SPECIFIC ABSTRACT	Mode of Production	Mode of Appropriation
GENERAL CONCRETE	Mode of Production	Observable examples of the specific abstract and possible subforms
SPECIFIC CONCRETE	Mode of Production	Empirical detail

Let's look at an example. If one were to study antebellum cotton plantations in the United States (specific concrete), one might begin by understanding them as an instance of a mode of production (general abstract). However, it would make more sense to extend this analysis under the category of class systems as a whole (specific abstract), given that the mode of production in 19th century Southern US states was based on the appropriation of surplus-labor, just as had been done in other slave systems, in feudalism, and in capitalism (general concrete examples of class systems). Next, one could re-abstract US slavery. Here, "class systems" becomes the general abstract category and "slavery" becomes a specific abstract category. Then, rather than subsuming slavery as one general concrete example of class systems as a whole, one could treat "slavery" as a specific abstract category, where "class system" was previously treated. In analyzing empirical observations through the specific abstract category of slavery, one could compare observations of US slavery to other historical forms, such as slavery in ancient Rome or agricultural societies as whole (general concrete). Their shared identities unite them as general concrete examples of a specific abstract form, i.e., slavery. However, if they have important differences, then the observations of general concrete examples should be broken down into subcategories for special analysis, e.g., slavery under agrarian societies versus slavery under capitalist systems. Thus, general concrete submodes are conceptually informed by a model of a specific abstract structure. However, abstract models and concrete empirical detail will often share some explanatory frameworks, but not others. Within such a framework, one would conclude that as general concrete forms of the specific abstract category, US slavery was different from previous forms because (1) its nature as a capitalist enterprise required that (2) the slave was reduced to property (3) whose work was exploited for private profits, which are three essential characteristics not present in previous slave systems in noncapitalist class societies, making US slavery "the meanest and most shameless form of man's enslaving recorded in the annals of history" (Marx 1984: 30).[5] Empirical reflexivity, knowledge of cross-cultural history, and controlled comparison assist this free movement in subject matter.

The determination of submodes can only be done after we have a model of the broader category. Once a working-model (the specific abstract) is conceptualized by using empirical observations, research uses this working-model in a controlled comparison across empirical cases to find similar structural relations common to broader realities, e.g., What makes all class societies class societies as such? This model can be used to help subdivide observations of the general concrete into new categories. Marx (1973: 108) depicted the unit of analysis in his study of the capitalist mode of production as focusing on the "categories which make up the inner structure of bourgeois society and on which the fundamental classes rest." Armed with such questions, research moves to an examination of those general concrete expressions indicative of the specific abstract model. In comparison to all class systems, what structural relations and essential structural components makes capitalism *capitalism*? Once this is answered, one may reconceptualize capitalism as a specific abstract category (or even a general abstract category) and continue research to see if its history can be divided into submodes and/or eras. This is the empirical detail at the level of capitalism in general but in its newly emergent specific concrete forms (e.g., agricultural, merchant, entrepre-

---

[5] "But as soon as people, whose production still moves within the lower forms of slave-labour, corvee-labour, &c., are drawn into the whirlpool of an international market dominated by the capitalist mode of production, the sale of their products for export becoming their principle interest, the civilized horrors of over-work are grafted on the barbaric horrors of slavery, serfdom, &c. Hence the negro labouring the Southern States of the American Union preserved something of a patriarchal character, so long as production was chiefly directed to immediate local consumption. But in proportion, as the export of cotton became of vital interest to these states, the over-working of the negro and sometimes the using up of his life in 7 years of labour became a factor in a calculated and calculating system. It was no longer a question of obtaining from him a certain quantity of useful products. It was now a question of production of surplus-labour itself" (Marx 1992j: 226–227).

neurial, industrial, monopoly, transnational, and global forms of capitalism; on periodization see Fine and Harris 1979, Weeks 1985–1986).

A series of structures that many general concrete formations possess can be sifted out in this manner by cross-cultural comparison in order to estimate their various dynamics (identity/difference). After a series of concrete cases are examined, they can be abstracted in terms of the common functions their shared structures perform, as well as the unique characteristics of their era. Thus, these forms can be taken as specific abstract models to inform continued investigation of specific concrete data. Subforms of capitalism have elements unique to them that deserve inspection, of course, but the important point is that what they share, the basic structure of capitalism, allows the analysis to move forward with a useful, empirically generated interpretive-analytical model. Moreover, when researching the here-and-now, one can use the model of capitalism and compare it to empirical observations of the present to discover if it is changing into something new, asking, What might foster this change? What is likely to replace this system? These conceptual-empirical relations – moving from production in general through class systems, capitalism, forms of capitalism, and specific capitalist spheres of action – are listed in Tables 5.3, 5.4, 5.5, and 5.6, and Figures 5.1, 5.2, 5.3, and 5.4.

Tables 5.3.–5.6. Moments Composing Marx's Determinant Abstractions

Table 5.3
Determinant Abstractions Beginning with Production in General

	ABSTRACT	CONCRETE
GENERAL	Production in general	Tribal, slave, Asiatic, feudal, and capitalist modes of production
SPECIFIC	Class systems in general	Capitalism

Table 5.4
Determinant Abstractions Beginning with Class Systems in General

	ABSTRACT	CONCRETE
GENERAL	Class systems in general	Capitalism in particular
SPECIFIC	Tribal, slave, Asiatic, feudal, and capitalist systems	Capitalism's historical subforms: agriculture, manufacture, industry, joint-stock companies, and monopoly

Table 5.5

Determinant Abstractions Beginning with Capitalism in General

	ABSTRACT	CONCRETE
GENERAL	Capitalism	Western Europe / US post-1800
SPECIFIC	Industrial Capitalism	England, post-1800

Table 5.6

Determinant Abstractions Beginning with Industrial Capitalism in General

	ABSTRACT	CONCRETE
GENERAL	Industrial Capitalism	England, France, Spain, Holland, etc.
SPECIFIC	Colonialism	English policies in Ireland

Now we can see how the method of successive abstraction unfolds. Although Figure 5.1 provides a general overview, these steps should be broken down into constituent moments. Starting with production in general (Figure 5.2), research begins with an inquiry into observable activity representing many different production systems (1). Patterned behavior and relationships indicative of a particular type of production system are conceptualized (2), e.g., class systems. Additional data is analyzed and compared across history (3). Provisional models are created for the categorizing the general concrete (4a). More inquiry and analysis are undertaken using the model of the general concrete for interpreting the specific concrete (4a, 1, 2, 3, 4a). In this process, the model and its attendant data are compared with specific concrete data and concepts at the specific abstract to check for utility and quality. If successful, a model of a specific abstract system, its basic structure and laws of motion established and finalized, is used to study the general concrete further (4a, 1, 2, 3, 4a, 4b, 1, 2, etc.), e.g., class analysis (exploitation, method of appropriation, the role of the state) informs the study of capitalism (e.g., the wage-system, the working-day, colonialism, legislation). Finally, models are modified when necessary, i.e., when historical developments produce changes in the system.

In turning from comparative historical research to that of political economy (Figure 5.3), Marx begins (1) with an abstract model of class society in general developed through the procedures shown in Figure 5.2. Inquiry

Marx's Conceptual Doublets • 171

Figure 5.1

The Movement of Successive Abstractions

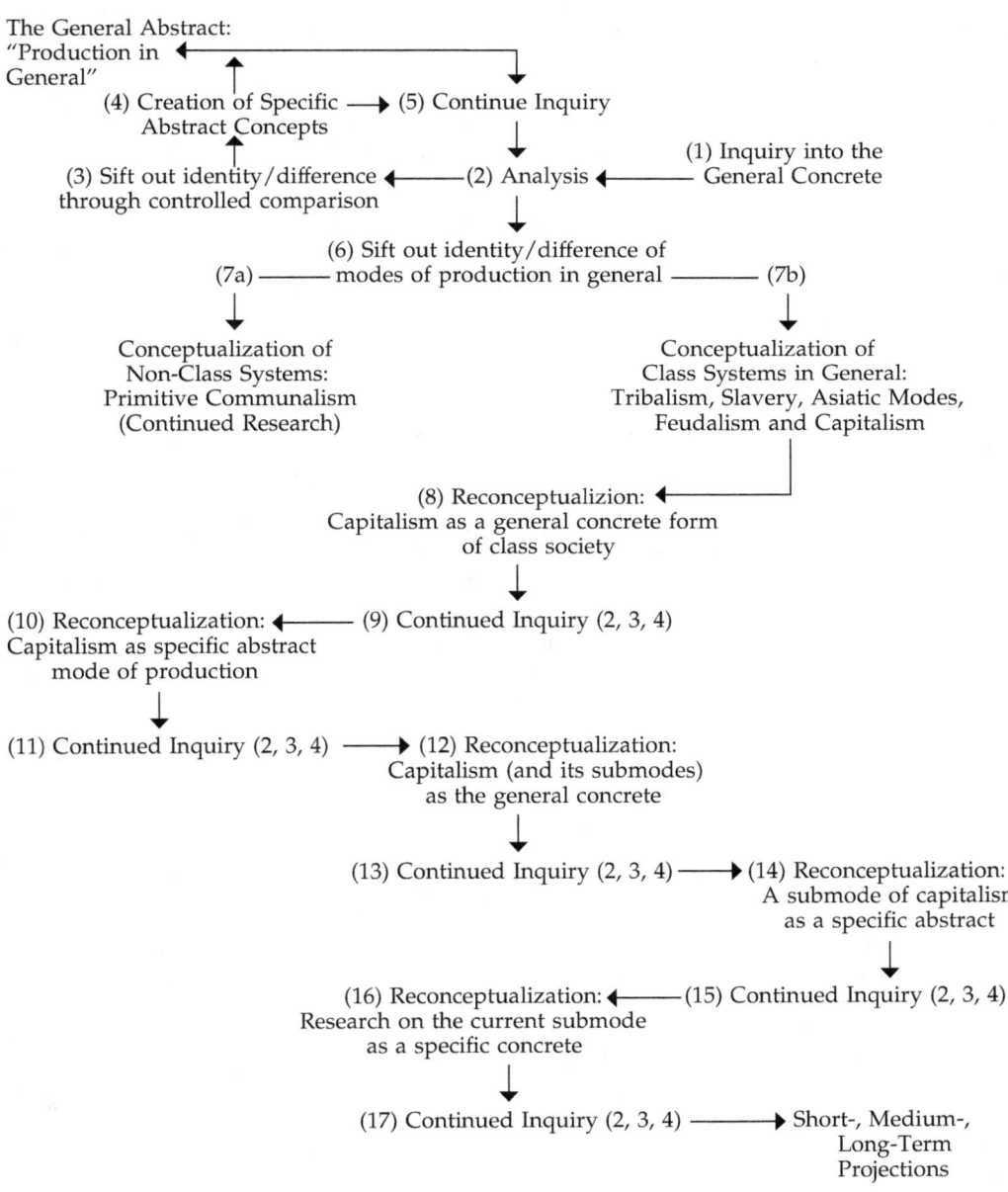

172 • Chapter Five

Figure 5.2

From Modes of Production to Class Systems in General to their Submodes

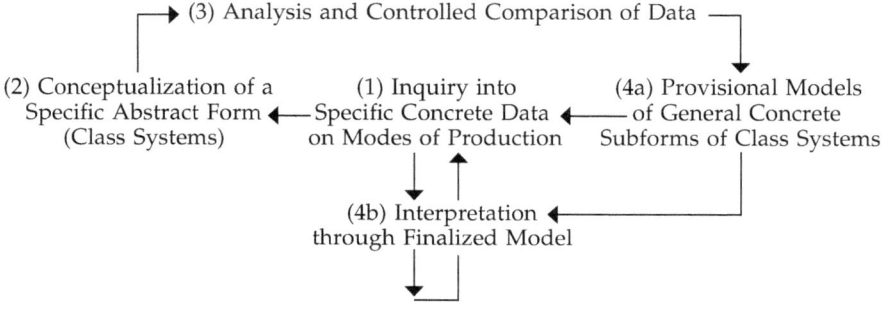

Figure 5.3

Conceptualizing Capitalism out of All Class Systems

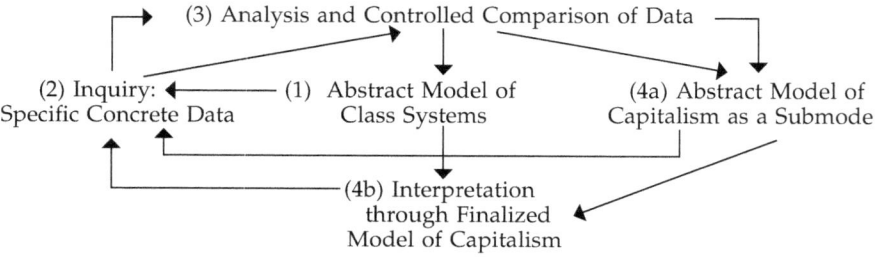

Figure 5.4

Conceptualizing Submodes of Capitalism

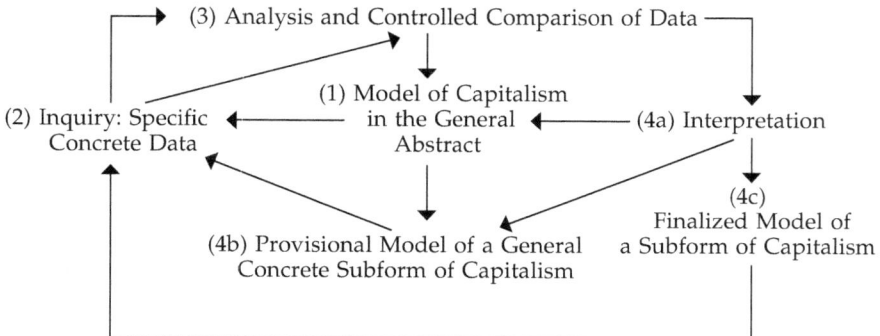

continues as data on the specific concrete is collected (2). Data is analyzed and compared cross-historically (3), and provisional models are created for the general concrete subforms of class systems (4a), e.g., capitalism. As more inquiry and analysis are undertaken, they are compared to a model of class systems (2, 3, 1) and the provisional model of capitalism as a submode (2, 3, 4a). As the identities and differences are sifted out, a model of capitalism in general is finalized for interpretation and used in continued inquiries into specific concrete developments (4b, 2, 3, 4a–b) (see Figure 5.3). If successful, the result is a specific abstract model of a general concrete system, with its basic structure and laws of motion established and finalized, i.e., the structure and laws of motion unique to capitalism. The model is modified when historical developments produce qualitative change in the system, discovered through continuing inquiry.

In conceptualizing the submodes experienced in capitalist history (Figure 5.4), one starts (1) with the abstract model of capitalism in general already developed (Figure 5.3). Next, inquiry (2) examines specific concrete data representing recent history, which is compared to other periods of capitalism (3). Next, one must interpret (4a) whether these periods are commensurable; if not, a provisional model of a general concrete submode is created (4b). After more inquiry and analysis, the model of a subform as a unique general concrete category is compared to capitalism in general (2, 3, 1, 4b), inquiry and comparison continue, and interpretation helps conceptualize a finalized model of the subform of capitalism (2, 3, 4a, 4c). After models of submodes of capitalism in general are settled upon, inquiry, analysis, and interpretation continue using these subforms as specific abstract categories if and when necessary (4c, 2, 3, 4a). This allows for further inquiry, observation, analysis, description, explanation, and presentation of the multi-segmented levels of determination capitalism and its history contain.

Marx's Conceptual Doublets: Moments of Conceptualization

Marx used several conceptual doublets to name the abstract relations he uncovered (List 5.2). After making distinctions between appearances and essence as well as wholes and parts, he broke these down and conceptualized them in terms of time and space, form and content, and quantity and quality. Within this set of concepts, Marx distinguished between relations and processes that were necessary and contingent across both history and structure. This is where Marx's method of successive abstractions and abstractions of determination do their analytical, interpretive, descriptive, and explanatory work.

List 5.2

Marx's Moments of Conceptualization

Appearances and Essence
Wholes and Parts
Time and Space
Form and Content
Quantity and Quality
Relations and Processes
Necessity and Contingency
History and Structure

The relation of appearance to essence, discussed previously as an onto-epistemological problem, is likewise a conceptual strategy. Because "man always contrives to neglect the things that are nearest to him" (Marx 1985i: 510), and given that "capitalist society presents itself as something other than it really is" (Geras 1971: 79), appearance is not "simply...a cognitive problem" (Rovatti 1972: 88). Capitalism's material conditions dominate social action, where calculative self-interest, an abstraction of our own making, appears to be the human essence, when for Marx (1973: 255) this is "pure semblance. *It is the phenomenon of a process taking place behind it*" (emphasis in the original). Traditional political economy, similarly, "proceeds from the fact of private property, it does not explain it to us" (Marx 1988a: 69). Surface appearances – data there for all to see – need explanation; they are not the point of analysis from which data is explained. But without the proper tools we are left with poorly carved abstractions and a conceptualization that distorts our view. If abstraction is done well, the analyst will be in a position to grasp and critique illusory appearances and what had once seemed chaotic and happenstance takes on coherence and meaning.

The distinction between appearance and essence does not allow Marx to instill his concepts with any content he wishes. He is not imputing invisible realities only he can name. An essence contains the common elements that phenomena share in reality and, therefore, what allows them to be united in thought. Surface-level data may provide indications about the essential relations and processes accounting for them, but these are nevertheless uncovered through inquiry, observation, conceptualization, and analysis. Moreover, the distances between essences and appearances are not static but dynamic (Zeleny 1980: 21). Nonetheless, some stability in the relations of social structures is required for laws to be operative (and therefore discoverable). Change occurs over the lifespan of a system, although systems retain enough of their essential properties to still be identified as systems. As capitalism matures, its essence becomes more apparent as the world is increasingly transformed by commodity fetishism and its material relations take on an increasingly determinative character. Marx intended to bring people to a radical consciousness before this development was so mature that it resisted efforts to change it and/or it reached the point of its implosion, carrying society along with it.

When wholes (or totalities) are abstracted from, their constituent parts are isolated for study. It follows that such parts express, to varying degrees, the properties of the whole, i.e., conglomerations of parts and their interrelations make up structures and combined structures make up wholes. The concentration of forces of determination among the parts express overall essential relations and this allows structural features to be uncovered.[6] Because inner-relations among parts are reciprocally shaped by the ever-changing dynamic of the whole, within its relative stability, qualities of wholes continually emerge out of the relations among their parts. Thus, neither the identity of the part nor the whole remains permanently fixed. Modernity is composed of the capitalist mode of production – itself containing methods of production, distribution, circulation and appropriation – as well as state, familial, religious, and artistic forms. However, though some parts express the whole, we cannot assume *a priori* that any part will express all of its features – e.g., a factory may contain capitalist rules, bureaucratic norms, a gendered division of labor, but there may be religious, artistic, or culinary norms that are not expressed, or perhaps only temporarily or just at certain times. With

---

[6] Foucault (1991: 76) argues that "the further one breaks down the processes under analysis, the more one is enabled and indeed obliged to construct their external relations of intelligibility. (In concrete terms: the more one analyzes the process of 'carceralization' of penal practices down to its smallest details, the more one is led to relate them to such practices as schooling, military discipline, etc.)."

capitalism's development many social forms converge. Thus, interconnections within one cluster of parts extend to other parts, i.e., "a radical change in the mode of production in one sphere of industry involves a similar change in other spheres" (Marx 1992j: 362). The result is that capitalism, as a dynamic system made up of interconnected parts, is marked by both stability and change, one of which will tend to predominate at different points in time. During analysis, the stability of a whole can be held temporarily constant; when this is done, abstractions should be carved cognizant of this temporary condition, abstracted at varying degrees of maturity.

Two important parts of wholes are time and space. Marx (1976a: 111) criticizes Hegel for positing civil society and estates as an "antithesis of *space* and *time*, etc., of conservative and progressive," a form of analysis which "turn[s] the *mobile* part of society also into a *static* one" (emphasis in the original). For his part, Marx sees double movements of both internal mechanisms and development over time/space depending upon how large or small, fluid or stable, the data he investigates. Thus, if "time and space are not two separate categories but one," then what is available for analysis is dependent on "how we conceptualize and therefore how we measure structural space" (Wallerstein 1991: 139, 143). Temporal location is necessary for understanding a phenomenon's location within a system but also for understanding that system's history (past, present, and future). Wallerstein (1991: 146) continues:

> Insofar as they are systems, they persist via the cyclical processes that govern them. Thus, as long as they persist, they have some features that are unchanging; otherwise, we could not call them systems. But insofar as they are historical, they are constantly changing. They are never the same from one instant to the next. They are changing in every detail, including of course their spatial parameters. This tension between the cyclical rhythms and the secular trends is the defining characteristic of a geohistorical social system. That is, they all have contradictions which implies that they must all at some point come to an end (also see Albritton 1999: 45–48).

However, there is no predetermined conclusion about historical development. Explanation and interpretation must be ready to incorporate unpredicted events, and this "disallows linear causality" (Martin 1998–1999: 515).

More abstract than concrete in research practice, neither time nor space can be completely removed from analysis, though one must be abstracted out of view so a part of a whole, or a cluster of relations, can be captured in thought. Abstracting time out of view leaves focus on space and research here is mostly of a structural form or an institutional configuration. Abstracting structural relations out of view provides for examining how past events account for outcomes in a particular period. By manipulating time/space abstractions, various historical/structural relations and processes can be pieced together in their order of importance. "It is also a perfectly correct method: being in fact the first method given above with this difference, that instead of being applied to space, in which the different parts of the completed product lie side by side, it deals with time, in which those parts are successively produced" (Marx 1992j: 214–215).

Form and content operate in conjunction with controlled comparison, where making poor distinctions between their identity/difference is of great concern.[7] Form is a regular institutional structure (in the abstract) within the configuration of a social formation as a whole, e.g., "When money leaves the home sphere of circulation, it strips off the local garbs which it there assumes, of a standard of prices, of coin, of tokens, and of a symbol of value, and returns to its original form" (Marx 1992j: 141). In each form, a series of structural regularities shape and encourage certain outcomes as the form extends its determining power, usually understood in terms of ranges of probability and possibility, limits and developments. Bureaucratic organization exerts its powers of determination within different institutional arenas that adopt this form of organization, whether they are a military apparatus, legislature, or school. Certain social forms tend to be regular within certain social structures, e.g., capitalism cultivates a general property form, a general state-form, a general family form. Thus, given that forms can exist at varying stages of development, the magnitude of their causal forces will vary over time too – e.g., although "the riddle represented by money is but the riddle represented by commodities," in capitalism "it now strikes us in its most glaring form" (1992j: 96).

---

[7] "Hegel separates *form* and *content*, *being in itself* and *being for itself*, and brings in the latter externally as a *formal* element. The content is complete and exists in many forms, which are not the forms of this content; whereas clearly the form which is supposed to be the actual form of the content, has not the actual content for its content.... The separation of the *in itself* and the *for itself*, of substance and subject, is abstract mysticism" (Marx 1976a: 62; emphases in the original).

Content, the companion to form, refers to characteristics a form manifests in specific concrete instances. Respecting the relationship between form and content is apparent in Marx's analysis of commodities – e.g. "Whether...a commodity assumes the relative form, or the opposite equivalent form, depends entirely upon its accidental position in the expression of value – that is, upon whether it is the commodity whose value is being expressed or the commodity in which value is being expressed" (Marx 1992j: 56). Research attempts to determine if a specific content warrants extended focus because of its ability to complement and/or override the regular effects of the form in question and/or its ability to add insight to the analysis as a whole.

Processes of change are a central concern in description and explanation. Here, the Hegelian doublet – quantity and quality – becomes appropriate. The commodity, for example, "may be looked at from the two points of view of quality and quantity" (Marx 1992j: 43). Internally related and regularly paired, quantity and quality allow social phenomena to be abstracted temporarily in time/space and described in terms of either one of them at a single moment. Just as ice and water turn into each other with quantitative changes in temperature, the growth in the magnitude of a sum of money "must...be capable of expansion and contraction." And, with enough expansion and investment the "conversion of money into capital" is made possible (1992j: 134, 189). Such a conversion of function leads to a shift in the properties and, therefore, in the structural role a relation serves in the system, resulting in the need for a new and appropriate conceptualization, i.e., money's quantitative growth transforms it into potential investment capital.

Forms of Determination: History/Structure, Relations/Processes, Necessity/Contingency

In Chapter Four, we saw how a synthesis of historical and structural analysis was central to Marx's studies and how the concepts of presuppositions and preconditions played important roles in these researches. Earlier in this chapter we also saw how the method of successive abstractions played a role in how Marx uncovered various moments of determination at multiple vantage points. Bringing these issues together with his concerns of flexibility and precision and his conceptual doublets, we are now ready to reveal in more detail the *forms* of determination (as opposed to determinant abstractions discussed above) that Marx uncovered through inquiry and observation and mobilized in analysis and presentation. Here, necessary and contingent historical and structural relations and processes take center-stage.

Once a structural analysis is undertaken, the researcher is in a better position to discover what happened in order for the system to take shape. Marx's corresponding language refers to certain presuppositions as necessary conditions, e.g., "The division of labour is a necessary condition for the production of commodities, but it does not follow, conversely, that the production of commodities is a necessary condition for the division of labour" (Marx 1992j: 49). Modern slavery is similarly understood as integral to the historical development of the capitalist system because it allowed for the initial accumulation of wealth later invested in industrial production and mass agriculture (Marx 1982b: 101–102). Structures in a social formation at hand are therefore presupposed by social phenomena that made them possible, e.g., the full development of capital necessarily presupposes the development of a money economy (though not vice versa). Presuppositions and preconditions help Marx's research explain the causal factors of historical development and systemic change, as we have already seen.

Key relations and processes – replacing the commonsense notion of externally related things – make a system what it is, e.g., "Magnitude of value expresses a relation of social production" (Marx 1992j: 104); in the "conversion of money into a commodity[, t]he apparently single process is in reality a double one" (1992j: 110). In moments of inquiry versus conceptualization, if the analysis of change is the focus, then the central units of analysis are the relations among empirical phenomena. As relations of social production, commodities' values are established in relation to each other, especially the cost of reproducing the requisite labor-power for the next round of accumulation. The realization of surplus-value is dependent on the relation between constant and variable capital and the process of appropriation, especially as these are related to the dynamics of the class struggle at the time. Thus, processes are regularized activities that characterize the functions within/ between relations of the overall structure and over its history. Engels (1941: 44) previously explained:

> The great basic thought that the world is not to be comprehended as a complex of ready-made *things*, but as a complex of *processes*, in which the things apparently stable no less than their images in our heads, the concepts,

go through an uninterrupted change of coming into being and passing away, in which, in spite of all seeming accidents and of all temporary retrogression, a progressive development asserts itself in the end (emphases in the original).

In conceptualizing processes that capture the functions of, and account for the changes in, social relations, the researcher examines how activities involved in some social relations set off changes in others. Thus, research of either relations or processes entails research into the other, though their relationship is often "established by a social process that goes on behind the backs of the producers" (Marx 1992j: 51–52). This is why processes are studied as internally related to relations, i.e., with relations come processes, and with processes are those that determine the characteristics of the relations in question.

Relations and processes can be subdivided into necessary and contingent forms. In a necessary relation, one dimension of a relation necessarily entails another (male/female, parent/child). In Marx's political economy, there are necessary internal relations between humans and labor-power, ruling classes and exploited classes, capital and labor, and the state and civil society. Although some internal relations are specific to systems, others are more general across space and time – e.g. labor and production in general. Like relations, there are also necessary processes. In the ongoing functioning of a system, necessary processes exist that must be actualized for social production and reproduction to occur. In capitalism, labor must be bought, commodities must be made and sold, surplus-value must be appropriated, and capital must be accumulated. Without these processes, capitalism ceases to exist. With them there are but periodic though regular crises because of the necessary process of the appropriation of surplus-value, capitalism's central contradiction.

Structured regularities of generalized social forms do not always assert themselves as immutable obstacles to the social relations and processes in and through which they develop. An element of contingency is always (and necessarily) in tension within a set of social relations. For instance, contingent internal relations exist between feudalism and capitalism, or between the techniques of production and the patterns of the distribution of wealth. While class struggle is a necessary process involved in capitalist class relations in general, the actual forms it takes in specific cases are contingent processes. Class struggle might be expressed by the actions of either class: by capitalists through shaping the political process, keeping the wage-bill low, fighting against redistribution policies, skirting workplace safety practices, etc.; by workers through forming unions, engaging in slowdowns, going on strike, forming political parties, demanding higher wages and benefits, etc. None of these are guaranteed to occur and "the 'accident' of the character of those who first stand at the head of the movement" can be explanatorily important (Marx 1989b: 137). Further, there is no guarantee that class struggle will take on a revolutionary-socialist character, a contingency that motivated Marx and Engels to write "The Communist Manifesto" and involve themselves in the political struggles they did.

If both structural and historical moments are synthesized with the doublets of relation/process and necessity/contingency, then we find that eight possibilities result within and between levels of generality.

(1) *Necessary historical relations* such as relations of production occur across societies in general. (2) *Contingent historical relations* are the outcome of emergent relationships, such as the evolution of the nuclear family. These types of relations, however, become (3) *necessary structural relations* when a specific mode of production is abstracted into view, e.g., the capital-labor relation in capitalism. (4) *Contingent structural relations* in capitalism include the lumpenproletariat and the petit bourgeoisie, groups not necessary to the essential capitalist structure. Capitalism's necessary process of growth has also brought into being new structural relationships now indigenous to the system, but in contingent ways, e.g., management and professional classes. All societies have (5) *necessary historical processes* they must accomplish if they are to survive, such as the production of goods and services and all the relations that extend therefrom – reproducing the means of production (e.g., the physical plant must be regularly reinvigorated) and labor-power (e.g., a new generation of workers must be born and raised). If neither of these two processes (which reproduce relations) is accomplished, then a social formation changes qualitatively. The emergence of a new particular social structure is not a necessary event. (6) *Contingent historical processes* (e.g., "rounding of the Cape") are often involved in societal-wide changes in social structures. Relations of contingency – or accidents – are recognized because of unpredictable events in history and/or the emergence of structural properties (for discussion see DeMartino 1993). Moreover...

> World history...would...be of a very mystical nature, if 'accidents' played no part. These accidents themselves fall naturally into the general course of development and are compensated again by other accidents. But

acceleration and delay are very dependent upon such 'accidents' (Marx 1989b: 136–137).

Once a system emerges, it possesses (7) *necessary structural processes* that must be realized if it is to reproduce itself as a system, e.g., the appropriation of wealth in class systems. Marx was also often concerned with understanding (8) *contingent structural processes*, e.g., the realization of value – that money invested produces commodities that are then sold, profits repatriated, and portions reinvested and other portions taken as personal gains. Given conditions of the class struggle, the distribution of wealth, availability of buyers, and the conditions of the world market as a whole, there is no guarantee of value's realization. It is contingent on various circumstances and capitalists' attempts to control these results in the prevailing "politics" of modern society.

The tables below depict how Marx put these forms of determination to work. After his long period of inquiry, Marx's analysis and conceptualization developed an understanding of social systems in general (Table 5.7), moved to using these abstractive-conceptualization practices to grasp production in general (Table 5.8), applied them to feudalism as a class system in order to interpret capitalism's preconditions and presuppositions (Table 5.9), and then used these frameworks as a comparative-historical model for developing research on capitalism as a specific abstract and general concrete mode of production (Table 5.10). These frameworks, and the data used to demonstrate them, were then presented in *Capital* (Table 5.11). In these models, Marx combines the necessary and contingent relations and processes that account for the rise and development of capitalism from the vantage point of historical and structural analysis. Here, we discover one moment of how Marx accounts for causal properties within and between systems.

Summary

In an early critique of Hegel's abstract terminology, Marx (1976a: 16) pauses and states, "Now let us translate this whole paragraph into plain language." In his later political economy, Marx (1992j: 95, note 2) sarcastically disparaged one author with, "What clearness and precision of ideas and language!" Clarity and precision are requirements of observation, analysis, and presentation in any scientific effort. Marx's acquaintances testify to the power with which he wielded his theoretical ideas, one of whom believed "nobody possessed

Marx's Conceptual Doublets • 183

Table 5.7
Social Systems in General at Moments of Abstraction

SOCIAL SYSTEMS IN GENERAL		NECESSARY	Observation, Analysis, & Description	CONTINGENT	Observation, Analysis, & Description
VANTAGE POINT OF HISTORICAL ANALYSIS	RELATIONS	Necessary Historical Relations	Abstract & Concrete Examples	Contingent Historical Relations	Abstract & Concrete Examples
	PROCESSES	Necessary Historical Processes	Abstract & Concrete Examples	Contingent Historical Processes	Abstract & Concrete Examples
VANTAGE POINT OF STRUCTURAL ANALYSIS	RELATIONS	Necessary Structural Relations	Abstract & Concrete Examples	Contingent Structural Relations	Abstract & Concrete Examples
	PROCESSES	Necessary Structural Processes	Abstract & Concrete Examples	Contingent Structural Processes	Abstract & Concrete Examples

Table 5.8
Production in General and Moments of Abstraction

PRODUCTION IN GENERAL		NECESSARY	Observation, Analysis, & Description	CONTINGENT	Observation, Analysis, & Description
VANTAGE POINT OF HISTORICAL ANALYSIS	RELATIONS	Necessary Historical Relations	Labor and nature	Contingent Historical Relations	Means and forces of production
	PROCESSES	Necessary Historical Processes	Transformation of nature, individuals, and social relations	Contingent Historical Processes	Mode of appropriation
VANTAGE POINT OF STRUCTURAL ANALYSIS	RELATIONS	Necessary Structural Relations	Relations of production	Contingent Structural Relations	Producers and non-producers
	PROCESSES	Necessary Structural Processes	Production, Distribution, Consumption	Contingent Structural Processes	Rate of exploitation and forms of class struggle

Table 5.9
Feudalism as a Specific Abstract and General Concrete Mode of Production

FEUDALISM AS A MODE OF PRODUCTION		NECESSARY	Observation, Analysis, & Description	CONTINGENT	Observation, Analysis, & Description
VANTAGE POINT OF HISTORICAL ANALYSIS	RELATIONS	Necessary Historical Relations	Combined community and landlord ownership of means of production	Contingent Historical Relations	Configurations of class and state power
	PROCESSES	Necessary Historical Processes	Conquering of land and competing monarchs	Contingent Historical Processes	Rate of wealth appropriation
VANTAGE POINT OF STRUCTURAL ANALYSIS	RELATIONS	Necessary Structural Relations	Lord-serf; guild master and apprentice	Contingent Structural Relations	Political relations: loyalty of army, relations with clergy
	PROCESSES	Necessary Structural Processes	Appropriation of wealth	Contingent Structural Processes	Intensity of class struggle

in a higher degree the quality of expressing himself clearly. Clearness of speech is the fruit of clear reasoning, a clear thought necessitates a clear form" (Liebknecht 1901: 69). Even if this was so, this clarity is not always observed in Marx's written work.

Marx's reputation for confusing readers stems from two main sources. First, such problems are partly a result of the relation between Marx's moment of inquiry and his method of presentation. Like the inversion of our knowledge about life under capitalism, Marx's ideas and concepts can come to a reader in a way that is the reverse of what is expected. Second, problems in understanding Marx are also the result of his failure to publish systematic instructions on his method of abstraction. It does not follow from either of these that his concept formation was imprecise or that it cannot be made more understandable.

Table 5.10

Capitalism as a Specific Abstract and General Concrete Mode of Production

RISE OF CAPITALISM IN GENERAL IN EUROPE			NECESSARY	Observation, Analysis, & Description	CONTINGENT	Observation, Analysis, & Description
VANTAGE POINT OF HISTORICAL ANALYSIS	RELATIONS	Necessary Historical Relations		Private ownership of the means of production	Contingent Historical Relations	Configurations of classes and states – e.g., old monarchies, new bourgeoisie, and states in transformation
	PROCESSES	Necessary Historical Processes		Expropriation of original producers	Contingent Historical Processes	Rate of primitive accumulation of capital in various regions
VANTAGE POINT OF STRUCTURAL ANALYSIS	RELATIONS	Necessary Structural Relations		Capital-Labor; states-citizens	Contingent Structural Relations	Organic composition of capital; size of petit bourgeoisie and proletariat
	PROCESSES	Necessary Structural Processes		Production of surplus-value, appropriation of wealth, and its conversion into capital	Contingent Structural Processes	Intensity and forms of class struggle

Marx's approach to conceptual precision strives to overcome the obfuscatory function capitalist material relations enact on the knowledge systems that accompany them. In prior attempts at clarifying Marx's central concepts, his discourse sometimes is reduced to a series of over-precise formulas of a metaphysical or Kantian variety that invests reality with an entity dubbed "the dialectic" that is supposed to *explain* worldly and natural events (De Koster 1964). Other dictionary definitions of Marx's terms do not always clarify *how* Marx's terms are rooted in his philosophy of science, which means they fail to explain why a social science requires flexibility in order to be precise (Stockhammer 1965, Bottomore et al. 1983, Carver 1987). Marxist dictionaries also go far in making it seem as if his work is presentable as a

Table 5.11
Capitalism in the Specific Concrete

CAPITALISM AS SPECIFIC CONCRETE: INDUSTRY IN 19TH CENTURY ENGLAND & EUROPE		NECESSARY	Observation, Analysis, & Description	CONTINGENT	Observation, Analysis, & Description
VANTAGE POINT OF HISTORICAL ANALYSIS	RELATIONS	Necessary Historical Relations	Growth of capitalist agricultural labor in Cork, Limerick, Kilkenny, etc.	Contingent Historical Relations	Class relations between old monarchies and new bourgeois in Spain, France, etc.
	PROCESSES	Necessary Historical Processes	Expropriation of landed peasants in 18th century England, e.g., enclosure laws in Herftfordshire, Northamptonshire, and Leicestershire	Contingent Historical Processes	Colonialism: the Dutch in Africa, Spanish and English in the Americas, etc.
VANTAGE POINT OF STRUCTURAL ANALYSIS	RELATIONS	Necessary Structural Relations	Labor discipline in workshops in Devonshire, Bedford, etc.	Contingent Structural Relations	Size of landless and unemployed in Derbyshire, Nottinghamshire, etc.
	PROCESSES	Necessary Structural Processes	English use of Ireland's labor and land in the appropriation of wealth; exploitation in English workshops and factories	Contingent Structural Processes	Rebellions and revolutions in Spain, Germany, France

finished philosophy and a seamless whole, though this is misleading. Other times dialectical treatises use a conceptual framework that is so abstract and unfamiliar in reference to both popular and scientific discourse that a dictionary of terms is necessary, many of which the reader is unlikely to have encountered before (Bhaskar 1993). This leads inquiry away from conceptual knowledge that corresponds to everyday life and is more likely to cultivate a smaller and more limited circle of interested inquirers than the wider audience for which Marx strove. This is not unique, however, because most Marxist-dialectical treatments (including this one) are difficult to translate into everyday language. It is between formulaic, determinist readings of Marx and confusing and obscure Marxisms where ground must be found for advancing both our understanding of his terminology and the philosophy of science justifying it.

Contrary to claims made by Analytical Marxists (see Chapter One), Marx's dialectical framework did invoke a language of causal mechanisms and effects. This can be shown in how his concepts and terms have internal relations with one another and address multiple determinations at varying levels of generality. The extent of his success relies on three interrelated onto-epistemological suppositions: (1) The extent to which he is correct to assume that social phenomena are best abstracted as internally related; (2) the degree to which the significant internal relations in question find expression in a system's parts, including the manner and magnitude of the expression; and (3) the extent to which abstractions successfully capture the internal relations between both concepts and a system's parts. Marx's precision increases when these criteria are met. In making his language and concepts commensurate with his initial assumptions and the explanatory frameworks he used for the quality of data collected, Marx, as this chapter has shown, achieved a more elegant science than he is usually credited with.

It now should be clearer why Marx agreed that he moved with rare freedom in the material as well as why *Capital* is difficult for many readers. The process of abstraction through observation and inquiry, engaging in analysis and interpretation, and developing descriptions and explanations are complex procedures. Marx's broad view of history allowed him to catch a glimpse of the possibility of general laws. Materialism, the centrality of labor, and the importance of class struggle were three general historical law-like social facts he concluded existed. As a result, in his research and conceptual development, Marx went from production in general to data indicative of various modes of production to the contemporary mode of production at specific stages of development. Differentiating these categories provided Marx the freedom to grasp history, to carve the general structure of capitalism out of its present inner-connections to the social systems it was developing within and overtaking, and to name the newly emergent social forms. The method of successive abstractions with which Marx carves up reality and the concepts he mobilizes for his studies are the sources of the flexibility and precision with which he could grasp duality, identity, and difference. It is with this set of abstractive practices that Marx went from historical and structural research to the study of contemporary institutional structure projected into its many possible futures – short (e.g., the rise of joint-stock companies), medium (i.e., crises and the possibility of proletarian revolution), and long term (i.e., the potential for a socialist transition out of capitalism and the rise of communism). What occurs through this process is a reorienting of the inquirer's thinking, the *re-inversion* in thought for which Marx was striving. What some experience, however, is cognitive dissonance rather than revelation. However, this point should not be over-exaggerated, because others often experience a mobilizing and liberating reorientation.

Researchers must keep in mind that, on the one hand, the language and concepts used must be abstracted in a way that is of sufficient qualitative similarity with the ontological presuppositions in force. On the other hand, these presuppositions are rooted in empirical observations. Material reality is internally related with the knowing subject in a way where material life is guided by rules, rules that language construction must acknowledge and conform to if quotidian life is to be made practicable. Though mental abstractions and analytical concepts are tools used to refer to and study concrete referents, their construction is not *the* goal of social science. Its goal is the successful description and explanation of the reality under consideration. Despite the fact that Marx is not available to explain himself, the goal of clarifying his meaning should not be assumed to be unattainable. It is in understanding the relationship between the terms he uses and the special meaning he attaches to them where one key to grasping Marx's scientific method and overall view lies. Once we have grasped Marx's conceptual practice, we are in a better position to see how he put his models of society, capitalism, and social change together.

Chapter Six
Marx's Models

What is the relationship between Marx's explanations of historical-structural dynamics and the language he used to describe them? Paul Lafargue (1890) reported that Marx "saw a highly complicated world in continual motion" and, simutaneously, believed that "in higher mathematics [there exists] the most logical and at the same time the simplest form of dialectical movement" (cited in McLellan 1981: 73, 70). So it is telling that in explaining his work on *Capital*, Marx (1936i: 222) told Engels that "the matter [of dialectical development] is too decisive for the whole book," while, in a letter to Sorge (June 21, 1872), he described this work as "rigorously and mathematically scientific" (cited in Rubel 1980: 97).[1] About his conceptual strategy, Robert Paul Wolff (1988: 80–81) remarks:

> To talk about this world, Marx...needs a language whose syntactic and tropic resources are rich enough to permit him...to [1] represent the quantitative relationships that actually obtain in capitalist production and exchange...[2] articulate the structure of mystification that conceals the exploitative and self-destructive character

---

[1] It should be noted that this passage is not found in the letter printed in *Karl Marx and Frederick Engels: Collected Works*, Volume 44 (New York: International Publishers, 1989).

of capitalism...[3] capture linguistically the way in which the mystification of value and equal exchange serve as the necessary surface appearance of the underlying structure of exploitation...[and, 4] implicate the speaker in the very patterns of mystification that are being exposed.

If these observations are correct, we should find Marx using dialectical and mathematical language to reflect a world in motion, one in which he captures the relative stability of social structures, their motion, and their historical development. This brings a double-edged problem to Marx's readers. They have both his terms and methods with which to contend as well as an entire upbringing in capitalist society that encourages a mystified form of thought. However, if we understand Marx's internal relations view, his assumptions about the social world, and how his epistemological methods are internally related to each (see chapters Two through Four) in conjunction with his method of concept construction (see Chapter Five), we are in a better position to understand Marx's dialectical presentation of his political economy.

David Riazanov (1927: 213) noted that "In the last days of his prime, [Marx] had created the essential contours of a model, a draft, in which the basic laws of capitalist production and exchange were expressed. But he had not the strength left to transmit this into an organism as living as the first volume of *Capital*." Although we can only surmise what this model might have looked like, there are avenues to pursue. Viewing Marx's mature work in the context of his research practices can help us deduce his synthesis of models of change within and between social systems with a model of capitalism and its historical development. His conceptual strategy animates these descriptive and explanatory models. In using dialectical and mathematical terms within historical materialist and political-economic frameworks as connective tissue, Marx's "concepts and ideas...are demonstrated in *Das Kapital* as much as any other place in his work" (Beamish 1992: 175). Here, Marx's scientific dialectic reaches its zenith.

Model Building within Dialectical and Mathematical Reason and Language

As complex as it is, *Capital* is not a hieroglyph: It has a method, a rhyme, and a reason. Before writing it, Marx's historical materialism had adopted a view of the structural-systemic parts in all societies considered in the abstract (i.e., the base-superstructure model; see chapters Three and Four). In transitioning from this framework to the study of a particular system, "the multiplicity of observed phenomena are [reduced] to a limited number of properties constituting a model" (Zimmerman 1976: 72). In modeling the capitalist mode of production, Marx used political economic terms (e.g., commodities, production, exchange, profit), terms associated with Hegelian thought (e.g., contradiction, negation, metamorphosis), and terms associated with mathematics and analytical philosophy (e.g., multiplication, variables, constants) to capture the system's qualitative and quantitative realities. He also employed a flexible set of terms to *bridge* such realities – as well as historical and structural realities – with each other (more below). This flexibility in shifting vantage points is informed by several questions:

- What systems have unique configurations of base-superstructure elements?
- What structural relations tend to dominate in specific systems?
- How many structural relations are involved in this or that system?
- To what extent are they involved?
- Are these relations necessary or contingent?
- How did they come to be?
- How do they function as wholes?
- How do they develop over time?

Examining how contradictions and interactions between parts in a system determine limits and/or enhance possibilities for other phenomena helps Marx answer such questions and construct models on their basis.

Abstract models function as descriptive and explanatory frameworks. Models are amended as data are collected, and, therefore, the reconstruction of the overall structure is advanced. This process may range from describing new structural conditions to explaining historical developments. In such an approach, "models of the social world and its lawful structure" should target not specific events but "the enduring social relationships that structure interaction" (Isaac 1987: 50–51). The researcher breaks down (through abstraction) the chaotic empirically "given" into its institutional structure, which in turn is broken down into constituent parts. Forms abstracted are analyzed with special attention paid to the characteristics manifested in specific concrete instances in relation to the abstract model. The researcher pieces together the logical relations between the parts, attempts to uncover how they function in tandem, and tries to discover how their structures shape social practices in concert with/against themselves and each other, as well as the part they play in the reproduction of each other and the whole. Elemental forms are understood as regular structural relations within the overall social formation. With such relations constructed to organize one's thoughts, the model "permits foreseeing how the model will react in case of a modification of one of its elements" (Zimmerman 1976: 72).

Modeling Change Within and Between Modes of Production

Causal models strive to capture concrete relationships that produce change within and between social systems. Wright (1978: 15–26) explains how social structures produce, shape, and change one another through six "modes of determination." *Structural limitation* (1) refers to the way "some social structure establishes limits within which some other structure or process can vary," where certain forms will be excluded and others are more possible or likely – e.g., capitalism's class structure limits its extension of democracy. Within a structurally limited range of possibilities, *selection* (2) is "the setting of limits within limits," where social structures select from among the various institutional forms available. Mechanisms excluding possibilities involve "negative selection," whereas mechanisms that encourage outcomes among those possible involve "positive selection" – e.g., capitalism's material base has selected the nuclear family while undermining the extended family. Structures may also assist in the *reproduction* (3a) of others, or, conversely, they may hinder them from changing in certain ways, i.e., *nonreproduction* (3b) – e.g., the credit-system has facilitated the reproduction of commodity exchange while assisting in the non-reproduction of barter. Some institutional forms have *limits of functional compatibility* (4) with others and/or the overall formation – e.g., earlier religious rules against usury had a limited compatibility with capitalism's rise. Other structures are determined by the *transformation* (5) of others and/or are often transformed as well – e.g., the transformation of the feudal economy transformed corresponding forms of governance. *Mediation* (6) expresses the idea that "a given social process shapes the consequences of other social processes," especially the other modes of determination – e.g., state regulation can mediate the determining influence of markets and class dynamics. These modes of determination shape interaction between the parts of organic wholes, and thus wholes themselves.

List 6.1 summarizes the different moments of determination discussed thus far. First, there are the *abstractions of determination*: general abstract, specific abstract, general concrete, and specific concrete (see Chapter Five). Marx does not always deploy these terms in his political economic analyses (though he does at times) but they appear more often in his various meditations on method. These are used in the moment of analysis that undertakes conceptualization. This is also in part true for his *forms of determination*, which involve necessary and contingent historical and structural relationships and processes (also see Chapter Five). Although his methodological discussions do not elaborate these systematically, his work contains extensive evidence on how he used such forms of determination in his descriptive and explanatory practices. Finally, we have seen how Marx's models employ several *modes of determination*: structural limitation, selection, reproduction/nonreproduction, limits of functional compatibility, transformation, and mediation. All of these moments of determination – from abstractions to forms to modes – sit within an internally related hierarchy of relations stretched between two poles, with analysis in the abstract on the one side and more concrete descriptions and explanations on the other. Epistemologically, abstractions of determination inform inquiry, analysis, and conceptualization. Analysis, description, and explanation relate to forms and modes of determination, where empirical material is pieced together, interpreted, and presented.

List 6.1
Summary of Abstractions, Forms, and Modes of Determination

Abstractions of Determination	Forms of Determination	Modes of Determination
General Abstract	Necessary Historical Relations	Structural Limitation
Specific Abstract	Contingent Historical Relations	Selection
General Concrete	Necessary Structural Relations	Reproduction /
Specific Concrete	Contingent Structural Relations	Nonreproduction
	Necessary Historical Processes	Limits of Functional
	Contingent Historical Processes	Compatibility
	Necessary Structural Processes	Transformation
	Contingent Structural Processes	Mediation

Abstract Ideas ⬅——————————————➡ Concrete Analysis

Determinism is not a construct of teleological inevitability. The approach to "'determinism' Marx so famously propounded refers in his actual analysis only to a *limited range of possibility* posited by a given material productive base." Although a range of possibility sets "determining limits of possibility within which events occur, [it does] not decide the results of the historical decisions and struggles *within* them" (McMurtry 1992: 311; emphases in the original). Among possible outcomes, some institutional forms fit better with others, whereas other outcomes are more or less likely given different time/space variables. The organizations or institutional forms – at first sporadic, rudimentary, and/or isolated – that are the most functionally compatible with the most powerfully determinant structures are likely to be encouraged or made more possible, e.g., the rise of private property, the state, and the nuclear family. Some struggles are more successful than others. Once established, however, institutional forms are not static but are transformed by contradictions and the forms and modes of determination within the structure of the whole.

Social formations are presupposed and determined by other social phenomena that make them possible. As one system meets its demise, the possibilities of new ones are born. Once born, they change, mature and, ultimately, die. To understand the structure and dynamics of an emergent form, researchers must identify the historical preconditions and presuppositions that account for it. This research uses a model of historical analysis, which identifies the signposts that indicate change, unified with structural analysis, which pinpoints the processes by which change most often happens. Marx, for his part, identifies "the preconditions for systemic transformation: (1) when forces of production can develop no further under existing relations of production, and (2) when revolutionary class struggle occurs" (Horvath and Gibson 1984: 15–16). Understanding Marx in this way allows us to model the relation between historical and structural change – or *change within and between systems* – implied in his work (Figure 6.1).

In this model, one begins with the present and grasps its historical preconditions and presuppositions but also projects the structure of the present as the preconditions and presuppositions of future change. One examines the relationship between forces and relations of production and how forms and modes of determination influence the behavior of the structures involved. New conditions emerge and evolve, conflicts and contradictions beset the

Figure 6.1
Model of Social Change within & between Systems

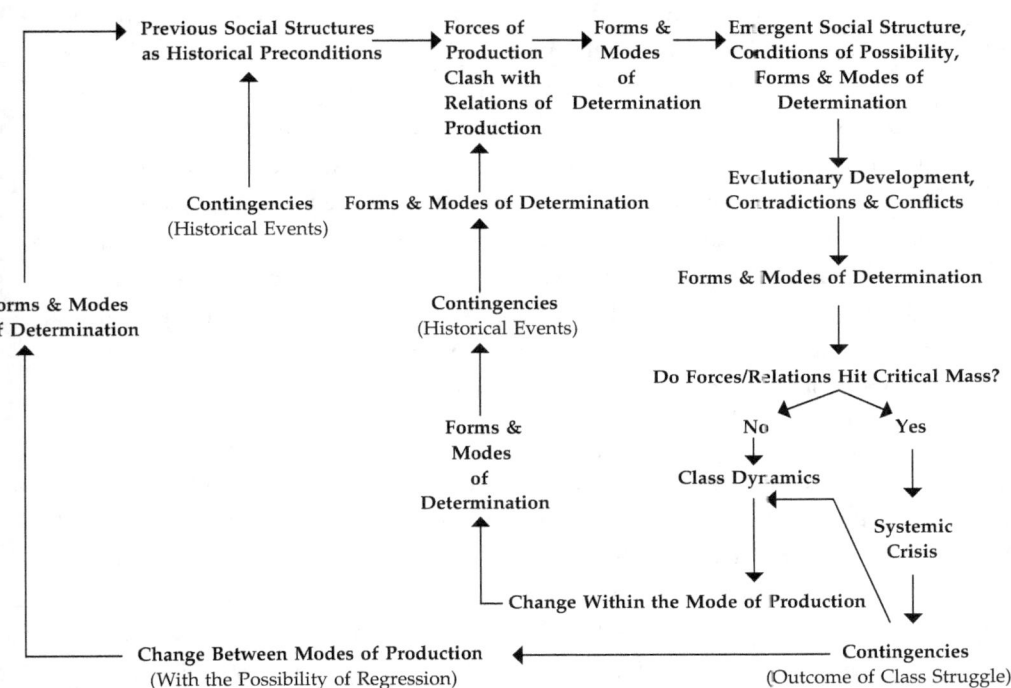

system, and as forms and modes of determination continue to influence outcomes, research is sensitive to if/when the forces and relations achieve a critical mass. If they do not, then class dynamics will usually shape change within that mode of production in conjunction with the forms and modes of determination. If they do achieve critical mass, the system goes into a systemic crisis, whose result is determined by contingent developments (e.g., outcomes of class struggle, forms and modes of determination, and unforeseen events). Often, however, the result is a change between that system and a new one, which itself becomes a new set of preconditions and presuppositions for future change.

Each historical mode of production has a configuration of tendencies, some more crystalline and deterministic than others; some tendencies it may share with others (e.g., class struggles). Thus, using controlled comparison in his analysis of change between and within modes of production, Marx subdivided systems by distinguishing between relations and processes indicative of all social systems from those fundamental to class systems based on whether or not they had a mode of appropriation, and out of those he identified the mode specifically associated with capitalism. However, as Ruben (1979: 68) reminds us, "a mode of production can be exemplified in a plurality of historical societies separated both in time and space." Agrarian production, slavery, and barter can and have existed side-by-side, across temporal and spatial distances, and within feudalism and capitalism. Marx sorted out this complexity by starting with the capitalist present and then arranged the most relevant modes of production that explain its rise in a conceptual order. In this work, only those formations involved in the development of capitalism, or those that share structures that inform our knowledge of it, serve as useful markers for doing a history of the present:

> Capital is the all-dominating economic power of bourgeois society. It must form the starting-point as well as the finishing point, and must be dealt with before landed property. After both have been examined in particular, their interrelation must be examined.... It would therefore be unfeasible and wrong to let the economic categories follow one another in the same sequence as that in which they were historically decisive. Their sequence is determined, rather, by their relation to one another in modern bourgeois society, which is precisely the opposite of that which seems to be their natural order or which corresponds to historical development (Marx 1973: 107).

Therefore, because feudalism determined the emergence of capitalism more than other production systems, it receives priority treatment. Only afterward do slavery, the Asiatic mode, and primitive communal societies require attention. Nevertheless, because the maturation of value into the commodity form under capitalism is a process that penetrates other systems, capitalism's expansion unites this hitherto loosely connected set of systems, increasingly absorbs its competitors, and becomes the world's future by transforming it. This casts an appearance of evolution on all systems as a whole, though it is capitalism that drives this process forward. We will see below how this approach allows for models of history and capitalism's place in it.

A model should contain the most essential structural relations that account for the central tendencies of the social totality in question. Marx models the capitalist mode of production through piecing relations between its parts together and uncovering how these relations shape real social practices in concert with (or against) each other and the role of these structures in the reproduction of each other, the whole, and the societies in which this whole develops. Because capitalism's "rationalization of isolated aspects of life results in the creation of – formal – laws" (Lukács 1971: 101), grasping of how its structure distinguishes it from other systems and modifying the model's parts in conjunction with a controlled comparison of empirical evidence helps expose its systemic relations and laws for discovery (Tilly 1984). Because one tendency, or law, is the outcome of one or more parts at different moments in their mutual development, various moments of analysis stress different relations. During research, by "moving to a lower level of abstraction we can consider how the laws of capital operate in a modified way" (Horvath and Gibson 1984: 20). As models more closely reflect relationships in the data, the more abstract laws assist in uncovering real causal mechanisms, whose outcomes are understood as central tendencies (see List 6.2). The extent to which laws and tendencies explain patterns that have hitherto been only described is the extent to which appearances are eschewed and causal forces behind the concrete are uncovered (Fraser 1997).

Tendencies are internally related to both the structures of the system and to each other. How such tendencies work themselves out sets the whole in motion and accounts for historical change. Marx, however, concludes that capitalism's laws and tendencies have built-in limits that, when transgressed, result in system-wide instability: The appropriation of surplus-value results in an increase in the concentration of wealth, a primary cause of economic crises. With an eye on its increasingly intense crises, Marx sees a limited future for capitalism, though this conclusion is tempered: "There must be some counteracting influences at work, which thwart and annul the effects of this general law, leaving to it merely the character of a tendency" (Marx 1909: 272). Thus, Marx's (1909: 272–283) analysis of the tendency of the rate of profit to fall includes countervailing factors such as increasing the pace of machinery and lengthening of the working-day (thus intensifying the rate of exploitation), finding cheaper raw materials and technology, relative overpopulation, and world trade. As a result, though relations of necessity, contingency, and asymmetry create central tendencies and laws of motion, "such laws cannot be read as predictions or claims about the actual or probable occurrence of events in the life of [a specific case], although clearly they

198 • Chapter Six

List 6.2
Structural Components of the Capitalist Mode of Production and its
Central Tendencies

<u>Structural Components</u>
Proletarians and Bourgeoisie
Labor-Power, Abstract Labor, Wage-Labor, and Surplus-Labor
Commodity Production and Systematic Commodity Exchange
Privately Owned Means of Production and Competitive Markets
Money, Private Property, and Profit (Interest, Rent, Capital)
The Organic Composition of Capital
Production of Surplus-Value: Absolute & Relative, Rate & Mass
Method of Appropriation & Rate of Exploitation
Systematic Accumulation of Capital
The Bureaucratic State
The World Market

<u>Central Tendencies</u>
Capitalists' Compulsion to Accumulate Wealth
The Tendency Toward Technological Revolutions
Capitalists' Unquenchable Thirst for Surplus-Value Extraction
The Tendency Towards Growing Concentration and Centralization of Capital
The Organic Composition of Capital Tends to Increase
The Rate of Profit Tends to Decline
The Inevitability of Class Struggle
The Tendency Towards Growing Social Polarization
The Tendency Towards Increasing Socialization of Labor
The Inevitability of Economic Crises
Constant Expansion of the System's Geographical Borders
Ongoing Commodification of Increasingly Large Parts of Social Life

Source: Marx 1992j, Mandel 1990: 24–31, Wallerstein 1982, 1999

can be used in a *mediated* manner.... Necessity pertains not to actual occurrences, but to the *tendencies* inherent in things, how they can be represented as behaving when studied at a certain level of abstraction at which both nonessential impediments to those tendencies and tendencies which pull in the opposite direction are disregarded" (Ruben 1979: 69–70; emphasis in the original). Counteracting tendencies may thwart the expected development of the system when its core variables are modeled alone. "The law therefore shows itself only as a tendency, whose effects become clearly marked only under certain conditions and in the course of long periods" (Marx 1909: 280). The work of political economy is to piece together, using empirical observations, the structure in the abstract, the central tendencies it exhibits, and the counteracting influence of other relations. Grasping these relationships between abstract models and concrete observations helps estimate a system's short-range, medium-term, and long-range projections. Thus, in political-economic work, "Where science comes in is to show *how* the law of value asserts itself" (Marx 1988b: 68, emphasis in the original).

Now we are in a better position to understand the specific terminology Marx used to depict the changing time/space relations and processes he modeled as essential to the laws of motion of the capitalist mode of production.

Change Within Capitalism: The Conversion of Quantity and Quality into Each Other

In conceptualizing structures within historical change (e.g., identity/difference and dynamics versus stability), the following things must developed: (1) a way to settle on periodization, i.e., establishing the essences of structures via controlled comparison; (2) which analytical signposts indicate change, i.e., dynamism of identity/difference; and (3) the processes by which social change most normally occurs, i.e., modeling inner-structure. A language must be used to indicate such dynamics.

In one approximation, Marx uses two types of special terminology to set his model of capitalism in motion. First, there are quantitative and mathematical terms analytically appropriate for a study of a society based on systematic commodity exchange. Capitalism involves the transformation of nature by labor into value and value into money through exchange and money into capital through profit accumulation and re-investment. When material relationships reduce themselves to mathematical relationships, there are corresponding changes, albeit alienating ones, in social life. This requires the use of mathematical terminology, at least in part, to an extent that is empirically and theoretically appropriate.

But Marx goes a step further. Marx's onto-epistemology assumes quantitative and qualitative realities are internally related. Because capital (something abstract and qualitative) has its origins in labor (something concrete but also qualitative), this origin needs to be traced though the conversion of nature into a commodity through labor (a qualitative transformation), the commodity's sale for profit and profit's appropriation as wealth (a quantitative transformation), and wealth's metamorphosis into capital (a qualitative transformation). Such work required tracing the inner-connections between these changes of quality into quantity and back again, but now as a distorted, alienating social relation of exploitation. Given this process is an "ever-changing reality," in any terminological strategy needed to capture it, "our concepts must reflect this continual change; they must flow into each other" (Pilling 1980: 128). Thus, second, we find Marx using a set of terms to *connect* his analysis of capitalism's quantitative realities with its qualitative ones. Capitalism transforms social and natural qualities by producing and exchanging quantities. The profits realized in these exchanges are transformed into capital. Analysis and presentation must provide a way of observing, conceptualizing, and making plain such transformations.[2]

Quantitative Terms: Basic Mathematical Functions. A system can be initially modeled outside of its temporal context and examined "in the abstract, apart from its historical forms" (Marx 1992j: 476). As a pure structure, capitalism involves systematic commodity exchange, wages, prices, and profits. Such relationships can be studied through "equations" calculated as a "mathematical rule, employed whenever we operate with constant and variable magnitudes, related to each other by the symbol of addition and subtraction only" (1992j: 55–56, 206). Some relationships – such as that between labor and value – can thus be examined in terms of the "magnitude" of their qualities (1992j: 47). Other times, variables' magnitudes are multiplied by other variables at other

---

[2] "You are quite right about Hofmann. You will also see from the conclusion of my chapter III, where the transformation of the handicraft-master into a capitalist – as a result of purely *quantitative* changes – is touched upon, that *in the text* I refer to the law Hegel discovered, of *purely quantitative changes turning into qualitative changes*, as holding good alike in history and natural science. In a *note* to the text (at that time I was just hearing Hofmann's lectures) I mention the *molecular theory* but not Hofmann, who discovered *nothing* exceptional of the affair, nothing *exceptional*, but instead I mention Laurent, Gerhardt, and Wurtz, of whom the last is *the real man*. Your letter brought a dim recollection of the thing to my mind and I therefore looked up my manuscript" (Marx 1936i: 223, emphases in the original).

magnitudes, such as the way "that the surplus-value produced by a given capital is equal to the surplus-value produced by each workman multiplied by the number of workmen simultaneously employed" (1992j: 305). Adding up the total amount of interactions and the total number of variables means that their "average" can be calculated (1992j: 307). Sets of outcomes numerically conceived and placed in relation with one another help us grasp their relative "proportions" (1992j: 45). These quantitative relationships – and others, such as compound ratios – provide for estimating the conditions endured by individuals.[3] Such basic mathematical terms and functions allowed Marx to combine and recombine different relations as a way to observe their interaction with one another.[4]

Quantitative Terms: Algebra and Statistics. Once quantitative relationships among variables were targeted, Marx combined and recombined them into more complex relations using algebraic and statistical reasoning.[5] Algebra allows for calculating time/space relationships by manipulating variables and constants. For example, in modeling the "rate of surplus-value," Marx (1992j: 207) first reduces "constant capital" from $c+v$ to v (i.e., from constant capital + variable capital to just variable capital), then recombines $v+s$ (i.e., variable capital + surplus-value, or the total labor expended in the process),

---

[3] For example, the rate of surplus-value "is determined by the compound ratio between the number of labour-powers exploited simultaneously by the same capitalist and the degree of exploitation of each individual labour-power.... [Thus d]iminution of the variable capital may therefore be compensated by a proportionate rise in the degree of exploitation of labour-power, or the decrease in the number of the labourers employed by a proportionate extension of the working-day" (Marx 1992j: 288).

[4] "Different operations take...unequal periods and yield therefore, in equal times unequal quantities of fractional products.... The division of labour, as carried out in Manufacture, not only simplifies and multiplies the qualitatively different parts of the social collective labourer, but also creates a fixed and mathematical relation or ratio which regulates the quantitative extent of those parts – *i.e.*, the relative number of labourers, or the relative size of the group of labourers, for each detail operation. It develops, along with the qualitative sub-division of the social labour-process, a quantitative rule and proportionality for that process" (Marx 1992j: 327).

[5] "These three factors, however, state of prices, quantity of circulating commodities, and velocity of money-currency, are all variable. Hence, the sum of the prices to be realized, and consequently the quantity of the circulating medium depending on that sum, will vary with the numerous variations of these three factors in combination.... While prices remain constant, the quantity of the circulating medium may increase owing to the number of circulating commodities increasing or to the velocity of currency decreasing, or to a combination of the two. On the other hand the quantity of the circulating medium may decrease with a decreasing number of commodities, or with an increasing rapidity of their circulation" (Marx 1992j: 123).

and thus concludes that the "relative quantity produced, or the increase percent of the variable capital, is determined...by the ratio of the surplus-value to the variable capital," or s/v. This relationship lies at the heart of the central dynamic between the bourgeoisie and the proletariat. The working lives of real people become a calculative measure inserted by capitalists into an equation along with the costs of natural resources and technology. This abstract relationship, played out in the real concrete, is the root of both alienation and class struggle in capitalist society.

In using variables, Marx conceived relations and processes as developing in one "period" versus another. Temporal variables are therefore crucial, for they allow a critical perspective:

> One thing, however, is clear – Nature does not produce on the one side owners of money or commodities, and on the other men possessing nothing but their own labour-power. This relation has no natural basis, neither is its social basis one that is common to all historical periods. It is clearly the result of a past historical development, the product of many economic revolutions, of the extinction of a whole series of older forms of social production (Marx 1992j: 166).

From the historical vantage point, "the process of capitalist production in the flow of its constant renewal...must have had a beginning of some kind" (1992j: 534). From the structural vantage point, when examining "antithetic and complementary phases," "Time is...required for the completion of the series" (1992j: 122, 121). Thus, in a historical-structural comparative analysis, general rules are made discoverable by depicting changes in variables over the "interval" of time "between two complementary phases" (1992j: 115), such as "between different due-days of the obligations" (1992j: 137) and longer intervals such as weekly, quarterly, or even longer intervals (1992j: 169, 259).[6] In this way one can estimate processes across various levels of time,

---

[6] Marx (1992j: 534) thus offers a "General Rule: The value of the capital advanced divided by the surplus-value annually consumed, gives the number of years, or reproduction periods, at the expiration of which the capital originally advanced has been consumed by the capitalist and has disappeared.... After the lapse of a certain number of years, the total value he has consumed is equal to the sum total of the surplus-value appropriated by him during those years, and the total value he has consumed is equal to that of his original capital."

all the way up to an "infinite series"(1992j: 70), using, if necessary, imaginary numbers, "like certain quantities in mathematics" (1992j: 105).

In modeling causal properties by changing temporal intervals, magnitudes in their quantity, and their spatial relations, Marx was concerned with discovering laws. These laws are not uniform at every interval measured.[7] Several cases at one point might be variable and/or the measurement for a particular case may fluctuate. The analysis of fluctuation therefore requires examining "deviations from the average level" (1992j: 123), as well as standard "errors," a method that allows Marx to recognize how cases collectively produce a regression to the mean.[8] Other statistical measures such as correlations also establish important relationships. We saw in Chapter Four how Marx (1992j: 574) established a correlation in the "organic composition of capital" – i.e., "the mass of the means of production employed, on the one hand, and the mass of labour necessary for their employment on the other" – which he found to be "the most important factor" in "the influence of the growth of capital on the lot of the labouring class." By extension, the "law of capitalist production...reduces itself simply to this: The correlation between accumulation of capital and rate of wages in nothing else than the correlation between the unpaid labour transformed into capital, and the additional paid labour necessary for the setting in motion of...additional capital" (1992j: 581). His establishment of such properties is also reflected in his explanation of corresponding changes, inverse ratios, inverse proportions, and fluctuations.

Analytical Terms: Qualitative and Quantitative Comparisons. Marx used several terms to depict and compare relations quantitatively and qualitatively. Some objects need to be sufficiently different in order to uncover their individual traits or functions – i.e., "The very polarity of these forms makes them mutually exclusive" (Marx 1992j: 56). Other times, variables compared

---

[7] "It has been seen that these minutiæ, which, with military uniformity, regulate by stroke of the clock the times, limits, pauses of the work, were not at all the products of Parliamentary fancy. They developed gradually out of circumstances as natural laws of the modern mode of production. Their formulation, official recognition, and proclamation by the State, were the result of a long struggle of classes" (Marx 1992j: 268).

[8] "In every industry, each individual labourer, be he Peter or Paul, differs from the average labourer. These individual differences, or 'errors' as they are called in mathematics, compensate for one another, and vanish, whenever a certain minimum number of workmen are employed together" (Marx 1992j: 306).

to one another must be "commensurable" – or, "qualitatively equal" (1992j: 65). Examining commensurable variables allows for an analytical extension of their causal properties. For example, labor of an "average quality" allows for conclusions reached at one moment of analysis to be generalized to other relations but only if the relations discovered in the former are commensurate with those in the latter (1992j: 191). To the extent that they are, it is assumed that at least a "partial" relationship exists (1992j: 121, 176). By changing the vantage point, research can shift its analysis to the "converse" of what it had been previously (1992j: 70). Viewing relations and effects from different vantage points helps in deducing conclusions about partial relationships in the whole. Thus, Marx often used deduction to extend inquiry into variables that are "equivalent forms" (1992j: 70, 73–75), e.g., as identities, rent, profit, and interest are forms of surplus-value. Targeting equivalent forms provides for legitimate comparisons so long as they are not formulated as tautologies, where qualitatively equal things have been mistaken for qualitatively unequal things, e.g., stating "the number of actual sales is equal to the number of purchases, it is mere tautology" (Marx 1992j: 114, also see 1992j: 104, 204).

Qualitative phenomena often have no common numerical bases, making an act of abstraction necessary to grasp them in their movements. In such an analysis, one must settle on a temporal frame in which an object of study will be examined. Variables, both caused and causal, may be thought of as having a point of origination, such as the idea, in structural analysis, that "The circulation of commodities is the starting-point of capital" (Marx 1992j: 145). Fluidity temporarily locked into place facilitates an adequate understanding before this artificial intrusion is relaxed. Patterns of interest must be isolated in thought – because they are "isolated processes of production" (Marx 1992j: 588) – in a way commensurate with their place in both the structure and their movement through mediating processes. Data extracted from momentarily frozen abstract moments are thus understood as within a set of structural relations "corresponding" (1992j: 455) to a historically developed form of society, which has a certain "distinction" and/or "difference" (1992j: 191) from other systemic processes or structures, other forms of society, or even workers within the same structure (1992j: 520). In analyzing such movements, the researcher must make sure that relationships among variables are controlled. In controlling variables, Marx (1992j: 359) freezes their movements and describes them as "fixed," often finding control variables to be internal to the system.⁹ At other times, grasping motion requires artificially imposing controls. Marx works both within capitalism's logic and within the logic of dialectical and traditional analytical terms. Quite a juggling act!

Conversion Terms. Marx never forgets the one-sided nature of mathematics, given that the quantitative relationships in capitalism have an alienating character. Capitalism converts the qualities of human life into quantitative relationships and these turn into alien qualitative forms, i.e., capital. Writing to Engels, he explained the solution to how profit rates could rise during inflationary periods but decline during deflation: "The whole difficulty arises from confusing *the rate of surplus-value with the profit rate*" (Marx 1936k: 239, emphasis in the original). Marx here warns against confusing a qualitative phenomenon (rate of surplus-value) with a quantitative one (profit rate). Shortly after this, Marx (1936l: 241) declares that he will examine the change of a qualitative reality into quantitative ones, explaining to Engels his plan for Book III is to trace "the transformation of surplus-value into its different forms and separate component parts." Conversely, Marx (1992j: 292) also understands that "merely quantitative differences beyond a certain point pass into qualitative changes." This transformation of qualitative phenomena into quantitative ones and *back again* (i.e., the nature-labor-wage-commodity-price-profit-capital circuit) requires a corresponding language to depict it.

One set of conversion terms capture what I refer to as *relations of composition*. Here, Marx's goal was to specify objects of inquiry, their number, range, connections, and internal relations. For example, abstracting things as "assemblages" (Marx 1992j: 43) reflects how social phenomena are the outcome of multiple crosscutting determinations. Some relations have a mutual attraction and may "congeal" to form new relations. In capitalism, commodities are a "congelation of homogenous human labour" (1992j: 46) and money "congeals into the sole form of value, the only adequate form of existence of exchange-value, in opposition to use-value, represented by all other commodities" (1992j: 130). As one of capitalism's historical preconditions, money conjoined with the rise of private property during its birth and, afterwards, individual labor-

---

⁹ "Just as in Manufacture, the direct co-operation of the detail labourers establishes a numerical proportion between the special groups, so in an organised system of machinery, where one detail machine is constantly kept employed by another, a fixed relation is established between their numbers, their size, and the speed" (for other examples, see Marx 1992j: 385, 518).

powers of a workforce were "combined" (1992j: 315) with one another. The "connexion existing between their various labours" (1992j: 314) resulted in a change where the "organised group, peculiar to manufacture, [was] replaced by the connexion between head workman and his few assistants" (1992j: 396). As capitalist industry increases in scale, there is the possibility of the "conglomeration" of a type of technology or work relation (1992j: 358) and "the old capital periodically reproduced with change of composition, repels more and more of the labourers formerly employed by it" (1992j: 589). In such changes, an "aggregation" is a conglomeration of relations where each contributing part does not lose its essential identity, is synthesized temporally and spatially into a whole, and serves an identifiable function.[10] Thus, it is necessary to know a congelation's "mass" if certain tendencies are to be observed and extrapolated.[11] But knowing the mass is not enough. The term "concentration" is used to describe when a set of relations is recombined with a greater magnitude and intensity but within the same (or lesser) spatial-temporal composition. "Hence, concentration of large masses of the means of production in the hands of individual capitalists, is a material condition for the co-operation of wage-labourer, and the extent of the co-operation or the scale of production, depends on the extent of this concentration" (1992j: 312). If a change in the quantity of a concentrated relation results in a change in its quality, then its composition has been converted into a new form, often a form that makes other changes possible.[12]

---

[10] "This is a superior class of workmen, some of them scientifically educated, others brought up to a trade; it is distinct from the factory operative class, and merely aggregated to it" (Marx 1992j: 396).
[11] "With the productive power of labour increases the mass of the products, in which a certain value, and, therefore, a surplus-value of a given magnitude, is embodied. The rate of surplus-value remaining the same or even falling, so long as it only falls more slowly, than the productive power of labour rises, the mass of the surplus-product increases. The division of this product into revenue and additional capital remaining the same, the consumption of the capitalist may, therefore, increase without any decrease in the fund of accumulation" (Marx 1992j: 566).
[12] "Greater outlay of capital per acre, and, as a consequence, more rapid concentration of farms, were essential conditions of the new method.... The continual emigration to the towns, the continual formation of surplus-population in the country through the concentration of farms, conversation of arable land into pasture, machinery, &c., and the continual eviction of the agricultural population by the destruction of their cottages, go hand in hand" (Marx 1992j: 633, 647).

Other conversion terms capture what I refer to as *general systemic relationships*. In political economy, Marx's central concern was the conversion of labor into capital and all that this entails. A partial relationship in this totality of conversions may be termed an "manifestation," "embodiment," an "incarnation," and/or a "crystallization." The same is true for concrete object. For instance, labor as the source of value may be crystallized (Marx 1992j: 48), incarnated (1992j: 96), or embodied (1992j: 190) in money or a commodity. Such terms refer to the coming into being of a concrete social phenomenon that is the expression a more abstract, general whole. Thus, these terms can be used verbs as well as nouns, e.g., embodies/embodiment, incarnates/incarnation, crystallizes/crystallization, manifests/manifestation. Further, the labor-value relation is in an on-going process of formation, which produces an accompanying "transformation" in social relations, e.g., "when they assume this money-shape, commodities strip off every trace of their natural use-value, and of the particular kind of labour to which they owe their creation, in order to transform themselves into the uniform, socially recognised incarnation of homogenous human labour" (1992j: 111). As a result, laborers are treated as embodiments of homogenous abstract labor and inserted into calculations for expanding capital's value, just as individual capitalists are incarnations of capital (1992j: 556).

Two of the most notable concepts Marx utilized to describe systemic relationships of negativity are the *inverted* and *antagonistic* metaphors:

> The law by which a constantly increasing quantity of means of production, thanks to the advance in the productiveness of social labour, may be set in movement by a progressively diminishing expenditure of human power, this law, in a capitalist society – where the labourer does not employ the means of production, but the means of production employ the labourer – undergoes a complete inversion and is expressed thus: the higher the productiveness of labour, the greater is the pressure of the labourers on the means of employment, the more precarious, therefore, becomes their condition of existence, viz., the sale of their own labour-power for the increasing of another's wealth.... Nowhere does the antagonistic character of capitalist production and accumulation assert itself more brutally than in the progress of English agriculture (Marx 1992j: 603–604, 630).

The features of capitalism that are both antagonistic and inverted allow for and demand an examination of "opposed tendencies" and "antithetical phases," such as the way the "establishment of the normal working-day is the result of centuries of struggle between capitalist and labourer...[which] shows two opposed tendencies," that of both lengthening and shortening of the working-day (1992j: 257–259). Marx (1992j: 115) also sees sales and purchases – in opposition to "the direct identity that in barter does exist" – as "two independent and antithetical acts" containing "antitheses and contradictions." This relation is expressed in the M-C-M' and the C-M-C circuit, where "Each circuit [C-M and M-C] is the unity of the same two antithetical phases" (1992j: 146–147). Because of antithetical phases and opposing tendencies, some relations – even if only temporarily – keep a system in "equilibrium" (1992j: 143), and this is often achieved through "mutual dependent" parts and their relations, such as buyers and sellers (1992j: 158). Rather than equilibrium, other antithetical phases and opposing tendencies produce instability: "The function of money as the means of payment implies a contradiction without a terminus medius.... This contradiction comes to a head in those phases of industrial and commercial crises which are known as monetary crises" (1992j: 136–137).

Recognizing relationships of antagonism and inversion, however, tells us nothing of their relative degree, such as the degree of the capitalist's "thirst for the living blood of labour" through extension of the working-day (Marx 1992j: 245), the "degree in which the agricultural labourer was a compound of wage-labourer and pauper, or the degree to which he had been turned into a serf of his parish" (1992j: 631). How strong do such relations sit with or against one another? Relations have variable "degrees of attraction." These often change in their "degrees of intensity."[13] The degree of attraction capital

---

[13] Marx (1992j: 133) tells us that "the value of a commodity measures the degree of its attraction for all other elements of material wealth, and therefore measures the social wealth of its owner"; "If the intensity of labour were to increase simultaneously and equally in every branch of industry, then the new and higher degree of intensity would become the normal degree for the society, and would therefore cease to be taken account of" (1992j: 492); "there is also an extension of the scale on which greater attraction of labourers by capital is accompanied by their greater repulsion; the rapidity of the change in the organic composition of capital, and in its technical form increases, and an increasing number of spheres of production becomes involved in this change, now simultaneously, now alternately. The labouring population therefore produces, along with accumulation of capital produced by it, the means by which it itself is made relatively superfluous, is turned into a relative surplus-population; and it does this to an always increasing extent" (1992j: 591).

has to surplus-value is perhaps of the greatest magnitude in capitalism. Over time, a "certain stage of capitalist production necessitates that the capitalist be able to devote the whole of the time during which he functions as a capitalist, *i.e.*, as personified capital, to the appropriation and therefore control of the labour of others, and to the selling of the products of this labour" (1992j: 292). Thus, within the capital-labor relation in the system of appropriation, "we see, that machinery, while augmenting the human material that forms the principle object of capital's exploiting power, at the same time raises the degree of exploitation" (1992j: 373). The concept of degree helps depict the extent of growth, diminution, and/or changes in the causal force of a qualitative relation.

Some relations are tied up or are "immanent" in some relations more so than in others. Contradictions are immanent in capitalist commodity exchange (Marx 1992j: 115) and the appearance of the natural productivity of capital outside of labor is immanent in the essence of capital, though it is labor that is "the immanent measure of value" (1992j: 315, 503). Some of "the laws, immanent in capitalist production, manifest themselves in the movements of individual masses of capital, where they assert themselves as coercive laws of competition" (1992j: 300). Immanent laws in the abstract translate into "inherent tendencies" in the concrete, e.g., "To appropriate labour during all the 24 hours of the day is, therefore, the inherent tendency of capitalist production" (1992j: 245). Other relations of the capitalist system are less marked by relations of immanence but rather operate as "circulating mediums" for other relations, such as the way money's function as a circulating medium results in its shape as coin (1992j: 125).

A final set of conversion terms attempts to capture *time/space relations* as systemic processes of motion, dynamics, and change. Not completely separate, structure and history, like quantity and quality, must be converted into one another as well.

Marx's conceptualization of temporal-spatial dynamics within the historical-structural relation is composed of three moments. The first are what I call *moments of indefiniteness*, where Marx extends time without a particular end. This sort of time might be "continual" (e.g., immigration to cities, growth of surplus population in the country, and the destruction of arable land [Marx 1992j: 647]) or "incessant" (e.g., how the labor-wage market continually throws the laborer back into a seller of labor-power [1992j: 542]). Second, there are *moments of finiteness*, where time is reduced to a single micro-unit. Here, things, such as co-operation in the industrial laboring process, happen "simultaneously" (1992j: 310–312). Third, there are *moments of periodization*, where the quality of a temporal moment requires an intrusion on the real concrete, such as dividing "the process of production" or the development of capital accumulation up into "stages" (1992j: 167, 714) and/or into "different economic epochs" (1992j: 175). These three moments together produce a view of "periodical changes...over the different spheres of production" (1992j: 590), such as "periods of prosperity" (1992j: 427).

Marx's (1992j: 715) analyses of spatial relationships acknowledge and examine the degree of a phenomenon's "centralization," such as relations between classes, institutions, or the distribution of wealth (also see 1992j: 586–589). When time/space relationships are combined, the term "sphere/s" is often used to indicate a crystallized set of relationships with attendant functions, such as "spheres of production" (1992j: 590). There are several spheres that produce and reproduce a whole or one of its structures: the "process of formation" (1992j: 588), the "process of organizing" (1992j: 588), and the "process of renewal" (1992j: 534, 589). Within such dynamics, the mutual interdependence of each part with others creates processes of "partial" (1992j: 121) and "constant metamorphosis" throughout the system as a whole (1992j: 590). The constant renewal of capital is therefore a continuous process of growth in "the amount of capital invested" and in "the industrial reserve army.... *This is the absolute general law of capitalist accumulation*. Like all other laws it is modified in its working by many circumstances" (1992j: 427, 603, emphasis in the original). Concentration versus dispersion, spheres of production and reproduction, and variable levels of metamorphosis mediate the explanatory power of laws through relations of necessity and contingency.

All processes and relations necessary to produce another process or relation are brought together and their mutual functions are brought to bear on each other toward an outcome through the process of "realization." The medium and process of "circulation" is how capital realizes profits through reproducing surplus-labor and actualizing surplus-value (1992j: 136). Mediation within/between processes of production, circulation, and distribution shape the range of other outcomes, especially labor markets, the working-day, and the rate of profit. For example, labor is converted into commodities, commodities into sales, and sales into profits, and the "conversion" of all of these entails the production of surplus-value that is a necessary component of capital in general (1992j: 115, 136). As the production of capital intensifies and/or accelerates, the laws of motion accompanying the capitalist mode of production correspondingly obtain concrete expression, eventually reaching their limits of functional compatibility.

Dialectical Movements of Negativity and Qualitative Change. One of the strengths of Hegel's dialectic was its grasp relations of negativity and processes of change.[14] Marx (1992j: 311, 427, 603) uses this logic to depict capitalism's structure and its history as marked by "oscillations," whether this is "the market-price above or below a certain mean" (1992j: 503), or the "movement of alternating expansion and contraction" that "take on the form of periodicity" (1992j: 593). However, Marx (1992j: 339–340) also tells us that "it is a law, based on the very nature of manufacture, that the minimum amount of capital, which is bound to be in the hands of each capitalist, must keep increasing; in other words, that the transformation into capital of the social means of production and subsistence must keep expanding." As an oscillating system, capitalism tends toward spatial growth over time, which is accelerated with the appearance of "competition and credit" as "the two most powerful levers of centralization" (1992j: 587). With its movements of alternating expansion and contraction, this growth occurs in a special way:

> But accumulation, the gradual increase of capital by reproduction as it passes from the circular to the *spiral form*, is clearly a very slow procedure compared with centralisation, which has only to change the quantitative groupings of the constituent parts of social capital.... The masses of capital fused together overnight by centralisation reproduce and multiply as the others do, only more rapidly, thereby become new and powerful levers in social accumulation. Therefore, when we speak of the progress of social accumulation we tacitly include – to-day – the effects of centralisation (1992j: 588, emphasis added).

---

[14] "Old Hegel made some very good jokes about the 'sudden reversal' of centripetal to centrifugal force, right at the moment when one has attained 'preponderance' over the other" (Marx 1987b: 185). Marx (1992j: 547, note 1) elsewhere notes that one author's analysis of how "the separation of property from labour has become the necessary consequence of a law that apparently originated in their identity" suffers because "the dialectical reversal is not properly developed."

Thus, capitalism is a system whose periodic cycles oscillate in spirals that grow in magnitude with the ever-growing amount of capital needed to realize a profit rate acceptable in a competitive market. Thus far in capitalism's history, this spiraling and oscillating movement of alternate expansion and contraction has been "ebbing and flowing" (1992j: 619). In this analysis, when combined with the abstract model of Marx's laws (see Chapter Four), his historical account of countries reaching capitalist development first (Spain, Portugal, Holland, France, and England) and the fact that "The particular task of bourgeois society is the establishment of the world market, at least in outline, and of production based upon the world market" (Marx 1936c: 117), what emerges is a model of the historical dynamics in the capitalist sphere (see Figure 6.2).

Figure 6.2 depicts how different countries come to capitalist development first. As various national bourgeoisies are successful and as other countries are brought into capitalist development, a world-market takes shape, absorbs its former progenitors, subsuming them, and throws all under the sway of its laws, to greater or lesser degrees over time/space. As the system expands,

Figure 6.2
Abstract Model of Expansion & Contraction in Early Capitalist Development

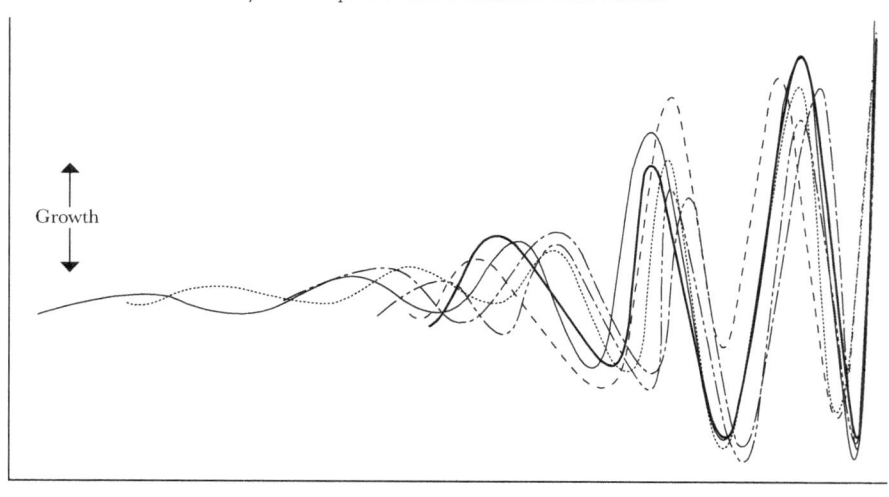

"Free competition brings out...coercive laws having power over every individual capitalist" (Marx 1992j: 257). With this law-like force of the market – something which "compels [the capitalist] to keep constantly extending his capital, in order to preserve it, but extend it he cannot except by means of progressive accumulation" (Marx 1992j: 555) – the tendency arises of increasing magnitudes of oscillation. As the size of the market – spatially and internally with the size of capital – increases, so do its fluctuating oscillations. This process has a "General result: by incorporating with itself the two primary creators of wealth, labour and the land, capital acquires a power of expansion that permits it to augment the elements of its own accumulation beyond the limits apparently fixed by its own magnitude, or by the value and the mass of the means of production, already produced, in which it has its being" (1992j: 566). Thus, for Marx (1992j: 570), "capital is not a fixed magnitude, but is a part of social wealth, elastic and constantly fluctuating."

However, this model as-of-yet does not incorporate counteracting tendencies, which would again modify its oscillations. For example, after several periods of crises, states have adopted policies to flatten out these oscillations. Further, the ability of capital to constantly expand and realize profits projects the trajectory of the oscillations asymptotically upward and keeps the system as whole above a level of net declines, though capitalists' ability to secure ever-cheaper labor and natural resources, find available markets, and commodify more features of social life have built-in limits (Wallerstein 1982, 1999). Any spiral whose revolutions increase in speed while requiring ever-greater levels of capital investment necessarily tends toward reaching such limits – this is not always true in mathematics, but it is true in capitalist society. That being said, the utility of the model is that it represents – in the abstract – potential outcomes should counteracting tendencies lose their power of influence.

Like any other growing system there are maximum and minimum limits on the extent to which capitalism's parts can grow, e.g., "the working-day is not a fixed, but a fluent quantity, it can, on the other hand, only vary within certain limits" (Marx 1992j: 223). The human body has a maximum limit to the number of hours it can physically work without rest before breaking down, money has a minimum level it must reach in order to function as capital, and profits have a minimum level they must not shrink below if they are to remain tolerable. This is also true for the system as a whole. When its limits are breached, they have met their "bounds" (1992j: 264), a "social barrier" (1992j: 285), and/or their "fetters" (1992j: 669) and a relation, a set of relations, or the system as a whole can no longer sustain themselves at that level.

The need for constant expansion of capital is internally related to growing levels of labor exploitation. The contradictory class relation immanent in capitalism thus presents its own fetter to the system: "The greater the social wealth, the functioning capital, the extent and energy of its growth, and, therefore, also the absolute mass of the proletariat and the productiveness of labour, the greater is the industrial reserve army. The same causes which develop the expansive power of capital, develop also the labour-power at its disposal" (Marx 1992j: 603). This relation necessarily leaves open the possibility that the economic crises that threaten capital become transformed into political crises for the capitalist class.

This possibility of capitalism's change extends from the logic of class relations and the study of class history. As the "economic structure of capitalist society [grew] out of the economic structure of feudal society" a process of "dissolution" emerged where "the latter set free the elements of the former (Marx 1992j: 668). The various time/space locations of "primitive accumulation" (i.e., Spain, Portugal, Holland, France, and England) required "a systematical combination" that included colonies, national debt, modern taxation, and protectionism, all of which required "brute force" and "the power of the State, the concentrated and organised force of society, to hasten, hot-house fashion, the process of transformation of the feudal mode of production into the capitalist mode, and to shorten the transition" (1992j: 703). Within capitalism itself, in processes of producing, realizing, and actualizing, some social objects undergo a "transmutation" (1992j: 112, 113) or a "transformation" (1992j: 206). Because of its contradictions, the system reaches a point of "crisis" (1992j: 231, 625), where it is threatened with collapse. Such is the general rule:

> At a certain stage of development it brings forth the material agencies for its own dissolution. From that moment new forces and new passions spring up in the bosom of society; but the old social organisation fetters them and keeps them down. It must be annihilated; it is annihilated (1992j: 714).

The seeds of capitalism's destruction grow within it and the "knell of capitalist private property sounds. The expropriators are expropriated" (1992j: 715).

Thus, in change within systems, one form of the system, or one of its relational clusters, displaces another, only for the process of renewal to reassert itself. In change between systems, one system produces the "negation" of another through a "process of transformation" (1992j: 714), only for new processes of formation, organizing, and renewal to begin.

Because capitalism was ready for a period of global expansion when *Capital* was written, its crises would be limited in time/space. In *Theories of Surplus-Value* (II), Marx's (1968b: 534) grasp of crises leads him back to the whole-part problem: "In world market crises, all the contradictions of bourgeois production erupt collectively; in particular crises (*particular* in their content and in extent) the eruptions are only sporadical, isolated and one-sided" (emphasis in the original). It is at this time/space interval – the system at the level of the world market – when a crisis within capitalism has the potential of being general and the system is most likely to meet its end. The interpretive framework, therefore, is that when the world market resembles the more closed system found in Volume One the laws Marx describes as internal to capitalism's social structure should not only find increasing expression but the countervailing tendencies that turn the back or modify them should become increasingly less available as an option to thwart them. The terminal crisis could happen through economic collapse, though it is a political revolution for which Marx's communist project strives.[15]

Models of System and History from the General and Abstract to the Specific and Concrete

With these stipulations in mind, capitalist history can be periodized by using models of historical generality at levels Two (recent capitalism) and Three (capitalism in general) in conjunction with a controlled comparison across the time/space relations of the system as a whole. This method assists in

---

[15] "The transformation of scattered private property, arising from individual labour, into capitalist private property is, naturally, a process, incomparably more protracted, violent, and difficult, than the transformation of capitalistic private property, already practically resting on socialised production, into socialised property. In the former case, we had the expropriation of the mass of the people by a few usurpers; in the latter, we have the expropriation of a few usurpers by the mass of the people" (Marx 1992j: 715).

specifying developments requiring new categorization. Political-economic models help identify if and when crises in the system are followed by periods of growth – and sometimes qualitative change – in the largest economic units, the characteristic by which capitalism's eras can be distinguished. In addition to capitalism in general, we can distinguish between agricultural, entrepreneurial, industrial, monopoly, transnational, and global capitalism (Marx 1992j; Sweezy [with Baran] 1966, Wallerstein 1974, Kolko 1988, Goldblatt et al. 1997). These forms can co-exist over time/space, e.g., the earlier capitalist eras contained a few large political-economic actors, such as the East India Company, although the most recent era still contains agricultural and entrepreneurial capitalists as it did before. The latter loci of early capitalism were outstripped by emerging industrialists, many of whom crystallized into monopolies, some of which were transformed into transnational corporations. Over this history, states, responding to the needs and conditions associated with capitalist development, rose, consolidated jurisdictions, jostled for position in the nation-state system, grew, engaged in class struggle, waged war, created institutions to settle disputes, learned to cooperate, and are now beginning to be colonized by transnational capital. In other words, today, the largest capitals are stronger than the majority of states. If the relation between capital and the state reverses itself in significant regions of the world-economy, then the hypothesis of an emerging new era, i.e., globalization, is perhaps justified (see Figure 6.3).[16]

In Figure 6.3, the left-hand column lists historically emergent forms, or epochs, of the system, which became structural presuppositions for the next form. The next column to the right represents prior periods as preconditions and presuppositions for the next stage. The third column from the left presents unique features emerging across capitalism's eras. The last column displays the central tendencies inherent to the system across its eras. This model should not be interpreted as presenting *the* or even *a* final understanding of this history but as an initial interpretive framework set up for additional empirical corroboration and to facilitate additional research, analysis, description, and

---

[16] "The same spirit which creates the corporation in society creates the bureaucracy in the state...if earlier the bureaucracy combated the existence of the corporation in order to make room for its own existence, so now it tries forcibly to keep them in existence in order to preserve the spirit of the corporations, which is its own spirit" (Marx 1976a: 45).

Marx's Models • 217

Figure 6.3
Model of Change within the Capitalist Mode of Production

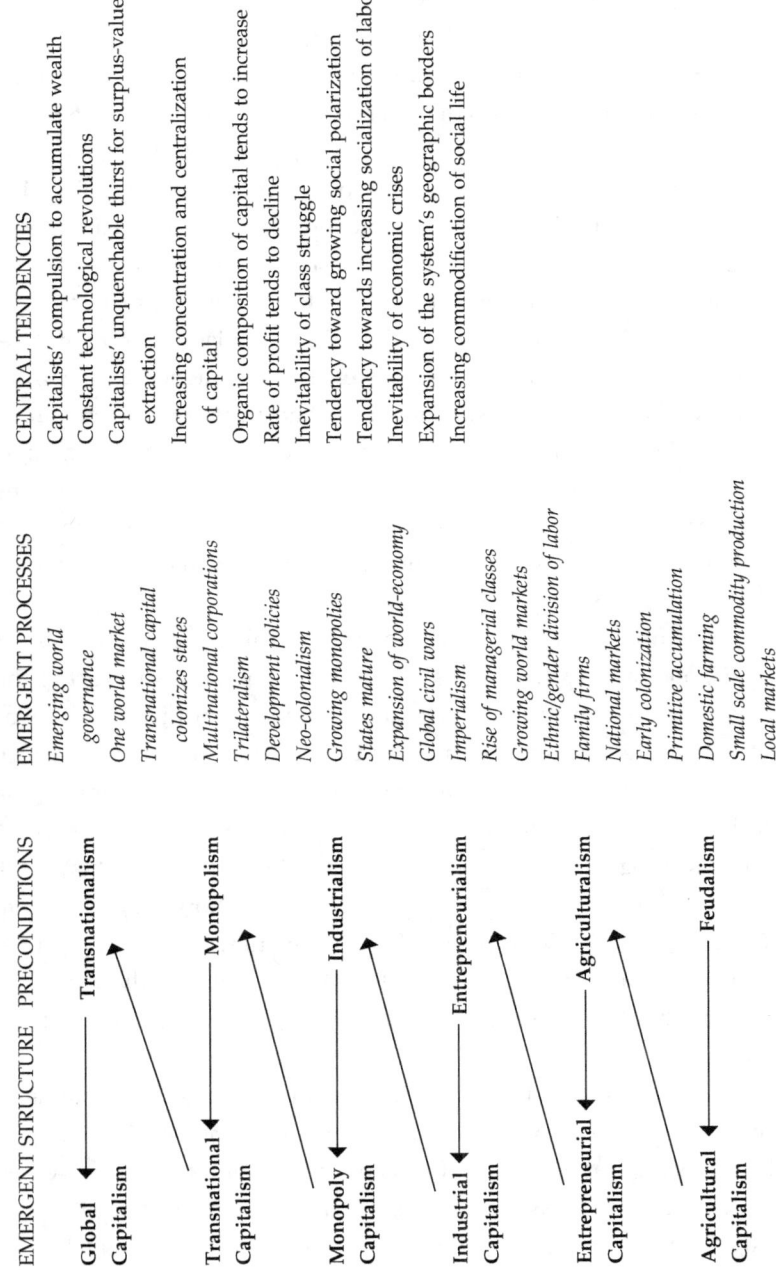

explanation. It also proposes to help us understand the limits capital may or may not be reaching on a world-scale.

We are now in a better position to understand Marx's general theory of how history and capitalism are internally related and how the present social structure will play a role in what emerges in the future. Figure 6.4 displays previous modes of production as developing though preconditions and presuppositions for the modes that follow and the moments of inquiry Marx used to describe each. Each mode of production can be abstracted either as (1) a precondition of what follows or (2) the result of what has come before. From this vantage point, the relation between historical development and modes of production can be modeled and studied. As capitalism matures and develops it overtakes other social systems. However, if abstraction begins in the past and moves forward, there is a danger of conceptualizing the relation of history to modes of production as linear and teleological. Marx's vantage point, however, is the present and he investigates the history of those things that made modernity possible. This is how capitalism gives history the appearance of a linear and universal evolution, i.e., its evolutionary tendencies absorb other systems and carry them along.

General models of historical systems – broken up along two poles of analysis, one ranging from the structurally general and abstract to the more structurally specific and concrete and another from historically general to historically specific – can be mapped across world history. A horizontal axis – from left to right ordering the most historically general to the most historically specific – contains abstract models of social systems, i.e., Marx's general sociological model. Vertical models start with broad historical social totalities and move to increasingly specific formations (levels of generality Five through One). Table 6.1 depicts the relationships concerning historical generality, abstract analytical categories, and the study of the concrete, portraying world history up to and including the various eras of capitalist development.

Organic and evolutionary models conceptually join the relations that are intrinsic to structures to what has happened over time, while sorting the central tendencies that exist in all forms of the modern system as well as those specific to different eras. These eras are understood to emerge and develop through the processes of change within and between systems. In analyzing such changes, research remains attentive to the necessary and contingent

Marx's Models • 219

Figure 6.4
The Relationship between Models of Production and Historical Change

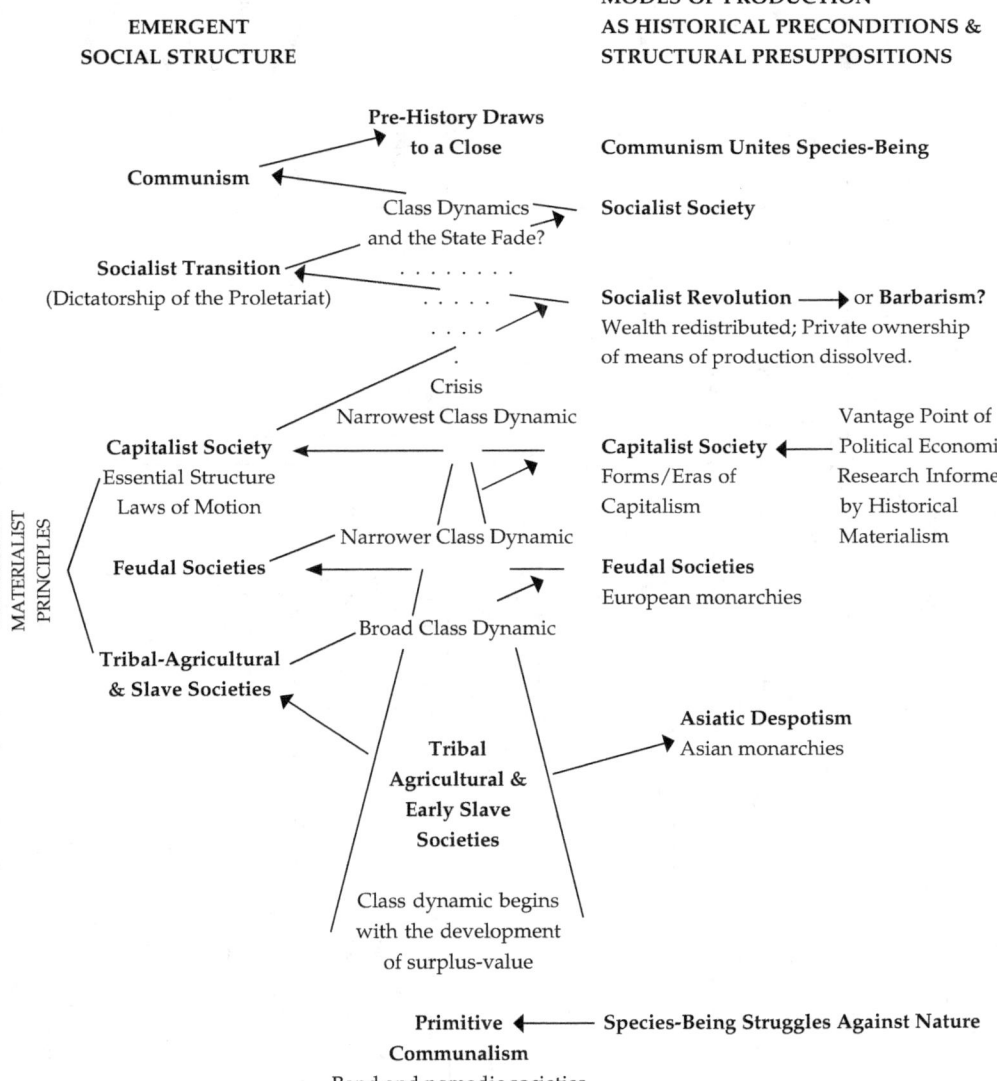

Table 6.1
The Abstract and Concrete in Historical Materialism and Political Economy

General Structure --> History

	MATERIAL BASE	SUPERSTRUCTURE		REAL CONCRETE
HISTORICAL MATERIALISM	**_Production in General_** Modes of Production (Means, Forces, and Relations of Production)	**_Political Relations_** Producers & Non-Producers Juridical & Legal Relations	**_Ideology_** Juridical, Legal, Religious, Artistic, and Scientific Knowledge.	World History
	Class Societies in General Abstract — Concrete Tribal Society — Utes, Celts, Zulus Asiatic Modes — Dynastic China Slave Systems — Rome, Greece Feudalism — Tudor England Capitalism — Western Europe	**_Ruling Classes in General_** Chieftains; Slave Masters; Monarchs; Capitalists **_General State Forms_** Warriors and Mercenaries Sovereign States Professional Militaries	**_Ideologies in General_** Traditional Authority Divine Right Noblesse Oblige Patriarchy Xenophobia Laissez-Faire Social Contract	**_Class Struggles in General_** Tribal Warfare Witch Hunts, Bread Riots, and Regicide Union Organization Strikes, Parties, etc. Proletarian Revolution
POLITICAL ECONOMY (THE ABSTRACT)	**_Capitalism as a Specific Abstract Structure_** Proletarians and Bourgeoisie; Labor-Power, Abstract Labor, Wage Labor and Surplus-Labor; Commodity Production and Systematic Commodity Exchange; Privately Owned Means of Production and Competitive Markets; Money, Private Property, and Profit (Interest, Rent, Capital); The Organic Composition of Capital; Production for Surplus-Value: Absolute & Relative, Rate & Mass; Appropriation Method & Rate of Exploitation; Systematic Capital Accumulation; Bureaucratic States; The World Market	**_Capitalist State:_** **_Specific Abstract Forms_** Bureaucratic States Professional Military and Policing Organizations Rationalized Law Surveillance & Discipline Market Manipulation through Policy Formal Democratic Freedoms	**_Bourgeois Ideology:_** **_Specific Abstract Forms_** Individualism Free Markets Social Contract Competition Meritocracy Profit Fetishism Bureaucratic Rationality	**_Specific Abstract Central Tendencies_** Capital Accumulation; Technological Revolutions; Capitalists' Thirst for Surplus-Value; Concentration and Centralization of Capital; Organic Composition of Capital Increases; Rate of Profit Tends to Decline; Inevitable Class Struggle; Social Polarization; Growing Socialization of Labor; Crises; Expansion of the System; Commodification of Social Life

Marx's Models • 221

	General Concrete Periods of the **Capitalist Mode of Production**	Capitalist State: **General Concrete Forms**	Capitalist Ideology: **General Concrete Forms**	General Concrete Class **Struggles in Capitalist History**
POLITICAL ECONOMY — THE CONCRETE	Agricultural Capitalism Entrepreneurial Capitalism Industrial Capitalism Monopoly Capitalism Transnational Capitalism Global Capitalism	Democratic-Republicanism Military Diplomacy Market Agencies Absentee Imperialism Police Apparatus Dictatorship Death Squads	Fascism Liberalism Conservatism Racism/Sexism Libertarianism	Primitive Accumulation; Small-Scale Commodity Production; Family Firms; Colonization; Imperialists Wars; Monopolies & States Mature; Managerial Classes; Multinational Capital; Neo-colonialism; Capital Colonizes States; Emerging World Markets & Government
RECENT CAPITALISM	Specific Concrete Events *within Periods of Capitalism* Industrial Capitalism – Industrial Revolution Monopoly Capitalism – Rise of Fordism Transnational Capitalism – Post-WWII Expansion Global Capitalism – World Trade Regimes, 1990s–2000+	Specific Concrete **State Actions** Military-Keynesianism Marshall Plan Bretton Woods System Welfare-State Policy Pentagon System CIA/National Security State	Specific Concrete **Political Ideologies** Laissez Faire Keynesian Economics Cold War / Detente Truman Doctrine Thatcherism Neo-Conservatives	Specific Concrete Events WWI-II, Korean & Vietnam wars; Genocide in East Timor & Central America; Structural Adjustment Programs. Prison-Industries; NAFTA, WTO; Wars in Iraq; 9/11; Wars in the Middle-East; etc.

Specific

See also Harvath and Gibson (1984)

relationships and processes involved within and between these systems, a recognition of the internally related character of history and structure. Marx's synthesis of the abstract and the concrete over time/space deciphered the contours of both world history and capitalist development and "provided a methodological gauge against which theoretical propositions pertaining to human society could be measured in order to determine their relative explanatory power" (Horvath and Gibson 1984: 14). One realizes Marx's justification in believing a coherence in world history can be grasped.

Marx's models are not set up to be "tested" except in a limited sense: Do they "fit" the facts and "work" as interpretive-analytical frameworks? Do the structural and processural variables targeted by the model correspond to empirical observations? If not, can countervailing influences be pinpointed? Do the models describe and explain systemic regularities, ranging from new structural conditions to current historical events? Do they do this better than competing models? Do models provide for continual refinement and analytical utility? This temporal flexibility distinguishes Marx's methodology, which is meant to retain its usefulness for studying ongoing capitalist development on a global scale as well as its transition into a new era. The "test" is whether Marx provided a method fit for studying both his and our conditions of life. What give Marx's analyses their power are not new facts, but new analyses of facts that reveal hitherto unseen mechanisms.[17] Thus, his dialectical models of history and system guide research in the same way a map helps one finding one's way through a physical landscape.

If unarmed with an internal relations approach to dialectical thought, in conjunction with an unclear understanding of Marx's approach to scientific method, we are left with diminished chances to recognize the form and

---

[17] "The vulgar economist has not the slightest idea that the actual, everyday exchange relations and the value magnitudes *cannot be directly identical*. The point of bourgeois society is precisely that, *a priori*, no conscious regulation of production takes place. What is reasonable and necessary by nature asserts itself only as a blindly operating average. The vulgar economist thinks he has made a great discovery when, faced with the disclosure of the intrinsic interconnection, he insists that things look different in appearance. In fact, he prides himself in his clinging to appearances and believing them to be the ultimate. Why then have science at all?... But there is also something else behind it. Once interconnection has been revealed, all theoretical belief in the perpetual necessity of the existing conditions collapses, even before the collapse takes place in practice" (Marx 1988b: 69, emphases in the original).

content of his epistemological strategies. If Marx held an internal relations philosophy of science, believed that a science was not yet fully developed if it could not be expressed mathematically, and thought mathematics demonstrated dialectical properties, then the examples discussed here are possibly what he had in mind when he attempted to bring all of these methods to bear on the history of our present. One difficulty in recognizing this stems from intellectual traditions. Another stems from the material conditions of capitalism itself. In order to comprehend this environment conceptually, and thus the possibility of analyzing it scientifically, Marx needed a way to bring Hegel to materialism, materialism to history, history to structure, and structure to mathematics, all the while avoiding the pitfalls of formula or rigidly determined inevitabilities. This is one key way we must understand the importance and meaning of Marx's words and models, their form, their interrelations, and their justifications.

Part III: From Marx's Science to His Politics

Several things have colored receptions of Marx. Many have been influenced by misconceptions about Marx's ideas as well as the problems reading his work, not the least of which is understanding the scientific basis of his dialectical method. Adding to this problem, the historical movements that used Marx's name and ideas as their inspiration often ended up discrediting him. If what has been called "really existing communist societies" are taken as representative of Marx's ideas in action, then Marx is deposited in the historical dustbin along with them. There are ample reasons to question this assumption.

Chapter Seven
From Political Economy to the Communist Project

With dialectics helping him grasp the logic of change, and with historical materialism directing his attention to class struggles and material contradictions, political economy brought Marx to at least three conclusions about his present moment. First, capitalism's demise will come through economic crises and/or a social revolution. Second, capitalism's technological complexity, material abundance, democratic potential, and class structure produced the presuppositions of communism. Third, for a communist political project to succeed it would need to cultivate the preconditions necessary for its emergence. This chapter examines this project and how it was later transformed under Marx's name.[1]

Prior to Marx's conversion to it, discourse on communism centered on the utopian theories of Robert Owen, Charles Fourier, and Henri de Saint-Simon. Wanting no part in utopian schemes in his early career as editor, Marx (1975d: 220) wrote that "The *Rheinische Zeitung*, which does not admit that communist ideas in their present form possess even *theoretical reality*, and therefore can still less desire their *practical realisation*, or even consider it possible,

---

[1] An earlier version of arguments in this chapter can be found elsewhere (see Paolucci 2004).

will subject these ideas to a thoroughgoing criticism," though he conceded that his paper's disagreements with such ideas meant that they "must take such theoretical works the more seriously" (emphasis in the original). It was only later that Moses Hess converted Marx (and Engels) to the communist program (see Bottomore 1973: 30, Hobsbawm 1982a–b, Jones 1982).

It is inaccurate to see Marx's approach to communism as based upon ethical considerations. Rather, assuming that "mankind is not beginning a *new* work, but is consciously carrying into effect its old work" (Marx 1975k: 144, emphasis in the original), he believed capitalism creates "a movement which itself produces the *material conditions of emancipation*" (Marx 1985l: 29, emphasis in the original). This was true regardless of any opinion, theory, or moral desideratum. For him, the "working class...have no ready-made utopias to introduce" (Marx 1968a: 61). The communist program was to be – unlike the Utopians – practical, never ossifying into a preformulated theorem. Moreover, as opposed to "crude communism" based on "envy" and "leveling-down," Marx's (1988a: 101–102) vision was a "fully-developed humanism" that involved the "re-integration or return of man to himself, the transcendence of human estrangement...accomplished within the entire wealth of previous development."

It is misleading to understand this vision as a "prediction." Like a weather forecaster, Marx operated with a set of key variables. In meteorology, laws of condensation and evaporation, atmospheric pressure, etc. allow for short-, medium-, and long-term forecasts of changes in the weather. In Marx's case, class configurations and technological capacities also allow for a science, a forecast of the inevitability of change, and an extrapolation of political strategies for shaping that change. Although both approaches allow for a relatively good, though not always perfectly clear, vision of the short-term picture, given the relationship between necessity and contingency, the further in space/time the inquiry extends, the more difficult it is to know what the results will be. Thus, Marx was necessarily less detailed the further from the present his discussion of communism moved. With communism's conditions of possibility rooted in capitalism, the question was: How to get there from here? Bringing with him scientific dialectic's concerns about precision and flexibility, stability and change, Marx's vision encompassed four interconnected stages, each further into the future and each more contingent on unforeseeable events. These stages include the revolutionary program, a socialist transitionary period, the emergence of early communist society, and long-term communist development (for additional discussion on this topic, see Ollman 1979b).

At the heart of Marx's *revolutionary program* are a flexible but radical politic, a nonformulaic vision of the communist turn, the encouraging of the elements of communism created by capitalism, and revolutionary action.

Mid- to late 19th century Europe displayed a variety of political-economic conditions: a combination of feudalism in retreat, capitalism in ascendance, historical relationships to external societies, as well as forms of knowledge functioning as a connective tissue between these social relations. During Marx's youth, Germany was backward relative to Europe's most capitalistically developed regions. France under went wars and revolutions between monarchies, bourgeoisie, peasants, and working classes in multiple configurations. England was being transformed into an industrial machine. The United States was increasingly announcing its presence as a power. All had colonies to a greater or lesser extent. This uneven development produced various conditions with which to contend and, therefore, it would be wrong to assume that "the ways to achieve [revolutionary] goal[s] are everywhere the same" (Marx 1978e: 523). "What is to be done, and done *immediately* at any given, particular moment in the future, depends, of course, wholly and entirely on the actual historical circumstances in which action is to be taken" (Marx 1992e: 66, emphasis in the original). Depending on the situation, Marx advocated forming parties, unionizing industries, wholesale strikes, and armed resistance.

In advocating resistance to capital, violence was one variable to take into account. Although workers in democratic societies such as United States, England, or Holland might achieve their goals "by peaceful means" (Marx 1978e: 523), they needed to make clear to the bourgeoisie that "we know that you are the armed force opposing the proletariat – we shall act against you peacefully wherever possible – and take up arms when that is necessary" (Marx 1986b: 618). Of the Jacobins in the French Revolution, Marx thought their use of violence as a first choice was "a sign of the weaknesses and immaturity of that revolution, which had to impose by violence what was not yet inherent in society" (McLellan 1975: 68). Nevertheless, any such "historical development can remain 'peaceful' only for so long as its progress is not forcibly obstructed by those wielding social power at the time.... However, the 'peaceful' movement might be transformed into a 'forcible' one by resistance on the part of those interested in restoring the former state of affairs" (Marx 1989e: 248). Because of what the bourgeoisie was willing to do, workers must be prepared to defend themselves and their revolution, violently if necessary. "Of one thing you may be sure," Marx (1992e: 66) warned, "a socialist government will not come to the helm in a country unless things have reached a stage at which it can, before all else, take such measures as will so intimidate the mass of the bourgeois as to achieve the first desideratum – time for effective action."

Marx's program of permanent revolution combined radical politics with flexible strategies for achieving practical goals. Capitalism's free laborers and democratic potentials come into conflict with its appropriation of wealth, resulting in systemic instability. These may be expressed as financial crises for capitalists and/or political crises of them and their state. During these struggles, workers must make "it clear to themselves what their class interests are, by taking up their position as an independent party as soon as possible and by not allowing themselves to be misled for a single moment by the hypocritical phrases of the democratic petty bourgeois.... Their battle cry must be: The Revolution in Permanence" (Marx and Engels 1978b: 287). With the transitional aims of widening access to wealth, increasing democratic participation, and overturning bourgeois property relations, revolution would be permanent so long as alienation, class divisions, and capitalist exploitation remained. Organize workers and force a revolution upon the bourgeoisie, this was the program (see Engels 1971).

The transition from capitalism to socialism involved encouraging the elements of communism "not as it has developed on its own foundation, but, on the contrary just as it emerges from capitalist society; which is thus in every respect, economically, morally, and intellectually, still stamped with the birthmarks of the old society from whose womb it emerges" (Marx 1978g: 529). In this period, revolutionary politics "will have to pass through long struggles, through a series of historic processes, transforming circumstances and men. They have no ideals to realize, but to set free the elements of the new society with which old collapsing bourgeois society itself is pregnant" (Marx 1968a: 61–62). Once consciousness of how communism was being made possible was attained, political action could cultivate that which best served the proletariat's interests, including efforts to "constitute itself as a political party," "go on strike," "combat the State," and "prohibit the employment in factories of children under age of ten" (Marx 1988d: 392–393). Other strategies called for a national bank, free universal education, the abolition of inheritance, and universal suffrage (Marx and Engels 1978a). And if a "true realm of freedom" was to be reached, the "shortening of the working day" (Marx 1909: 954–955) was required, as was colonialism, "since capitalism had to encompass the whole world before it could be overthrown" (McLellan 1973: 289).

Some strategies to transition out of capitalism were temporary and strategically pragmatic. For example, *The Demands of the Communist Party in Germany*, listed only four of the ten points of "The Communist Manifesto": a state bank, nationalization of transport, progressive income tax, and free education. "The *Demands* were a plan of action for a bourgeois (and not socialist) revolution; they were designed to appeal to the petty bourgeoisie and peasants as well as to the workers" (McLellan 1973: 194). In Marx's view, even temporary advantage had its limits, e.g., if a "critical conjuncture for a new international working men's association has not yet arrived ... [then] all labour congresses and/or socialist congresses, in so far as they are not related to the immediate, actual conditions obtaining in this or that specific nation, to be not only useless but also harmful. They will invariably fizzle out in a host of rehashed generalised banalities" (Marx 1992e: 67). Marx's politic evinces signs of a pragmatic, if bold, strategic program. Neither liberal reform nor anarchist tactics demonstrated similar sensibilities.[2]

Marx's view of the non-elitist nature of revolutionary change can be seen in several places. His questionnaire's Introduction declared that "it is the workers alone ... and not some providential saviours – who can energetically apply remedies to the social miseries that they undergo" (cited in McLellan 1973: 443). In a critique of the tactics of the Social-Democratic Party, Marx and Engels (1989: 264, 269) summarily stated that "we cannot co-operate with men who say openly that the workers are too uneducated to emancipate themselves, and must first be emancipated from above by philanthropic members of the upper and lower middle classes." Intellectuals' role is as intellectual laborers,

---

[2] According to Marx's friend, Dr. Kugelmann (1928), "he spoke very disapprovingly of Bakunin. He said his motto was 'Everything must be ruined' and that it was absolute nonsense to destroy values, to pull down one's own and other people's houses and then run away without knowing where and how to build another one" (cited in McLellan 1981: 89).

not as party leaders.[3] While the "first phase of the proletariat's struggle... is marked by a sectarian movement... [because] the proletariat has not yet developed sufficiently to act as a class... fantastic solutions thereof, which the mass of workers is left to accept, preach and put into practice [should be rejected]. The sects formed by [the likes of Bakunin, Owen, Fourier, Saint-Simon] are... alien to all real action, politics, strikes, coalitions, or, in a word, to any united movement" (Marx and Engels 1988: 106). For Marx (1985k: 14), "the emancipation of the working classes must be conquered by the working classes themselves" and their struggle is "not a struggle for class privileges and monopolies, but for equal rights and duties, and the abolition of all class rule." The goal was for the proletariat as a class was to win back the means of production, which would provide them the means for destroying class antagonisms once and for all. This only they alone could do.

In the *socialist transition to communism*, communism would not arrive as if a metaphysic. Capitalism's collapse would handover a world, "if certainly not idyllic," whose conditions for socialism "will also be there" (Marx 1992e: 67). In this "period of revolutionary transformation of the one into the other" (Marx 1978g: 538), the remnants of the old society must be handled properly and the maturity of the elements of the new one nurtured. Seizing and retaining possession of the means of production and redistributing wealth were necessary. But doing this through attaining state power would present problems. On the one hand, the "working class cannot simply lay hold of the ready-made state machinery, and wield it for its own purposes" (Marx 1968a: 54). In a revolutionary situation, the state has its own material force and must be dealt with appropriately. On the other hand, workers could not disregard the state. Why? First, the state apparatus needs to be used to make goods and services available. The proletariat not only needs such services but also must win general legitimacy from the population as a whole. Second, because private property must be dismantled, the state has the capacity to transfer

---

[3] On his and Engels's association with radical parties, Marx (1991b: 288) explained: "Neither of us cares a straw for popularity. Let me cite one proof of this: such was my aversion to the personality cult that at the time of the International, when plagued by numerous moves – originating from various countries – to accord me public honor, I never allowed one of these to enter the domain of publicity, nor did I ever reply to them, save with an occasional snub. When Engels and I first joined the secret communist society, we did only on condition that anything conducive to a superstitious belief in authority be eliminated from the Rules."

social wealth from the ruling class to society as a whole. Third, a victorious proletariat can be defeated by a bourgeois counterrevolution, which must be prevented. If a revolutionary proletariat could gain the allegiance of the middle classes, and if they could gain the assistance and solidarity of other working classes, then the revolution stood a chance of success. However, an unprepared proletariat could easily be ousted. Immediately destroying the material state apparatus provided by capitalism would make such problems more intractable.

Given these considerations, one socialist strategy was "a political transition period in which the state can be nothing but the revolutionary dictatorship of the proletariat" (Marx 1978g: 538). For some, by advocating the dictatorship of the proletariat Marx held less-than-democratic commitments. Indeed, Annenkov said of a younger Marx: "This tone expressed the firm conviction of his mission to dominate men's minds and prescribe them their laws. Before me stood the embodiment of a democratic dictator such as one might imagine in a daydream" (cited in McLellan 1975: 8). Whatever this characterization might tell us about Marx, it tells us little about his thoughts on the political conditions likely to obtain after a workers' revolution. According to McLellan (1975: 74), "the word 'dictatorship' did not have the same connotation for Marx as it does nowadays. He associated it principally with the Roman office of *dictatura*, where all power was legally concentrated in the hands of a single man during a limited period in a time of crisis." The term was meant to reflect how the proletariat would reorganize society after a revolution, including both political defense and economic reorganization. Socialism, that is, would help the working class "win the battle of democracy" (Marx and Engels 1978a: 490). Eventually, however, the state should be dismantled and/or it would fade away with the dissolution of class distinctions.[4]

In this socialist work, the dictatorship of the proletariat, "as the necessary transit point to the abolition of class distinctions generally," has several tasks, including "the abolition of all the relations of production on which [the classes] rest... [and] the abolition of all the social relations that correspond to these

---

[4] "If you look at the last chapter of my *Eighteenth Brumaire*, you will find that I say that the next attempt of the French revolution will be no longer, as before, to transfer the bureaucratic-military machine from one hand to another, but to *break* it, and that is essential for every real people's revolution on the Continent" (Marx 1989a: 131, emphasis in the original).

relations of production ... [all of which results in] the revolutionising of all the ideas that result from these relations" (Marx 1978c: 387–388). The demands in "The Communist Manifesto" offer such transitional measures that, though "different in different countries," "will be pretty generally applicable" for "the most advanced countries" (Marx and Engels 1978a: 490).[5] This list includes national ownership in land, steeply graded taxation, abolition of inheritance, confiscation of property of rebels opposing the proletariat, nationalized banking and transport, planned industrial and agricultural development, compulsory work for all, interpenetration of agriculture and industry, free education, and abolition of child labor. In contrast to the Utopians, the goal is not the abolition of individually owned property but the private ownership of the means of production. "Hence the abolition of private property and communism are by no means identical, and it is not accidental but inevitable that communism has seen other socialist doctrines – such as those of Fourier, Proudhon, etc. – arising to confront it because it is itself only a special, one-sided realisation of the socialist principle" (Marx 1975k: 143). The "immediate aim" of these strategies is the "formation of the proletariat into a class, overthrow of the bourgeois supremacy, and conquest of political power by the proletariat" (Marx and Engels 1978a: 484). In other words, undermine the political-economic power of the capitalist class and increase that of the proletariat. Socialism, only temporary if communism was to be, had to protect the revolutionary victory and to unleash forward the liberating possibilities seized from capitalism as well as those newly cultivated.

Taking the analysis of the Paris Commune as "Marx's view of the shape of the future communist society after the revolution" rather than seeing this as an actual socialist revolution or even an accurate reflection of those events, McLellan (1973: 397) explains that, for Marx, socialism required a communal regime to be established and connected to secondary centers of power, removal of the standing army, abolition of class property, expropriation of

---

[5] Engels (1988: 206) explained: "when it was written, we could not have called it a 'Socialist Manifesto'" because at that period socialists "were understood, on the one hand, the adherents of the various utopian systems: Owenites in England, Fourierists in France ... mere sects ... the most multifarious social quacks." Also: "[Henry George's *Progress and Poverty's*] basic tenet is that *everything* would be *in order* were rent paid to the State. (You will also find payment of this kind among the transitional measures included in the *Communist Manifesto*.)" (Marx 1992h: 99–100; emphases in the original).

expropriators through "united co-operative societies," planned economics, and the distribution of social wealth to the masses. These tactics are meant as preparation for dissolving the state and establishing the basic structure of communist society. As capitalism is increasingly a thing of the past and as socialism transitions to communism, "the 'present-day state,' in contrast with the future, in which its present root, bourgeois society, will have died off" (Marx 1978g: 537). That is, "the proletariat, instead of fighting against the economically privileged classes in each individual instance, has acquired sufficient power and organization to use the general means of coercion against them; however, it can use only such economic means as abolish its own character as wage worker, hence as a class; so its complete victory coincides with the end of its domination, for its class character has come to an end" (Marx 1978f: 544). Informed by past praxis, modern scientific knowledge, and a rationally planned vision of the future, the dismantling of capitalism transforms its state. In this view, alienation, ideology, and superstition give way to reason with the increasing success of socialism's transition of the world to communism.

In *early communist society*, the state, something that was erected by the bourgeoisie as a coercive force over civil society, would become increasingly unnecessary as distribution of socially produced wealth equalized. As the state "fades away" it will leave behind the economic and technical apparatus created by capitalists but now staffed by classless, decentralized communists. Although envisioning bureaucratic domination as indicative of bourgeois states (see Marx 1976a: 46–47), Marx (1978g) also equated a bureaucratically organized society with "vulgar socialism," a vision primarily focused on distribution. Marx's vision of communism in this era included the self-government of producers, a national militia (with a short term of service), rural areas divided into districts (each district would have an assembly of delegates in the central town) and each district would send deputies to the national delegation (such delegates would be confined to the formal instructions of constituents and they could be revoked at anytime). A central government reduced in power but still possessing functions would exist at first, the unity of the nation would not be broken, and universal suffrage would be established. There would be no permanent hierarchical offices. At this stage, Marx surmised that a person's social contribution could be represented in a certificate, which would establish that they had provided so much labor to society and with which they could draw from a commonly produced and held pool of social value. But, rather than perfect equality, each person's draw would vary with the quantity and quality of labor provided.[6] Without the goal of profits and surplus-value via surplus-labor, this would be, presumably, qualitatively different from wages.

As the distribution of goods and the access to social wealth becomes commensurate for individuals in relationship to communist society's level of technological, cultural, and political development, there should be a corresponding a reduction in the worse abuses attendant with capitalism, such as swindlers, speculators, agents, crimes against property, and standing armies (see Carver 1983: 63). Presumably, social problems unrelated to class exploitation would not be eliminated. Marx predicts no universal peace, no universal riches, no complete freedom from labor, and no guarantee of perfect social tranquility. In other words, there is none of the utopianism of which he is often accused. But he does believe that class exploitation and all that comes with bourgeois property relations are things that can be overcome.

After the initial stage of communism, i.e., *communism's long-term development*, "the enslaving subordination of the individual to the division of labour, and therefore also the antithesis between mental and physical labour [vanishes and labor] becomes not only a means of life but life's prime want [where

---

[6] "But one man is superior to another physically or mentally and so supplies more labour in the same time, or can labour for a longer time; and labour, to serve as a measure, must be defined by its duration or intensity, otherwise it ceases to be a standard of measurement. This *equal* right is an unequal right for unequal labour. It recognises no class distinctions, because everyone is only a worker like everyone else; but it tacitly recognises unequal individual endowment and thus productive capacity as natural privileges. *It is, therefore, a right of inequality, in its content, like every right*. Right by its very nature can consist only in the application of an equal standard; but unequal individuals (and they would not be different individuals if they were not unequal) are measurable only by an equal standard in so far as they are brought under an equal point of view, are taken from one *definite* side only, for instance, in the present case, are regarded *only as workers* and nothing more is seen in them, everything else being ignored. Further, one is married, another not; one has more children than another, and so on and so forth. Thus, with an equal performance of labour, and hence an equal share in the social consumption fund, one will in fact receive more than another, one will be richer than another, and so on. To avoid all these defects, right instead of being equal would to be unequal.... But these defects are inevitable in the first phase of communist society as it is when it has just emerged after prolonged birth pangs from capitalist society. Right can never be higher than the economic structure of society and its cultural development conditioned thereby" (Marx 1978g: 530–531, emphases in the original).

there also increases] the all-round development of the individual, and all the springs of cooperative wealth flow more abundantly [after which] the narrow horizon of bourgeois right [can] be crossed in its entirety and society inscribe on its banners: from each according to his ability, to each according to his needs!" (Marx 1978g: 531). People will not want for their needs and individuals' abilities will be able to be freely directed toward the labor of their own choosing, without coercion. The alienation resulting from class exploitation will allow humans to interact once again as a whole species, rather than as a divided one. Laboring activity could return to expressing the creative capacities of our species in a way where there is a "re-integration or return of man to himself... the complete return of man to himself as a social (i.e., human) being... [and a] resolution of the conflict between man and nature and between man and man – the true resolution of the strife between existence and essence, between objectification and self-confirmation, between freedom and necessity, between the individual and the species. Communism is the riddle of history solved, and it knows itself to be this solution" (Marx 1988a: 102–103).

History's riddle is yet to be solved by what Marx envisioned, though it would be a mistake to assume his vision has been denied and an even a larger mistake to assume that the best-known self-proclaimed attempts to enact Marx's program are commensurate with that program itself.

The Discursive Transformation: From Marx's Communist Project to Soviet Diamat

Marx's communist program was transformed into a field of associated statements found in Soviet Union's official state ideology of *Diamat*. As an object of knowledge, an "associated field turns a sentence or a series of signs into a statement, provides them with a particular context, a specific representative content, forms a complex web" (Foucault 1972: 98). The meaning of an associated field is established through "all the series of other formulations within which the statement appears... all the formulations to which the statement refers (implicitly or not)" – including repetition, modification, adaptation, or opposition – "all the formulations whose subsequent possibility is determined by the statement, and which may follow the statement as its consequence, its natural successor, or its conversational retort," and "all the formulations whose status the statement in question shares, among which it takes its place without regard to linear order, with which it will fade away, or . . . be valued, preserved, sacralized, and offered, as a possible object, to a future discourse." The identity of an associated field "varies with a complex set of material institutions" (Foucault 1972: 98–99, 103). "Marxism" became such an associated field and the Soviet Union was often seen as the set of material-institutions that expressed it. How did this happen?

The post-World War II intellectual milieu in the West, influenced by the rise of the Cold War, held the somewhat standard view that the Soviet Union was a reflection of Marx's vision of philosophy, history, politics, science, and the future, e.g., "Marx's conception of freedom . . . leads inevitably to an antilibertarian, that is to say, an authoritarian view of man and of the individual's relation to society" (Harris 1950: 79–80). Popper (1966: 333), after acknowledging Marx's commitment to scientific method and his fight against utopianism, goes so far as to accuse Marx of allowing "romantic" and "irrational and aesthetic sentiments usurp, in places, complete control of this thoughts," resulting in mystical "wishful thinking." However, as discussed in Chapter One, reading the romantic traits of the early Marx very far into his intellectual biography and political program is unjustified. And, as we have just seen, Marx's approach to communism is not simply an article of faith nor is it a teleological-metaphysic, but rather a program requiring struggle, strategy, and work. Finally, Marx's understanding of communism's potential was founded upon a democratic proletarian revolution in *capitalist societies*. The Russian Revolution was a peasant-worker revolt in a primarily agrarian-czarist society. Historical materialism tells us a revolution *may* occur in a noncapitalist society but the conditions there should not be confused with the requisites for a revolution with socialist-communist potential. However, many Marxists, Soviets, critics of each, and sociologists in general have interpreted Marx as advancing the teleological idea that all social forms evolve from the less economically developed to more advanced stages, ultimately leading to capitalism and a socialist-communist revolution.[7]

---

[7] "Soviet Marxists under Stalin insisted on a simple, unilinear view of rigidly similar evolution in all societies. In a typical Soviet Marxian text, Keiusinen claimed: 'All peoples travel what is basically the same path. . . . The development of society proceeds through the consecutive replacement, according to definite laws, of one socioeconomic formation by another' (1961, 153). In the Soviet view, every society must pass through slavery, feudalism, capitalism, socialism, and communism" (Sherman 1995: 59).

These interpretations are inconsistent with the Marxian archive. Marx (1979c: 321) warned that one should be careful not to "transform my historical sketch of the development of capitalism in Western Europe into a historical-philosophical theory of universal development predetermined by fate for all nations, whatever their historical circumstances." When asked whether capitalism was Russia's inevitable future, Marx (1992f: 71) explained:

> Hence the 'historical inevitability' of this process is *expressly* limited to the *countries of Western Europe*.... In this Western movement, therefore, what is taking place is the *transformation of one form of private property into another form of private property*. In the case of the Russian peasants, *their communal property* would, on the contrary, have to be *transformed into private property* (emphases in the original).

The possibility of Russia bypassing capitalism and building socialism outside of the capitalist political economy was not likely to succeed. Russia's material situation was not ideal for building socialism because it did not contain the prerequisite mix of class, political, technological, and ideological conditions that make up its necessary and sufficient presuppositions and preconditions.[8] However, a peasant revolution in noncapitalist Russia is often used as a testing ground for Marx's theory of proletarian revolution in capitalism, a problematic interpretation and evaluation of Marx's approach to both class struggles in capitalism and the possibilities of a socialist transition out of it (for discussion, see Godelier 2000).

The rise of the Soviet Union was accompanied by several discursive shifts in Marx's ideas. In one, Engels's approach to dialectic – more codified than Marx's – established a framework in which Marx's views would later be read. In another, the ideas of proletarian revolution and a socialist transition to

---

[8] In his critique of Bakunin, for instance, Marx (1978f: 544) claims that he "understands absolutely nothing about social revolution; all he knows are its political phrases. For him its economic requisites do not exist. Since all hitherto existing economic formations, developed or undeveloped, have included the enslavement of the working person (whether in the form of the wage worker, the peasant, etc.), he thinks that a *radical revolution* is possible under all these formations. Not only that! He wants a European social revolution, resting on the economic foundation of capitalist production, to take place on the level of the Russian or Slavic agricultural and pastoral peoples and not to overstep that level; although he does see that *navigation* creates a difference between the brothers, but only *navigation*, for that is a difference all politicians know about! *Will power* and not economic conditions is the basis of his social revolution" (emphases in the original; also see Engels 1989: 39–40).

communism were read through "scientific socialism" and transformed into a project called "Marxism." In a third transformation, what Marx and Engels had called "our outlook" was turned into a world-philosophy of "dialectical materialism." The general outcome was the conflation of Marx's dialectical method with his communist project, a materialism shaped into a mechanist pseudo-science, and the dictatorship of the proletariat was transformed into the dictatorship of the Communist Party.

Soviet intellectual discourse came to see Marx's dialectic as a sensuous force and general organizing principle in all of nature, social life, and history. Over time, the associated field of statements organized within Soviet institutions morphed Marx's dialectic into justifications for the political leadership's aspirations. Engels's application of dialectical categories to natural science, where the dialectic was a law-like force in nature, "reached its apogee in Soviet textbooks on dialectical materialism. It was this trend that presented Marxism as a philosophical world-view or *Weltanschauung* consisting of objective laws and particularly laws of the dialectical movement of matter taken in a metaphysical sense as the basic constituent of reality" (McLellan 1973: 423). With evolutionary theory used to give them a stamp of legitimacy, armed with economic determinism in one hand, and asserting "the dialectic" as a force in reality in the other, the Soviets demanded doctrinaire uniformity in thought and saw their system as the representative of transcendental historical forces (for comment, see Bellamy Foster 1996: 16). It is through this lineage that an over-precise approach to dialectical reasoning evolved from Engels's systematizations and was transformed posthumously into an *apologia* for Leninism and Stalinism. Some historical background will help us understand how this happened.

Though socialist ideas in Russia predate the introduction of Marx's thought, from 1870 onwards Marx's "doctrine was taught and discussed in Russia ... by many professors of economics" (Bochenski 1963b: 26). A central lesson early Russian radical intellectuals took away was "that the development of Russian capitalism was inevitable [and that] Russia would be forced along a similar path to that followed by countries of Western Europe" (Smith 1996: 52–53). But Russian political culture was not uniform. By the time of the Russian Revolution (October 1917), radical, anti-czarist, pro-socialist sentiment existed among professionally trained academics who opposed the

Bolsheviks. Another group, the *narodniki*, was devoted to overthrowing the czar and bringing socialism to the people while bypassing capitalist development. However, because the cultural-political conditions of czarist Russia allowed little or no freedom of inquiry, the intellectual class was ill-prepared for the sort of inquiry required by science and democracy, being more familiar with the jailing of dissenters, police spies, and a general lack of open social-political debate (Ball and Dagger 1999: 158). In the process, academic approaches to Marx became increasingly suspect as remnants of bourgeois culture and a class of thinkers seeking abstract, general principles made for a milieu preconditioned toward creating doctrines.[9]

Those in control of the aftermath of the October Revolution would have to work with and within larger sociological and historical variables. Russia was a mix of czarist oligarchy and peasant agrarianism with a rising capitalist class but was hardly industrially developed compared to Europe. Russia's relationship to the global society brought it external threats. It was surrounded (primarily) by capitalist societies and global markets and was invaded by the West shortly after the Revolution. "By 1920", writes Smith (1996: 28), "international isolation of the Revolution was already beginning to have its dire effect on the theory of the communist movement. Lenin and Trotsky, as well as other leaders of the International, struggled to find a theoretical framework within which to tackle the terrible economic and social issues facing the Soviet state." This struggle for theoretical clarity created pressure to establish a "true" orthodoxy over Marx's thought. Engels's work provided the groundwork for the transformation that Marx's ideas would suffer at their hands.

---

[9] "Russian writers characterize the mentality of the Russian intelligentsia of the second half of the nineteenth century and of the early twentieth century as [embracing] that special love of the abstract and that interest in great principles which also goes hand in hand with a deep distrust of the concrete.... A mental attitude of this kind is almost always allied to an extreme form of dogmatism which lacks the critical and skeptical spirit characteristic of the West" (Bochenski 1963b: 23).

Transforming Marx's Dialectic: From Engels to Kautsky, Bernstein, and Plekhanov

The ideas of Marx and Engels should not be reduced as if they were a single person.[10] Nor was "Marxism" something either had in mind.[11] Engels (1936b: 472) remarked that in observing the French "Marxists" of the late 1870s, Marx would retort, "All I know is that I am not a Marxist." After Marx's death, nevertheless, it was Engels's works that provided for several discursive shifts that would become official "Marxism." What Engels and Marx called "our outlook," "the theory," and "the materialist conception of history" was transformed into something called "dialectical materialism." When Engels (1941: 43) wrote that "Marx was a genius; we others were at best talented. Without him the theory would not be what it is today. It therefore rightly bears his name," this was in a chapter ostensibly titled "Dialectical Materialism." However, a footnote by the editor in the first chapter explains, "Titles have been added to the numbered sections." This was an important step in creating a supposedly coherent "Marxist" political-philosophy. But as Rubel explains, Engels was "disinclined to use such terminology [because it] risked corrupting the essential meaning of Marx's writings," and only then using quotation marks or speaking of "so-called Marxists" in an attempt "to forestall the risk of confusion and ideological corruption." Following this line of evidence, Rubel (1981: 22) concludes that Engels "would have vetoed the use

---

[10] "L. Blanc, Blanqui, Marx and Engels are his Unholy Quadrinity which he never forgets. We two – the propounders of 'equal economic rights' – are said *inter alia* to have advocated '*Armed* (!) appropriation of capital.' Even the jokes we cracked about Switzerland in the *Revue* 'fill him with indignation.' 'No Civil List, no standing army, no millionaires, no beggars' – 'Marx and Engels hope that Germany will never sink to such depths of degradation.' It is exceedingly odd, the way he speaks of us in the *singular* – 'Marx and Engels *says*', etc." (Marx 1983d: 63–64, emphases in the original).

[11] "When Engels decided to adopt the designations 'Marxist' and 'Marxism', forged by his and Marx's adversaries as pejoratives for use in moments of polemic, and to put the followers of 'scientific socialism' in defiance of their adversaries by transforming these terms into titles of glory, he hardly expected that his act of defiance (or was it resignation?) made him the spiritual father of a mythology destined to dominate the history of the twentieth century.... The word 'Marxism' seems to have been used for the first time in the title of a writing in 1882... which attacked the 'Marxists' in the French socialists party. Gradually, however, the French disciples grew accustomed to the new titles, which they had done nothing to create, and helped develop them from sectarian labels to concepts with political and ideological content" (Rubel 1981: 20–21).

of Marx's name." Still, Marx's ideas soon became "Marxism," an abstraction only possible after his death.[12]

In "Socialism: Utopian and Scientific" (a central source for Russian revolutionary intellectuals), Engels (1978a) claims in his 1892 Introduction that he provides Marx's approach "a more connected form." Thus, it is of consequence that Engels (1978c: 681) believed that "Just as Darwin discovered the law of organic nature, so Marx discovered the law of development of human history." According to Callinicos (1983: 61), in his efforts...

> Engels...present[ed a] picture of a universe governed by certain objective, generally applicable, scientifically ascertainable laws. Historical materialism, he argued, had discovered the specific version of these laws operative in the realm of human society. Knowledge of these laws is essential to the success of a socialist party, since the proletarian revolution will be the outcome of an objective social process.

For Marx (1978f: 546), however, the term "scientific socialism" was "used only in opposition to utopian socialism, which tries to impose new hallucinations and illusions on the people instead of *confining the scope of its knowledge to the study of the social movement of the people itself*" (emphasis added). For Marx, "scientific socialism" was not a world-philosophy. Lacking a work on dialectics from Marx, the Soviets turned to Engels's (1934, 1939) dialectical vision of science and the natural world. His works implied that history contains a law called "dialectics," much the same way natural selection is a law of natural processes. Over time the Soviets adopted "dialectical materialism" as a philosophy that bridged materialism, natural science, and the movement of history. The linking of "Marxism," "dialectical materialism," and "scientific socialism" with Engels's dialectic helped reduce what Marx thought required political struggle down to a universal metaphysical law. These shifts conflated the historical, materialist, and dialectical features of "the theory" with its

---

[12] An author's name "performs a certain role with regard to narrative discourse, assuring a classificatory function. Such a name permits one to group together a certain number of texts, define them, differentiate them from and contrast them to others. In addition, it establishes a relationship among the texts.... [The] name seems always to be present, marking off the edges of the text, revealing, or at least characterizing, its mode of being. The author's name manifests the appearance of a certain discursive set and indicates the status of this discourse within a society and a culture" (Foucault 1984: 105–107).

political program, making these levels of discourse increasingly intertwined. The linguistic turn set the terrain of debate, i.e., Marx apparently had developed a finished "system" and all that was required was to discover what that philosophical system was and then implement it.

Although contemporary theorists differ on whether Engels and Marx shared the same dialectical outlook, it is clear that Engels did not present and explain dialectical reason as effectively as Marx used it. The "more connected form" of Engels's dialectic resulted in a less flexible, more operationally defined, Kantian-like set of principles that assisted in formalizing Marx's thought. For instance, after calling dialectics the "science of inter-connections," Engels (1934: 62–63) claimed that it . . .

> . . . is, therefore, from the history of nature and human society that the laws of dialectics are abstracted. For they are nothing but the most general laws of these two aspects of historical development, as well as of thought itself. And indeed they can be reduced in the main to three: The law of the transformation of quantity into quality and vice versa; The law of the interpenetration of opposites; The law of the negation of the negation. . . . [Hegel's mistake] lies in the fact that these laws are foisted on nature and history as laws of thought, and not deduced from them. . . . We are . . . concerned here with . . . showing that the dialectical laws are real laws of development of nature, and therefore are valid also for theoretical natural science.

In the same book, Engels (1934: 267) writes that his dialectic is "conceived as the science of the most general laws of *all* motion . . . its laws must be valid just as much for motion in nature and human history as for the motion of thought" (emphasis in the original). Thus, Engels accepted the premise that reality can be reduced to matter in motion and "that all matter obeys dialectical laws" (Graham 1972: 33, also see Ball and Dagger 1999: 154). "Neglecting the interrelation of praxis and cognition suggested by Marx," Grier (1978: 55) writes, "Engels also re-invoked the copy theory of knowledge in a manner reminiscent of Lockean representative realism. In the . . . *Anti-Duhring*, Engels asserts that 'an exact representation of the universe, of its evolution and that of mankind, as well as of the reflection of this evolution in the human mind, can therefore only be built up in a dialectical way'." Engels's laws are not simply analytical frameworks for studying the world but are also *forces in* nature and society. This change allowed theorists to claim history was predetermined and as predictable as matter.

Marx does not treat dialectical laws as forces in nature (e.g., as gravity) and given his rejection of metaphysics, assumptions of multiple determinations, and his rejection of the use of "a universal key to a general historical-philosophical theory" (Marx 1979c: 322), he never would have claimed that "historical development" could be reduced to three dialectical laws. According to Callinicos (1983: 63), "In the hands of the Second International's chief theoreticians, Karl Kautsky, Georgi Plekhanov, and Antonio Labriola, the dialectic, conceived as an ontology, an account of the fundamental nature of being, served to justify a version of Marxism in which social change is an organic process whose outcome is determined in advance." As self-labeled "Marxists" emerged, they selectively published and edited (Marx's and) Engels's works "while freely making use of his prestigious name... [and] cut him off from the International" and turned Marx's dialectic into "tablets of stone" (Smith 1996: 52, 49). This "allowed later writers, and finally Stalin, to rush headlong into an ontology" (Althusser 1977: 14).

How were Engels's shifts passed on in and through Soviet ideology? In terms of dialectics, Engels's works – *Anti-Duhring*, "Socialism: Utopian and Scientific," *Ludwig Feuerbach and the End of Classical German Philosophy* – were "the most influential works of the Marxist tradition" and "whose tenets were passed on in lectures, primers and handbooks, down to official Soviet dialectics" (Carver 1983: 96–97). This "most influential" character was facilitated by the idea that Engels's discussion of dialectic was a reflection of Marx's dialectical method. Thus, Engels's work "was decisive in forming the philosophical orientation of the Second International, which the Lenin of 1908 for the most part accepted" (Sayer 1979: 106, citing Colletti 1972). "Official Soviet Marxism stated unambiguously that [the] three laws [the unity of opposites, change of quantity to quality, and negation of the negation] were laws of the movement of the universe as well as statements of method" (Sherman 1995: 236). The discursive path to Leninist-Stalinist philosophy leads away from Marx's works but through Engels's.[13]

---

[13] A note on this claim. Some are skeptical about targeting Engels in this transformation (see O'Neill 1996). The central objection is that Engels cannot be accurately framed as forwarding an anti-democratic politic. However, it is not that Engels pushed Marx's revolutionary program in an authoritarian direction politically. Rather, the point is that Engels's version of dialectics changed the form of Marx's dialectical thought and that this would be influential in the type of intellectual philosophy the Soviets would eventually adopt.

Kautsky (1854–1938) (with Engels) was a theoretical leader of the German Social Democratic Party (GSDP). Marx (1992g: 82), though, did not think highly of him, believing him to be "a mediocrity, narrow in outlook, overwise (only 26 year old), a know-all, hard-working after a fashion, much concerned with statistics out of which, however, he makes little sense, by nature a member of the philistine tribe, for the rest, a decent fellow in his own way; I unload him onto to *amigo* Engels as much as I can." As a popularizer of Marxism, Kautsky...

> ... placed great emphasis on the 'scientific' character of this orthodoxy. He saw the movement to socialism as being guaranteed by the operation of 'laws of history.' These resembled laws of nature, in that they operated independently of human will and consciousness. They applied universally and used human beings as their instruments.... Armed with the scientific doctrines of 'Marxism,' the 'Marxist Party' had the task of bringing the truth to the masses. The socialist intellectuals would teach scientific socialism to the workers.... Socialism meant chiefly that industry would come under the centralised control of the state, a state he envisaged as a form of advanced parliamentarism (Smith 1996: 34–35).

Kautsky concluded that a socialist revolution would be an extended affair punctuated by political revolutions, not a sudden change in social structure or class configuration (Callinicos 1983: 63–64). Eduard Bernstein also endorsed gradualist ideas but saw a more peaceful transition than did Kautsky. Bernstein programmatically articulated the party's demands, particularly the democratization of the German state. In terms of method, Bernstein thought dialectics amounted to mysticism, materialism was simply a prejudice, and argued for a "return to Kant" (Smith 1996: 35). Revolution was seen as arriving from the election of socialist candidates and the installation of Bernstein's program. Attempts to speed up the pace of revolution, according to Kautsky, would only endanger its possibilities of success. Thus, the German Second International adopted gradualism, reformism, alliances with liberals over revolution, tactics of negotiation and compromise, and the belief in the inevitability of socialism outside of human agency. This associated field of statements served as the linchpin of debate and division amongst subsequent radical movements (Callinicos 1983: 64–65).

Plekhanov (1856–1918) and his colleagues, influenced by the German Social Democrats, were the first Russians to label themselves followers of Marx.

Though compelled to leave Russia by the 1880s, Plekhanov (1972, 1969) authored several works on Marxism and participated in its revolutionary politics and philosophical debates from the 1890s until several years after the 1917 Revolution. He was the scholar who...

> ... first analyzed the Russian situation as a whole in Marxist terms, thereby earning renown as the 'Father of Russian Marxism.' His writings became the school for a whole generation of revolutionaries.... For years his thought was regarded as true orthodoxy, so that any Russian Marxist had to side with or against it at least until 1905.... Kautsky, regarded Plekhanov 'as the most important of the younger Marxists' (Steila 1991: 1, 7).

Although Plekhanov argued against any "one-sided, economic determinist meaning" in Marx (Allen 1969: 7), his work had several lasting and distorting effects.

In his examination of the "fundamental problems of Marxism," Plekhanov (1969: 21) assures his readers that "Marxism is an integral world outlook," though Marx never offered any such thing. His work was unfinished at his death, his mature vision developed over time, he changed his opinion on several matters, and admitted past mistakes.[14] Plekhanov also claimed that Marxism's philosophical views "in their final shape... were quite fully set forth, although in polemical form, in the first part of Engels's book, *Herr Eugen Duhring's*." Plekhanov's transformation of Marx's ideas often came with no specific citation. Though Marx never used the concept, in 1891 Plekhanov coined the term "dialectical materialism" in an "article in Kautsky's *Neue Zeit* [thinking]... that he was merely adapting it from Engels's usage in *Anti-Duhring* and *Ludwig Feuerbach*.... Plekhanov [advanced a version that] installed a materialism which left no room for will at all and this is what he foisted on to Marx" (Smith 1996: 40, also see Macintyre 1980: 132). He assumed that Marx would have supported the "neo-Lamarckian" position that argued for the animism of matter (Plekhanov 1969: 23, 45–47), a metaphysic Marx certainly would have rejected. In Plekhanov's terms, humans are impassive, and matter and history direct human affairs outside of individual agency. Plekhanov (see 1969: 48–49) used Marx's views as a rigid doctrine to be followed in the course of social life, politics, and revolution rather than as a

---

[14] "Late in life, asked about his 'collected works,' [Marx] is said to have remarked, 'they would first have to be written'" (Nicolaus 1973: 53).

general theoretical, interpretive, and/or analytical framework for research and action. Viewing Marx's materialist dialectic equivalent to the theories of Copernicus and Darwin and as a key to all questions, natural or social, provided the theoretical background for the teleological idea that Soviet Communism represented an *inevitable* force in society and history – no doubt a politically serviceable view.

Marxism after the Russian Revolution

After the 1917 revolution, the Soviet state became increasingly insular, its leadership concentrated and detached from the populace, and its allegiance to Marx devolved into orthodoxy. Marx was believed to have discovered the objective and regulatory laws of all history and the Soviet state was viewed as history's most recent evolutionary development. By the time of the Communist International in 1919, Marx's concept of "dictatorship of the proletariat" had been transformed into a "workers' state." "Marxism" now meant the construction of a state, not its dissolution. Who was to control this state? Leon Trotsky answered this question in 1920, in "Terrorism and Communism," where he argued that "In the hands of the Party is concentrated the general control.... It has the final world in all fundamental questions.... The last word rests with the Central Committee" (cited in Smith 1996: 27–28). Rather than more democracy under a dictatorship of the proletariat, the "worker's state" under the Central Committee offered less.

At the Party congress in London, in a debate on the character of the party, Lenin's views on strict organization won the majority support (*bolsinstvo*). Others, such as the Mensheviks, the "criticists," the "economists" (who believed economic work alone was necessary; the revolution would arrive on its own), and the "empirio-criticists" (influenced by European thought), went to the other side and were purged or disappeared (Bochenski 1963b: 26–27). Left in control of the Party, Bolsheviks adopted the following plank at the Third International:

> As leader in the struggle of the proletariat for emancipation, there is only the new, the Third, the Communist International, to the ranks of which belongs the Russian Communist Party. This International has in fact been created by the organisation of communist parties out of the genuinely proletarian elements among the socialist parties in various countries, and especially Germany; it was

formally constituted in March, 1919, and its first session held in Moscow. The Communist International, receiving more and more support from the proletarian masses in all lands, has returned to Marxism not merely in the name it has adopted, but also in its ideological and political tenets; and in all its activities it realises the revolutionary teaching of Marx, cleansed of bourgeois-opportunist perversions (Bukharin and Preobrazhensky 1966: 378–379).

Marxism had also been cleansed of much of Marx. To understand this process better, a review of the positions and the changes made from Plekhanov to Bukharin to Lenin is instructive.

One of the central issues of philosophical debate at the time was whether materialism was mechanical and all-embracing or whether separate realms of knowledge existed. As a leading "mechanist," Nicolai Bukharin (1925: 61) believed that "mental life of society is a function of the forces of production." For mechanists, all phenomena could be deduced from organic matter and explained by mechanical laws in a unified science. Residue of this reflexively mechanical view of nature resulted in a mechanical view of dialectics, where material conditions produce their own knowledge systems – e.g. "*Dialectical materialism*, as a method of cognition applied to social development, *has created the theory of historical materialism*" (Bukharin 1931: 19, emphasis added). In opposition, the Deborinites held that there were distinct areas of knowledge, a view closer to Marx's. Deborin, however, like Engels, thought that the dialectic was fully exemplified in nature itself and not simply a method of research.

In 1925, the Deborinites used Engels's *Dialectics of Nature* and Lenin's "On the Question of Dialectics" as a way to defeat the mechanists. By 1929, the Deborinites controlled the dominant Soviet publications and intellectual institutions of learning. Bukharin and the Right Opposition were politically defeated and the Party condemned mechanism as a dangerous deviation, a fate that would befall the Deborinites shortly thereafter. The "defeat of the two groups ended free discussion of the general questions of science and philosophy, and of the particular issue of dialectics, for the mechanist and Deborinite debate gave way to the official version of dialectical materialism" and this official version, Stalin argued in 1929, could be shaped any way the current leadership cared (Macintyre 1980: 141–142). With the publication of Lenin's philosophical notebooks in 1929 and 1930, "it appeared that Hegel studies had secured a firm basis and serious intellectual leader in the Soviet

Union, [but just at this moment] Deborin's views on the relation of dialectical materialism to dialectical idealism were condemned" (Grier 1978: 61–62). With both Hegel's and Marx's concern with alienation exorcized, the political leadership found a serviceable interpretation of Marx and his flexibility was swept away.

Using dialectical materialism as a world-philosophy, the "dictatorship of the proletariat" was transformed into the rule of an elite under Bolshevik tutelage. Bukharin believed that "in the Russian revolution, the form of the dictatorship has become manifest as the Soviet Power. The Soviet Power is the realisation of the dictatorship of the proletariat, organised in its soviets as the ruling class, and, with the aid of the peasants, crushing the resistance of the bourgeoisie and the landlords" (Bukharin and Preobrazhensky 1966: 167–168). In his book, *The ABC of Communism*, "Bukharin recommends the dictatorship of the proletariat as the only way to make the transition to the classless society, and explains that 'dictatorship' signifies strict method of government and a resolute crushing of enemies" (Smith 1996: 30). The promise of worker democracy was soon transferred to the Communist Party as the moral, spiritual, and political authority of Soviet society. Lenin became the first genuine authority (Comtean high priest?) of this movement.

What Lenin concluded would be greatly influenced by the refined, specified and narrowed discourse on Marxism that preceded him. Lenin accepted Engels's copy theory of knowledge and was influenced by the rigid and elitist Plekhanov (Lenin considered it "impossible to become an intelligent, real communist without studying – precisely studying – all that Plekhanov wrote on philosophy, because that is the best there is in the whole international literature on Marxism" [see Lenin, in Wetter 1963: 100]; however, he later broke with Plekhanov over matters of doctrine). Lenin also split with Trotsky and Rosa Luxemburg over the issue of whether revolutionary consciousness must be brought to the masses through educated elites (Lenin) or if it comes spontaneously and organically to the proletariat (the latter two).

After the revolutions of 1905 and 1917 moved Lenin into power, this provided him the means to assert which vision of dialectics, historical materialism, and the communist project would rule. The stage was set: Criticism of Lenin or the Party became bourgeois and counterrevolutionary. Mechanical-historical evolutionism, dialectical materialism, dictatorship of the Communist Party, and matters of purity in ideological doctrine were the field of associated statements that provided the content, context, and discursive web Lenin inherited and would advance in three discursive shifts: (1) a Kantian formulation of dialectic; (2) codifying "dialectical materialism" as the official Soviet ideology; and (3) advocacy of a vanguard party over proletarian political control.

That Lenin developed his outlook close to the views of Kautsky and Plekhanov is instructive. Kautsky once wrote in a note to Plekhanov that "the economic and historical viewpoint of Marx and Engels is in the last resort compatible with neo-Kantianism" (in Smith 1996: 36, 34). Within this discursive horizon, it was no leap for Lenin's belief that the emergence of socialism-communism was an inevitable historical process. As a result of this influence, Grier (1978: 60) writes, the . . .

> . . . entire range of Lenin's philosophical writings would seem to embrace at one extreme a largely pre-Kantian response to what was essentially the epistemological problem inherent in Cartesian dualism, and at the other an endorsement of the Hegelian dialectic (modified in some appropriately 'materialist' fashion which has yet to be satisfactorily clarified) as the heart of Marxist philosophy. Between these two poles Soviet Marxist philosophy [was] free to oscillate.

Not only were these poles more narrow than Marx's, they were incommensurable with them. To the degree that Lenin was influenced by a Kantian philosophy of science, he moved further from Marx's flexible, non-metaphysical, non-*a priori* approach to scientific inquiry and helped transform his dialectic away from method and into a universal "fact."

An example of Lenin's (1952: 9) codifications is in the preface to his work, *Materialism and Empirio-Criticism*, where he claims "Marx and Engels scores of times termed their philosophical views dialectical materialism." This was untrue. Marx never used the term nor did he develop a general *a priori* theory of knowledge and society. Nevertheless, Lenin's *Materialism and Empirio-Criticism* became a central text in Soviet ideology and conceptualized Marx's thought as a closed and completed system. Lenin, as Bochenski (1963b: 30) tells us, "fused into one the doctrines of Marx and Engels" and "saw Marx only as reflected by Engels. . . . For him, too, the 'kernel' of this dialectic lay in the 'unity of opposites' . . . their struggle and the destruction of the thesis by the antithesis . . . [a] philosophy [that he] *linked to a party*" (emphasis in the original). Thus, Marx's political program and scientific dialectic were turned into a metaphysical and elitist politic that Marx explicitly rejected.

With dialectical materialism as a metaphysical world-view and scientific socialism an entanglement of scientific and political discourse, Lenin's advocacy of political strategy could hardly avoid a similar entanglement. Note how Lenin negates Marx's dialectical flexibility with a coded, rigid, and conflating vision of scientific analysis and political action's relationship:

> The fundamental task of proletarian tactics was defined by Marx in strict conformity with all the premises of his materialistic-dialectical world outlook. Nothing but an objective calculation of the sum total of all the mutual relationships of all the classes of a given society without exception and consequently a calculation of the objective stage of development of this society as well as a calculation of the mutual relationship between it and other societies, can serve as the basis for the correct tactics of the class that forms the vanguard (Lenin's pamphlet, *The Teachings of Karl Marx*, Little Lenin Library, No. 1 – cited in, Adoratsky 1934: 85).

Lenin shifted, reversed, and negated what by then had been termed "Marxism"; Marx's class-based view had become a "vanguard party."

Marx, on the contrary, had a fluid and nonpermanent organizational view of a worker's party. In being "firmly convinced that [his] theoretical studies were of greater use to the working class than [his] meddling with associations which had now had their day on the Continent," Marx (1985a: 82, 87) distinguished his earlier participation in the Communist League, which he saw as being one of many episodes "in the history of a party that is everywhere springing up naturally out of the soil of modern society," from his support of "the party in a broad historical sense." Lenin, on the other hand, believed the working class should be forced into allegiance to *his* political party. In *What is to be Done?*, for instance, he argued again that "the working class movement" should be brought "under the wing of revolutionary Social-Democracy" (Lenin 1929: 41). Marxist discourse became the political theory of elite party cadres that should be "small, exclusive, highly organized, tightly disciplined, and conspiratorial.... Democracy throughout society was not yet feasible, Lenin believed, because the masses could not yet be trusted to know their own real interests.... [W]ithout a vanguard party to tutor and guide them, the workers would make wrong or even reactionary decisions" (Ball and Dagger 1999: 158, 160). As a result, "Leninist theory ... went far beyond anything that Marx had suggested ... [where the] working class was merely the passive recipient, whether of doctrines or of five-year plans worked out 'above'" (McLellan 1975: 83–84). So-called "Marxism" was now not only far from Marx's ideas but was in fact in opposition to them.

Lenin's leadership sustained Russia's vertical vision of politics, inherited from czarist times, which set the stage for oligarchy under the ideology of "Marxism." A dynamic had been set in motion that represented a continual refinement of "dialectical materialism" into an official ideology of the Communist Party where the Party (especially its leader) had the final say in political and scientific life. Working to consolidate official ideology following consolidation of political power, Stalin (1942: 406) wrote:

> Dialectical materialism is the world outlook of the Marxist-Leninist party. It is called dialectical materialism because its approach to the phenomena of nature, its method of studying and apprehending them, is *dialectical*, while its interpretation of the phenomena of nature, its conception of these phenomena, its theory, is *materialistic*.... Historical materialism is the extension of the principles of dialectical materialism to the study of social life, an application of the principles of dialectical materialism to the phenomena of the life of society, to the study of society and of its history (emphases in the original).

Completing the transformation of Marx's discourse into Soviet political dogma, Stalin then cites Engels (*Dialectics of Nature*) and Lenin. Stalin advocated "socialism in one country," which Trotsky (and, it should be noted, Marx) vigorously opposed. This change in Marxist doctrine led to an even more insular and autocratic regime.

With these developments the tautological circuit of illusion of uniform agreement between Marx's work and Stalinism was complete. Marx's thought was streamlined into a mechanical, party-centered ideology called *Diamat*, dissenters were purged, and science was subordinated to the needs of a bureaucratic state, its secret police, and any policy the leadership settled upon. Stalin, as leading dialectical thinker, would not last, however. On February 25, 1956, he was demoted and lost his status as an authority on "dialectical materialism" (Bochenski 1963b: 1). Nevertheless, official state ideology had become detached from individual authorities. Party and bureaucratic authority became ensconced in self-perpetuating regimes, which contained their own contradictions and external pressures to be sure (for an overview of officials' theoretical positions, see Bender 1975). Limiting dissent and debate continued from the Internationals through the construction of the state ideological apparatus (see Steila 1991: 2–3, Glazov 1985: 53). Recantations, purges, and murders enforced dogma, restricted discourse, and guaranteed that the transformation of Marx remained relatively unchallenged. Later, obedient Soviet intellectuals concluded that "various shades of opinion are [not] always to be permitted in the Party.... [U]nity is achieved by fighting every deviation from revolutionary Marxism" (Adoratsky 1934: 77–78). Much of this discourse depicted adherence to Soviet Communism as a moral obligation: "Surely it was the duty of the Socialists of other countries to give every possible support to the Soviet revolution" (Coates 1945: 92). Marx's communist project had become a moral imperative of a state apparatus.

The Soviet Union was not a failed experiment in Marxism, no matter what its leadership or its critics said about it. That a society called itself Marxist or communist does necessarily make it a reflection of Marx's ideas about science and/or what would/could happen in/after capitalism. Marx was clearly an early anti-Leninist and an anti-Stalinist. Soviet ideology adopted a distorted and rigid pseudo-dialectical materialist metaphysic, denied Hegel's usefulness, and censored Marx's humanist motivations. In the process, his disdain for bureaucracy and authority were lost and his dialectic went from revolutionary and radical to reactionary and conservative. The Soviets became metaphysical conservative Hegelians, truly un-Marx in outlook.

The narrowing of Marx's ideas into dogmatism extends far back. Young Marxists' works were scoured deviations from orthodoxy. At the 1903 Congress, Plekhanov argued in favor of "dispersing" a parliament if elections turned out badly, "not after two years, but, if possible, after two weeks." Though "Lenin did not actually speak in this discussion," according to Smith (1996: 39), "he was completely unified with Plekhanov." After the October Revolution, purges of potentially troublesome intelligentsia began and "Moscow was to decide what was right and what was wrong" (Zirkle 1949: 73, see 67–80 generally). After Lenin's death in 1924, "Stalin emerged as master of the situation. He had almost all the proved Bolsheviks condemned and executed" (Bochenski 1963b: 34). David Riazanov, an early Marx scholar for the Bolsheviks, was purged and executed on Stalin's orders (Ball and Dagger 1988). These tactics also eliminated Trotsky. Georg Lukács was forced to recant his criticism of Soviet leadership and Karl Korsch made overtures of conciliation but eventually criticized Stalinism and was purged from both the International and the German Party (for summaries of Soviet dogma, see Blakeley 1961, 1975, Bochenski 1963a). By 1939, "the last of the Old Bolsheviks, those who had led the 1917 Revolution, had been humiliated in the Moscow Show Trials, and had been forced to 'confess' to the most fantastic of crimes. They were shot or sent to perish in the Gulags" (Smith 1996: 22). Between 1931 and 1947, approximately 6,000 anti-Stalinists were executed (Bochenski 1963b: 32). Circular reasoning backed by political power was used in silencing legitimate scientists, additional purges, forced collectivization, imprisonment, and killing millions (for general discussions, see Kolakowski 1978: 103, Solzhenitsyn 1974, 1975, 1978).

The Soviets tended to not just downplay Marx's early writings, they denied their informative value (Tucker 1964: 169). They often censored Marx. His commitment to humanism via Hegel's concern over freedom from alienation had to be denied if the Communist Party was to justify its domination. As of 1891, Engels was still trying to get Kautsky to publish Marx's "Critique of the Gotha Program" in *Neue Zeit*, fifteen years after it was written (Smith 1996: 52, also see: Callinicos 1983: 66). When Maximilien Rubel introduced documentation that "Marxism" was invented, not necessarily intentionally, by Engels after Marx's death, the Soviet representatives at the 1970 conference demanded that he and his paper be removed from the proceedings (O'Malley and Algozin 1981: 15–16; for Rubel's argument, see 17–25). A metaphysical "Marxism" stripped of humanism and married to authoritarian power was too valuable to be challenged. Even the most dogmatic Communist knew the incompatibility between the Party-line and Marx's humanist concerns.

Whatever the Soviet Union represented, it was/is not a testing ground for the validity of Marx's ideas, whether these be the application of dialectical thought to the study of history and society, the utility of dialectical and historical materialist principles in the study of political economy, the likelihood of proletarian revolution within capitalist society, and/or the potential for a working-class-based political movement to transform capitalism through a socialist transition into communism. This is not to say that Marx's political program for revolution is the sufficient and necessary politic and any deviations from it are therefore automatically subject to criticism. Rather, if we are to excavate what lessons Marx still has to teach us for our research, his method of studying the social world is not to be evaluated on the grounds of the Soviet Union's historical evolution-devolution. In relating what we can learn from Marx's mode of inquiry to social action today, his method of thinking about science and the study of capitalism shows us how to bring flexibility and precision to a progressive politic. This and keeping our eyes on capital as the central organizing force of modern life are a few of the lessons Marx still has for us today.

Final Remarks

> More than a century after Marx wrote his last word, and long after the industrial deluge of the nineteenth century and the military holocausts of the twentieth, the most difficult puzzle for us to solve may ultimately be whether we may now be at all capable of understanding Marx – Thomas Kemple, *Reading Marx Writing: Melodrama, the Market, and the 'Grundrisse'*, 1995, p. 23.

The preceding chapters have attempted to rise to Kemple's challenge. We can understand Marx and we can use his ideas and his work. It is now a question of how we might do so.

Chapter Eight
Recovering Marx

> I could not rest until I had acquired modernity and the outlook of contemporary science – Marx in a letter to his father, November 10, 1837 (see Marx 1975b: 19).
>
> In that case we do not confront the world in a doctrinaire way with a new principle: Here is the truth, kneel before it! We develop new principles for the world out of the world's own principles. We do not say to the world: Cease your struggles, they are foolish; we will give you the true slogan of struggle. We merely show the world what it is really fighting for, and consciousness is something that it *has to* acquire, even if it does not want to – Marx to Arnold Ruge, in Letters from the *Deutsch-Franzosische Jahrbucher*, 1844 (see Marx 1975k: 144, emphasis in the original).
>
> At any rate, I hope the bourgeoisie will remember my carbuncles all the rest of their lives – Marx to Engels, June 22, 1867 (see Marx 1936i: 221).

These three quotes tell us much about the intensity of Marx's work and how his relationship to it changed as he matured, from passionate young man of science struggling to capture in thought his historical moment and all its potential, to active revolutionary working to get out a message he thought could harness the energies of an oppressed world, to a more resigned scholar hoping his work would at least remain a thorn in the side of his adversaries. It would be a mistake to assume the latter man had jettisoned the commitments of his early days. Marx retained much of his energy and enthusiasm until he died, though his optimism and zeal were increasingly tempered by the sway of capitalism and history's movements. Nevertheless, there was much prescience in his desire that his work would become a weapon against the powers that be, though it was not as sharp and as effective as he hoped, nor has it been wielded in a manner he would have found ideal. But capitalism is still running its course and Marx's weapons are not out of reach. They remain sharp and available for us to use, if we will.

This book has presented a series of problems reading and using Marx's method and possible solutions to them. Its central thesis is that Marx's work provides us essential tools for understanding the world in which we live. And any effective action in this world is predicated on this understanding. Skeptics, viewing the history of "really existing" communist societies, are likely to find resonance in one of Marx's (1992j: 186) comments: "The way to Hell is paved with good intentions." It is ironic that though it is possible this axiom was in the public domain when he used it, today it is often assumed to come from Samuel Johnson, the Bible, or a bit of folk-wisdom. Finding a citable source in any text but *Capital* proves difficult. This is not the only wisdom and insight from Marx that audiences find appealing, especially when disassociated from his name. In 2002, for example, a study commissioned by Columbia Law School reported that two-thirds of Americans believed that Marx's axiom for communism – "From each according to his abilities, to each according to his needs" – comes from, or could come from, the US Constitution.[1] Among mature capitalist societies, Americans are, perhaps, among the few who require subterfuge in order to rate Marx's ideas favorably. In 1999, a BBC online survey picked Marx as the top thinker of the millennium, ahead of Albert Einstein (also a socialist), Isaac Newton, Charles Darwin, Thomas Aquinas, Stephen Hawking, Immanuel Kant, René Descartes, James Kirk Maxwell, and Friedrich Nietzsche.[2] In a 2005 public opinion poll in Britain,

---

[1] See http://www2.law.columbia.edu/news/surveys/survey_constitution/index.shtml
(Retrieved August 26, 2006)
[2] See http://news.bbc.co.uk/2/hi/461545.stm
(Retrieved August 26, 2006)

Marx was chosen as the world's greatest philosopher.[3] Marx and/or his ideas evidently retain an appeal to the everyday concerns of publics, save those at the upper echelons of the capitalist social structure. Should we rid ourselves of commonplace assumptions about what is or is not in his work, a discourse remains to assist us in understanding the world around us and acting upon that understanding.

Many assumptions about Marx's views are misplaced. Marx supported working-class democratic organization, was not in favor of censoring either the press or religion, supported free public education, and advocated tax revolts by workers. He offered no doctrines beyond a commitment to using scientific and dialectical reason for a ruthless critique of social relations, an embrace of progress over superstition, a repudiation of the exploitation inherent to class systems, and a commitment to actively encouraging the trends within capitalism that can help eliminate the domination of classes and states over society.

Marx had no blueprint for a future society nor did he offer any law-like political strategy to achieve it. His ultimate goal for life after capitalism was a functioning, technologically advanced society without repressive states or ruling classes. Even when this view is admitted, it is often rejected as utopian or at least impractical. To such skeptics, Marx offers the retort: If it was possible to rid society of aristocrats and slave drivers, then why not rid our world of capitalists and their state caretakers? Marx was animated by a vision of a future where this had been accomplished. His political strategy attempted to discover how to get there.

The practices of self-labeled Marxists have contributed to confusion over Marx's ideas and the targets of his criticisms. Revolutionaries in Russia, China, Cambodia, and elsewhere saw the road to communism as going through forced collectivization and equated market exchange with capitalism itself. Thus, they worked to abolish both markets and personal property, confusing the latter with the form of property that was Marx's target, i.e., private ownership of the means of production. Their practices mobilized a dictatorship of individuals and cabals underneath the banner of liberating the proletariat.

---

[3] See http://books.guardian.co.uk/news/articles/0,,1528336,00.html and http://observer.guardian.co.uk/comment/story/0,,1530250,00.html
(Retrieved August 26, 2006)

Wielding "Marxism" as an ideology, the rise of the Leninist-Stalinism nation-state system convinced observers that Marx's thought involved an authoritarian politic. Thus, critics of Marx and Engels laid the responsibility for the crimes against peasants and workers committed by various dictatorships at their feet. This was the experience with Marx in the East.

In the West, militaristic warfare ascendant in the 20th century provided the catalyst to overrun and destroy emerging working class movements in Europe and the United States. After World War II, radical foment was often centered in those societies newly absorbed into the world-economy. This resistance was often cast as Marxist and/or communist in character and provided political cover for the use of subversive tactics by core industrial powers to undermine working class solidarity in name, in form, and/or in content, especially true in the Southern Hemisphere. The growth of wealthy core sectors of the world capitalist system, accompanied by violent reaction from the conquered, displaced and/or otherwise increasingly desperate sectors in the periphery, became the fulcrum against which an anti-Marxian rhetoric could flourish. The occasional support of anti-systemic movements by the Soviet Union only fueled suspicions of Marx's nefarious designs.

The discourse Marx started became something called "Marxism," a construction that he rebelled against. Unfortunately, this term now threatens to be fatal to the practical use of his ideas and work. Why would (or should) anyone risk venturing down his path again given the appearances its history presents to us? Moreover, should we overcome the standard view, how can his work and politics inform our collective thought and action without becoming another *Marxism*, with all the cult-like trappings this implies? There is no easy or ready answer to this question. It is, and will remain, a dilemma.

But there is more to overcome than simply Marx's tarnished reputation. We must also overcome the way capitalism has shaped and formed our own thinking, even at its most private. In any society, not just ours, the external world – its institutional structure – becomes the terms and *nomos* of thinking and feeling. When our imaginations venture outside these safe confines, we experience unease and trepidation. This is because the "symbolic universe shelters the individual from ultimate terror by bestowing ultimate legitimation upon the protective structures of the institutional order" (Berger and Luckmann 1967: 102). This "anomic terror," a fear of fluctuating norms, unsure futures, and dissolving pasts, makes it difficult for Marx to receive a friendly audience, especially in those areas where capitalism has become institutionalized. It is true that imagining an alternative world and acting upon such imaginings is the root of religion, education, and politics. But Marx asks us to be radical and revolutionary, not safe and conservative. His vision often evokes reaction, concern, unease, and fear. Historically, working classes have evinced as much conservative sentiment as radical, maybe more. Thus, it is not at all certain that capitalism's crises will, or can, be resolved in the manner Marx envisions.

Capitalism produces a developed set of knowledge, distributed institutionally, that explain why capitalism equals human nature and why revolt is logically, politically, and morally unacceptable. This knowledge has never fully won the day but has waxed and waned with the results of real struggles. If there is a central message in his work, it is that, given the future capitalism has in store for us, we must overcome this fearful consciousness. His motivation is that his work will help us find a way to do this.

On the back cover of his book on Marx's method of abstraction, Derek Sayer (1987) reiterates Weber's assertion that Marx's work is not a "taxicab" one may take anywhere one pleases. This observation can be interpreted in a dual way. On the one hand, Marx does not offer theories as coverall assertions that can be applied in any direction; on the other hand, Marx's view does not point in only one direction. Marx is scientific and dialectical in his thinking; his work attempts to help us to do and be the same.

The taxicab that is Marx's work has been taken many places, with multiple interpretations of both its point of departure and its final destination. Marx (1992b: 30), for his part, wrote that the "royal road to science" was only for those who do not "dread the fatiguing climb of its steep paths." Travels on Marx's road – struggling with relating empirical observations to abstract explanations of them – aspire to bring popular and professional audiences with him to a more scientific understanding of modern society. On this path, Marx tried to present his work in a way that could be understood by these audiences. But understanding scientific discourse requires experience with the subject matter, as well as with the research process, something his audiences possess in varying degrees. Add in the normal ideological conditioning societies instill in their members, and the difficulties with which dialectical thinking is grasped, and we come to the conclusion that Marx perhaps overestimated the extent to which he would, or could, be understood. It is hard to imagine how this inherent tension can be overcome in a social – or any other – science. Thus, Marx's work is threatened not only with its historical bastardization but also the inherent difficulties contained within its own discourse and the material conditions in which it finds itself.

Marx was never afforded the opportunity to fully explain where his methodological royal road began, his trek up its paths, and the destination to which it brought him. After almost a century-and-a-half of Marxian scholarship, there still exists no general agreement about the stops on this journey. Marx's narratives contain a combination of the familiar, the unfamiliar, and (occasionally) the obtuse. When writing for self-clarification, Marx's words were more abstract and complex. When writing for a more popular audience, he often turned to more concrete analysis, especially as his political economy matured. But even here his meaning is by no means obvious. Marx's dialectical reasoning did not exist apart from his analytical thoughts. As a result, his work contains a variety of terms, ideas, and concepts upon which critics or supporters can focus, especially those elements more familiar to their intellectual backgrounds.

Those of a philosophical background often gravitate towards Marx's theory of alienation or the value of his materialist thinking. And, although some take Marx's dialectical reason very seriously, his social scientific methodology is often unfamiliar to philosophers, whose discourse often stays on a broad plane of abstraction, rarely touching the earth.

Sociologists are frequently attracted to how Marx examined social relationships, but sociology as a discipline rarely requires its practitioners to undertake a thorough study of the philosophy of science, which is necessary to fairly appraise how Marx handled these same social relations. For example, in sociology the question of class is often treated as a location and/or measured via income or wealth rather than as a structured social relation and dynamic within capitalist society. Moreover, much of sociology does not place its observations squarely within a capitalist analytical context. As a result, sociologists often misconceptualize their variables and the present in which they exist.

Economists often debate Marx's engagement with classical political economy, especially the labor theory of value and/or of the tendency for the rate of profit to fall. However, those who find Marx's dialectical framework anywhere from a nuisance to unfathomable do their best to ignore it. Given the importance he attributed *Capital*'s dialectical form of development, this tactic forces his political economy into an incommensurable framework and thus unintelligible in the terms in which he offered it.

We must be able to distinguish between the form and content of Marx's various analyses and strategies. The form of his inquiries – i.e., his dialectical thinking about change, his materialist onto-epistemological foundations, his approach to political-economic analysis, and how to view capitalism's dynamic conditions – remain tools we can acquire and take with us as we travel on the road capitalism is unveiling before us. The strength, power, and insight these provide are irreplaceable and necessary for any sociological inquiry. And they will enable us to translate the knowledge we win into practice. Marx's conclusions – the content of his findings – are contained in his historical research, his specific conclusions about the capitalist political economy, and his political strategies. This content – particularly his theories about capitalism's limits and tactics of revolutionary struggle – is not something we should accept as cast in stone. Capitalism's identity can remain intact while many of its surface features change.

Who or what is the proletariat today? For Marx's analysis in *Capital*, of course, the term "proletariat" applied most obviously to the industrial working class. But capitalism in much, if not all, of the world has moved past the point where industrialism is its signature feature. But it was not industrialism for Marx that marked capitalism as a *differentia specifica*. Rather, it was private ownership of the means of production (a class relation), the production of surplus-value based on the exploitation of free laborers (a productive relation, including the classes), and the ceaseless accumulation of capital (an economic process, including productive and class relations). Given that Marx thought that this system's essential classes were a capitalist class and a working class, and given that the structural view of the system prioritizes capital*ism*, we should perhaps start with asking whether or not there exists a class that is dependent upon the expansion of capital for its role and place in the productive system and if there is a class whose labor upon which this expansion is dependent.

Of course, both those who own and those who labor are dependent on this expansion, though dependent to different degrees and in different ways. Thus, the question arises of whether to define the classes based upon this legal *ownership* alone, upon the specific *control* of stocks of capital, or the *stewardship* of capital in general. Should ownership be the criteria, the result is a conglomerate, where clear capitalists exist but so do a greater number of states, organizations, individuals, and families who own stocks through pensions, investments, and the like. The picture that emerges here is unclear, overlapping, and contradictory (logically). Should the control of capital be the signature and distinguishing feature of the class structure, the result is a picture closer to the bifurcated class structure Marx posited. But capital comes in large and small configurations and there is also a class of managers who make investment decisions outside direct ownership. Are we to assume that small entrepreneurs are stewards of capital's interests in general and therefore an opponent of the proletariat, even though they too are dependent on the profitability of larger capitals and their own labor? The confusion comes in by trying to match up individual locations with class relations.

In Marx's examination of the modern mode of production, it is capital that is the standpoint the story telling subject, i.e., "Having transcended us and our limitations, capital now posses its own identity separate from us" (Sekine 1998: 436–437). In this view, the class structure is determined by the relationship capital in the abstract has to labor in the concrete. It is the private ownership of productive resources, and the control of society that stems from the process of accumulation, that must be targeted in political struggles. In such a view, it is the processes and relationships of capitalism that require struggling against and the fight against capitalists themselves as a class will extend from this.

By extension, the political strategies for fighting capitalist exploitation and state domination Marx saw as fit for his era may not be fit for ours. Some tactics, such as strikes, forming a flexible anti-systemic political force, redistribution of wealth in terms of universal education and health care, insuring pensions, increasing wages, and progressive taxation (especially on inheritance) are necessary and useful. However, other tactics, such as seizing the means of production, can only be useful and successful should they occur on a massive (i.e., global) scale. Further, the electoral process avails only limited options. Not only would truly Left candidates face massive and organized opposition from capital's representatives, but should they win majorities in any assembly, they face both a material infrastructure put in place (and owned) by the capitalist class and an entire global-system resting upon capitalist auspices. Thus, any revolutionary action isolated within the global system could be easily and quickly destroyed and/or undermined. This is not to say any of this is out of bounds for radical politics, but it *is* to recognize that there is no simple formula for achieving capital's transformation and any such tactic must have conditions ripe for it to succeed.

The view that emerges here is that class struggle must attack the stewardship of capital, undermining those relationships and processes that support capital's expansion: the monopolistic control of resources and the political control of the state by capital's caretakers. Extracting laborers from their dependence on wages and their lack of control of their own productive resources would be one crucial tactic in this struggle. Exploitation of laborpower is still the "name of the game" in capital's expansion, and this laborpower extraction extends beyond simply "wage" employees but includes everyone who has to work for a paycheck in order to secure the means of existence. In addition to traditional wage-workers, the working-class includes everyone except those who decide on the methods of capital's expansion. We are still living in capitalist society and grasping its essential character and our ability to connect its determining relations to empirical realities under concern must not be lost.

We must learn how to bring the form of Marx's precision and flexibility into our analytical and political thinking. Marx's communist program was for his time and place; looking to it as an unvarying roadmap would be naïve. Distinguishing between its form and content is useful. Marx's form of political thought contained flexible but radical strategies: changing capitalism from within and through revolutionary struggle. The content of specific features of his and Engels's project – i.e., armed class struggle – may not work well today. Such tactics must be understood in practical terms across the worldsystem. Peoples in peripheral societies have every right to defend their rights and their access to traditional lands. In core societies, an armed class revolution is likely to produce the destruction of the very basis upon which Marx believed a progressive future would rest, i.e., a vibrant working class and a technologically advanced apparatus. Today, the armed class struggle Marx and Engels thought useful for one period of European development could easily devolve into a mass civil war, bringing with it destruction of productive infrastructure (as one colleague noted metaphorically, "Today, we need hackers"). It is unlikely a progressive class consciousness animated by mass democracy and egalitarianism would emerge from this. I believe that Marx and Engels would admit this too. But their dialectical sensibility can still help us understand the continuing need for struggle in the present, to work for goals that target the fundamental aspects of political-economic domination, to gear activities for making the present ready for a future, of eschewing dogmatism, party doctrine, authority, and cults of personality.

Marx aspired to re-invent our traditional approach to politics. He believed that modern democratic theory creates the illusion of sovereignty of the masses over elite rule. Living in alienating conditions, people in capitalism tend to believe they control the state, although, in fact, the capitalist class remains largely in control. Throughout history, it has been the material forces leading social change and our belief in our control of contemporary society has masked what is going on behind our backs. However, by engaging in critical political discourse, education, and action, we can bring idealist forces into the forefront, animate the masses, and help reveal the way capital rules our lives. Thus, a Marxist politic today still needs to work toward helping the masses take control of social, economic, political, and technological resources but it must do this while explaining to them why this is necessary.

In pursuing such a program, we must disabuse ourselves of the mythos surrounding Marx's work. Across the history of his interpreters, one myth that has persisted is that Marx accepted a sort of eschatological-metaphysic.[4] Other myths skew Marx in different directions. One sociological primer on the "core concepts in sociology" informs new readers that Marx "is perhaps best known as the 'father' of communism," that "As far as Marx was concerned... only economics mattered," that "Marx's conception of the world was a singular one," that "the people of any society could be divided into two distinct classes: the bourgeoisie and the proletariat," that Marx believed "everything else – ideas, values, social conventions, art, literature, morals, law, and even religion – was 'epiphenomenal,' or secondary to and in the service of the economic realities of society" and that "religion... existed only

---

[4] Leonard Wessell (1984: 176–178) asserts that "Marx had already identified divinity with social collectivity" and that a "full understanding... will elude the student if he fails to grasp the religious nature of Marx's Hegelian part-whole problem." Reading his youthful, quasi-romantic commitments forward, Wessell (1984: 185, 188) proposes that Marx held to an implicit salvation model with the proletariat as the soul, the Christ figure and the purifier of mankind. Wessell also repeats other traditional claims: Marx was teleological; Marx worked with a *priori* frameworks; Marx used the triad of thesis-antithesis-synthesis.

to mask the inequalities and injuries of the economic system. Religion had no real importance in the overall scheme of things" (McIntyre 2006: 20–21). As we have seen, this view is wrong at every step but it is not atypical in mainstream academic depictions of Marx's ideas. Such descriptions often make their way into theory classes and steer new students away from Marx, or toward a Marx that did not exist. If we are to understand and use Marx effectively, we have to get a better handle on the basics in the way we read, teach, and communicate his ideas.

Liebknecht (1901: 32) believed that it was only "in London... the center of the world and of world trade and... whence [a place where] the trade of the world and the political and economic bustle of the world can be observed in a way impossible in any other part of the globe – here Marx found what he sought and needed: the bricks and mortar for his work. *Capital* could be created in London only." Though he was a man of his time and place, Marx's work and vision aimed to reach beyond both – i.e., he tried to develop a way of thinking about the world that could be applied to many other times and many other places. There is nothing out of the ordinary about this in comparison to other figures. But Marx was neither psychic nor soothsayer, neither philosophical-moralist offering wisdom to ponder nor utopian prophet presenting a universal model for others to emulate. Marx did not simply invent a system early on in life and then spend his life defending a malformulated theorem. Marx's thought evolved over time and was influenced by many other thinkers. He constantly worked through problems old and new as he continued his studies. What might be called "Marxist thought," or even "Marx's thought," therefore, did not spring forth from his mind like a preconceived set of doctrines and conclusions as if some multi-headed hydra. We should avoid treating his ideas this way.

There remains, nevertheless, a certain cohesiveness, at times precise and at others less so, flowing through Marx's analytical narratives. In the main, there are three strands of thought continuing over his work. First, although a student and already gravitating toward materialism, Marx returned to Hegel's dialectic to assist in moving away from metaphysical and ahistorical approaches to knowledge. Second, he converted to the communist program during his early years in journalism. Third, Marx's commitment to science was grounded in his assumption that observable material life is the place where human knowledge should begin. As a result, his materialist dialectic and a communist politic were radical stances. He later became a specialist in political economy. Marx mobilized his dialectical, historical, and political-economic inquiries toward his concern with real improvements in living conditions for the people of modernity. These interests – containing no elements of speculative philosophy – would be part of Marx's lifelong program. If we want to understand Marx's claims to science in the service of humanity, it is this certain cohesiveness that we must endeavor to understand. But because of the fragmented manner of the presentation of his central ideas, and because of the persistent myths repeated in the literature, one must have the tools necessary to see this cohesiveness and reconstruct it.

Another common persistent myth about Marx is that he is basically offering a moral critique of capitalism, or, at least, his study of capitalism, his condemnation of it, and his communist program are primarily morally driven. Marx, however, rarely wavered in his disdain for moralism, moralists, or the consideration of abstract moral dictums. Although his journalism sometimes ventured into value-laden language, this is something less apparent in his more analytical work (thought not completely absent). And, though Marx admitted he was likely to die before his favored political program would find its first real concrete successes, he did not he see himself as guide for what people might look like in a future society:

> Once a gentlemen asked him who would clean shoes in the future state. He answered vexedly, 'You should.' The tactless questioner understood and was silent. That was perhaps the only time that Marx lost his temper. When the visit was over, my mother said frankly, 'Herr Doktor, I don't wish to defend the man's silly questions, but I did think after your answer that it was better that he kept silent than that he should perhaps have answered that he did not feel he had the vocation of a shoeback.' When Marx agreed, she added, 'I cannot imagine you in an egalitarian age, since you have inclinations and habits that are so thoroughly aristocratic.' 'Neither can I', answered Marx. 'These times will come, but we must be away by then' (Kugelmann 1928, cited in McLellan 1981: 83).

Social criticism does not always jibe with what individual critics are apt to do, no matter what their vision of a more humane, egalitarian, and non-exploitative future is. Marx's personal biography is not *the* place where we should turn for insight about what to do with our own lives.

Where does this leave us? What does Marx have to teach sociology? What does Marx still have to tell us about capitalism and what to do about it? The distinctions and interconnections between Marx's levels of discourse mapped out in this study help answer such questions. Let us take a look at how they fit together.

Marx's contributions to social science have not always been overlooked or denied. Joseph Schumpeter (1954:12) held that Marx's materialist conception of history – encapsulated in "two propositions: (1) The forms or conditions of production are the fundamental determinant of social structures which in turn breed attitudes, actions and civilizations... (2) The forms of production themselves have a logic of their own; that is to say, they change according to necessities inherent in them so as to produce their successors merely by their own working" – offered "invaluable working hypotheses." Mainstream sociologists have agreed: "With increasing bureaucratization, it becomes plain to all who would see that man is to a very important degree controlled by his social relations to the instruments of production. This can no longer seem only a tenet of Marxism, but a stubborn fact to be acknowledge by all, quite apart from their ideological persuasion" (Merton 1957: 196–197). Today, more sociologists accept these ideas than before, but their general influence has been limited, especially as compared to other traditions in the discipline. The connections between historical materialist principles and the specific machinations of capitalism must remain central to our sociological inquiries. This is so even if it is only on the path to examining other social realities not strictly understood as being *of* capitalism. One cannot imagine how anthropology could have advanced without recognizing the *type* of society under consideration.

This having been said, it is also the case that both Marxist and non-Marxist sociologists have done little work on bringing statistical and dialectical thinking together. Conventional sociologists are not remiss in advocating the tools that statistical analyses contain. By the same measure, Marxist scholars are correct to emphasize that grasping the logic of change dialectically is a central priority in social research. Capitalism's material conditions offer each of these concerns ample empirical grounds for investigative work. Marx, for his part, clearly thought it possible, even necessary, to bring these approaches together. Nevertheless, one can imagine social scientists' response to the prospect of "dialectical statistics." Should such an inquiry proceed (and I believe it should), it must do so without metaphysical assumptions, tacit or otherwise.

We can now put the general Marxian vision – ontological, epistemological, and political – together. Table 8.1 and Figure 8.1 present a way to place these moments of inquiry into an interpretable picture. Table 8.1 displays the inner-connections between Marx's various moments of inquiry (DM, HM, PE, CP), the epistemological strategies and the levels of generality associated with each, and the analytical domains and the conceptual frameworks operative within and between these various moments. Figure 8.1 places Marx's four moments of inquiry into their relationship with history, social structure, and political strategy, where each level contains different relations of necessity and contingency. Understood in this way, we see how Marx's method strove for a complexity commensurate with the realities he studied, eschewed any metaphysical or otherwise teleological projection of the future, and grounded his politic in a scientific realism whose complex interrelations are unsurpassed to this day.

There are still other hurdles to overcome in communicating an accurate picture of Marx's view of capitalism and its future. Marx's supporters must overcome accusations that his views are animated by utopianism, naiveté, and dogmatism. Critics also suspect Marx's program holds designs for an authoritarian future, one where Marxist intellectuals maneuver themselves into positions of leadership and facilitate a new form of domination. Evidence for such accusations and suspicions is not simply imaginary. Leninist thinking (under the flag of Marx) still animates a number of schools of thought and apologetic accounts of "really existing Communism" can be found in the Marxist archive. Though not all Marxian scholars have walked down this path, this is nevertheless a history they are forced to bear. And even if not all of them view "really existing Communism" as something that can be placed on Marx's shoulders, a number of Marxists and their critics have adopted a stance that holds that Communism's collapse has implications for the viability of Marx's thought, even for his method and political economy. This *non sequitur* will be, perhaps, a burden that must be shouldered for the immediate future, no matter that it is empirically and logically false.

By extension, another hurdle to overcome is the widely held assumption that capitalism, freedom, and democracy are coterminous and (therefore) socialism and communism are necessarily anti-democratic. This view tends

Table 8.1
Interconnections between Marx's Moments of Inquiry & Levels of Generality

Moments of Inquiry	The Dialectical Method	Historical Materialism	Political Economy	The Communist Project
Critical Criticism	*Negative*: idealism, metaphysics, speculation, mysticism, inversion	*Negative*: ahistoricism	*Negative*: atomism, universalism	*Negative*: utopianism, reformism
	Positive: negativity	*Positive*: analysis of historical and structural development	*Positive*: study of social laws, labor theory of value	*Positive*: capitalism as progress but limited; grounding action in the material present
Scientific Naturalism informs...	Materialism, evolution, experimental model, mathematics	Controlled comparison of class systems, natural selection in change	Controlled comparison of variables within capitalism, statistical analysis, organicism	Humans as a species are a communal animal
Levels of Generality	*Naturalism at Seven and Six*	*Naturalism at Five through One*	*Naturalism at Three through One*	*Six and Five inform One and beyond*
Dialectical Method informs...	Study of contradiction, opposition, negativity, and transformation	Changes within & between modes of production	Changes within capitalism and its future potentials	Transformation of capitalism and a new structure's emergence
Historical Materialism informs...	NA	Base and superstructure, division of labor, conflicts between relations & forces of production, modes of appropriation, revolutionary change	Fetishism, commodity production, crises, conflicts between capital and labor, exploitation, the capitalist state, revolutionary class struggle	Political focus on class relations, material conditions must be mature for successful revolution, labor is and/or can be a revolutionary class
Levels of Generality	NA	*Five through One*	*Five & Four inform Three & Two*	*Six to Four inform One*

Table 8.1. (cont.)

Moments of Inquiry	The Dialectical Method	Historical Materialism	Political Economy		The Communist Project
			Political economic models in the abstract	Political economic realities in the concrete	
Political Economy informs...	NA	NA			Capitalists dominate politically, private property needs abolishing, proletariat needs organizing
Levels of Generality	NA	NA	*Five, Four and Three inform Two and One*		*Five through Two inform One and beyond*
Communist Project informs...	Critical criticism of historical relations of domination, master-slave relations	Early societies and species-being seen as communist	Elements of the future are contained within the present		Classless & stateless future, species-being united
Levels of Generality	*Six and beyond One inform Five through Three*	*Six informs Five*	*Levels beyond One inform Three and Two*		*Six informs those beyond Three through One*

Figure 8.1
Scientific Dialectics and the Communist Project

TIME ↓	← SPACE →			
	General Sociological Principles		Broadest and Most Speculative Levels	
	Human History →	← Onto-Epistemology		
	Scientific Dialectics		Increasing Necessity ↓	
Concrete: History in General	Historical Materialism	Abstract: Base and Superstructure	Science of Social Change, Focus on Class Systems	
Concrete: Capitalism in General Concrete	Political Economy	Abstract: Structure of Capitalist Society	Science of Capitalist Society	
Concrete: Recent Capitalism	Capitalism's Laws of Motion	Abstract: Central Tendencies	Present Activity in Capitalist Society	
Contingencies	Communist Project	Concrete Revolutionary Praxis	Increasing Contingencies ↓	
Contingencies	Proletarian Revolution	Seize Means of Production	Middle-Term Future	
Contingencies	Socialist Transitionary Period	Redistribute Wealth, Expand Democracy	Longer-Term Future	
Contingencies	Early Communist Society	Private Property Abolished	Emergence of New Social Structure	
Contingencies	Mature Communist Society	State Fades Away	Maturing Social Structure	

to dominate in the core centers of capitalist power, though it is not exclusive to them. However, it is in these centers where mass publics have a better opportunity to attack the foundations of the system. Overcoming these false assumptions will not be something to be done only at the level of discourse. In other words, any anti-systemic discourse that strives to overcome capital while increasing democracy and freedom will necessarily work within the gyrations of the global economy. As the contradictions of capitalism increasingly make conventional ideological assumptions unsustainable, those animated by Marx's program must be prepared to offer a palpable alternative, and this must necessarily include the promise and realization of expanding democratic participation and securing social freedom. Unfortunately, even academic audiences who understand that these goals are entailed in Marx's program often find such goals utopian in spirit and therefore impractical.

Even within sympathetic academic audiences there remain other myths and suspicions that will not be easily overcome. Marx's view of the relationship of capitalism to communism has been seen as succumbing to teleological reasoning. Many of Marx's supporters today still believe that he promoted a theory of uni-linear historical evolution and that he saw communism as a teleological force in history. In addition to the material already covered in this book, there are other reasons to challenge this view. Marx (1988a: 114) once argued that "communism is the necessary pattern and dynamic principle of the immediate future, but communism as such is not as such the goal of human development – the structure of human society." For Marx, communism is an outcome inherent in capitalism's trajectory, but history is not a goal-directed process steered by communism. His statements in private sometimes belied the optimism in his public statements. John Swinton recounted his dialogues with Marx about what he envisioned for the future: "It seemed as though his mind were inverted for a moment [and] while he looked upon the roaring sea in front and the restless multitude upon the beach, Marx's laconic reply came in a deep and solemn tone: 'Struggle!'" (cited in Boyer and Morais 1955: 84). This observation tells us as much about Marx's vision of the future as do his better-known pronouncements. Although Marx thought he saw a tendency toward communism in history, especially as it was squeezed through the capitalist present, this tendency should not be confused with a teleological process, as if a communist future shapes events in the present outside of human agency. Marx was trying to say, I believe, that history was tinged with a continuing struggle of the oppressed to free themselves. Marx's politic asks us to join this struggle, though we must remain agnostic on what the future might bring beyond continued struggle.

Defending Marx against teleological interpretations is not simply a question about forms of analysis. A political dimension comes with it as well. Teleological interpretations often lead critics to dismiss the goal of increasing social equality and diminishing the power of the ruling class, i.e., "If Marx's communism is a teleological fallacy, then so is any theory of politics that strives to eliminate social inequalities." This sort of dismissal is the ideological cousin of the critique of Marx's politics based on the history of Communist societies. Both forms of reasoning are *non sequiturs* and antithetical to any serious analysis of the present. Our agnosticism must not turn into pessimism or resignation.

We must animate our struggles with a vision of how human life can be organized in a way that cultivates and values equality and rejects and fights domination. At the same time, we must struggle against accusations of utopianism when we defend Marx's views. For example, what will socialism look like? At first, in Marx's vision, socialism will *look* much like capitalism, i.e., what capitalism has built for us today is what socialists will have to use in the future. It follows that early socialism will have cities, mass transportation and communication, suburbs, rural regions, etc. In other words, socialism is not the exact opposite of capitalism, its mirror image, such as collective living arrangements, destruction of personal property, or lack of mediums of exchange. Marx (1936j: 237) saw capitalism as containing many "unconscious socialist tendenc[ies]." These are things we should strive to develop on the way to overturning capital. What any system after socialism – communism in Marx's view – would look like is too difficult for most of us to imagine. That Marx understood this explains his less-than-specific account of what a communist future would be like. Thus, the point to get across to audiences is that Marx is not asking us to imagine a hypothetical society outside of our experiences, but that we should build a future based upon the features of the present while eliminating the sources of its main problems – the power of capital.

Pronouncements about capitalism being "the best system" – or other claims that come down in favor of capitalism over socialism – offer us false dilemmas. Being for or against capitalism is irrelevant. People have never imagined a system and chosen to construct it and live in it. However, Marx's inspiration is that capitalism is the first system in history that allows us to do this in the imagination because it is being made possible in reality. He wants us today to pick socialism out of capitalism's possibilities and struggle to forge it so that we may transcend capital and the domination of society by classes and states that comes with it. As a dynamic system, capitalism continually brings changes upon itself and, not being a reflection of human nature, it will at some point change into something else. If capitalism's change is inevitable – given a broad enough temporal lens – then the question becomes how we position ourselves in respect to this change. This is what Marx wants us to understand and grasp in all its possibilities.

We must keep our eyes on the realities of capitalism, its transformative character, its trajectory, and animate our imaginations with facilitating those features it produces that allow for its progressive transcendence as its moves through its final death throes, which are occurring all around us. These realities – capitalism's collapse and transformation – and the conditions necessary for a better future are everywhere. "Owing to a certain judicial blindness even the best intelligences absolutely fail to see the things which lie in front of their noses. Later, when the moment has arrived, we are surprised to find traces everywhere of what we have failed to see" (Marx 1936j: 235). We must work to not allow the appearances and illusions that capitalism produces fool us into undue pessimism about what sort of future we can make.

Immanuel Wallerstein reminds us that the world has no more geographical regions to incorporate for continued economic expansion and the scramble by corporations for more and more areas of human life to commercialize is reaching its limit. The threat of nuclear war is often used to justify increased government funding into space-based weaponry in order to help high-tech industry find an acceptable rate of return for their capital awaiting reinvestment, suggesting that the largest capitals are failing to find satisfactory levels of expansion. The number of banks and the increasing amount of wealth controlled by finance capital continue to concentrate. Cultures continue to draw closer together (though the loss of certain forms of traditional knowledge is perhaps not always a part of this process worthy of celebration; but sometimes it is, such as declines in xenophobia, racism, sexism, and others forms of traditional prejudice). Populations throughout the world increasingly suspect that their role in the global order is as pawns for corporate policy. States increasingly fail to serve the populations that fund and stand as subordinate to them. The growing awareness of ecological destruction forces the industrial and corporate source of the problem into public consciousness. Social movements in the global South, the European Left, and the worldwide anti-globalization movement are drawing closer together as well. The unilateralist war waged by the United States against Iraq was almost universally opposed outside the United States, and even there support was rarely very high. None of this calls for undue optimism, but rank pessimism is not called for either.

But there *are* reasons for pessimism. The caretakers of capital have placed one roadblock after another in the way of addressing ecological problems, while the industrial apparatus continues to belch out more toxins, chemicals, gasses, and particulates each year. A prison-industrial-complex continues to grow. Disease and famine grip large swaths of the world, mostly regions that were formerly colonies. Slavery and sweatshops continue. Culturally speaking, with "capitalism" largely left out of critical public political discourse, tensions, strife, and struggles often become directed toward one or another political party, nationality, ethnicity, gender, and/or religious group. Unable to grasp the moment they are in, publics turn to other establishment political parties, blame other ethnic groups, reinvigorate patriarchal norms of the past, embrace (often fundamentalist) religious doctrine, and encourage the slide toward barbarism, i.e., fascism. With "Marxism" now a mystified catchword, how to mobilize Marx's insights about capitalism in order to facilitate collective solutions to our social problems is doubly problematic. If he has anything to teach us, it is that failing to recognize the role capital plays in the ebb and flow of modernity is to fundamentally misconceive the moment we are in.

What is to be done? What Marx called "revolution" is something contingent on practical working class activity in the context of capitalism's crises. The failure of such class-based activity could allow capitalism to devolve into a rapidly declining phase of human history. Perhaps unstrapping ourselves with the burden of "communism" would help in forwarding a progressive politics of the present. Advocating social and economic democracy would advance similar themes. At the same time, calling the barbarism capitalism produces as the "fascist" reality that it is remains necessary. As the forces of reaction have increasingly learned, words carry power and this power animates the actions of publics, hence the blossoming of the public relations industry and its entry into commerce, education, and political discourse. "Here, therefore, it is completely in the interests of the ruling classes to perpetuate the unthinking confusion," a task they farm out to "the sycophantic babblers" who are paid to make sure that we "may not think at all!" (Marx 1988b: 69).

Radicals must speak to people in a way that allows them to see how the current system leaves them relatively powerless, how the ruling class, both its liberal and conservative factions, has manipulated democracy in order to augment their own power, how the system of production passes wealth up to the very few, and how all of this is currently threatening the ecological balance that allows human life to continue. These threats are real and become additionally intractable with the media monopoly, the manipulation of social discourse by the public relations industry, and the general servitude of the intellectual class. "The daily press and the telegraph, which in a moment spreads its inventions over the whole earth, fabricate more myths in a day (and the bourgeois cattle believe and propagate them still further) than could have previously been produced in a century" (Marx 1989c: 177). And, as repeatedly demonstrated, when eschatological rhetoric and its adherents make their way out of the pulpit and into the halls of political power, the synergy produced by the unification of religious leaders and the handmaidens of capital simply places another wall in the house of mirrors. It does seem, indeed, to be an uphill battle that we have on our hands.

When Marx spoke of praxis he did not mean only political revolution. He was pointing to methods of practice that helped shift capitalism away from capitalism. Directing energies away from economic reliance on capital and toward self-sustenance is one method of practice that can allow people to reclaim a piece of their labor and its products. Will a new working class party be effective? Perhaps. However, it is also likely that unless this party is both decentralized yet constituted on a global scale, it is likely that the forces of bourgeois reaction will keep people's movements tied up nation by nation in a distracting wrestling match, deflecting attention away from their need to secure their economic independence.

Flexible precision and precise flexibility were as much a part of Marx's political thinking as they were central to his scientific thinking. This way of thinking about the world remains scientifically vital and Marxist scholars continue to imagine what sort of socialism capitalism is now making possible

(Albritton et al. 2004). Nevertheless, many contemporary criticisms of Marx view him, and by extension his academic supporters, as irrational, romantic, and/or utopian. There is often the accompanying assumption that his political program of overthrowing capitalism through political struggle and scientific criticism is misguided. Thus, the common *non sequitur* holds, the search and struggle for a more equitable, democratic, and nonexploitative world is fruitless. The elimination of slavery was also once thought to be utopian, naïve, and fruitless. Marx accepted that such changes could emerge at a time many thought it impossible.[5] If the social problems associated with capitalism are to be transcended, and if Marx's work is a vital contribution to that process, then it will remain crucial to defend Marx's way of thinking about the world from those critics who foist misplaced and discreditable critiques upon him. As broadcast in "The Communist Manifesto," we must recover our "sober senses," grasp our "real conditions of life, and [our] relations with [our] kind" (Marx and Engels 1978a: 476), and persuade those who see his work as a quixotic and misguided effort to reconsider their views and refute Marx's (1989d: 206) accusation that "No one is so deaf as those who will not hear!"

---

[5] "When you reflect, my dear Uncle, how at the time of Lincoln's election 3½ years ago it was only a matter of making *no further concessions* to the slave-owners, whereas now the avowed aim, which has in part already been realized, is the *abolition of slavery*, one has to admit that *never* has such a gigantic revolution occurred with such rapidity. It will have a highly beneficial influence on the whole world" (Marx 1987a: 48, emphases in the original).

References

Acton, Harry Burrows. 1967. *What Marx Really Said*. London: Macdonald.

Adler, Max. 1983. "The Sociology in Marxism", in Tom Bottomore and Patrick Goode, eds., *Readings in Marxist Sociology*. Oxford: Clarendon Press, pp. 30–33.

Adoratsky, V. 1934. *Dialectical Materialism: The Theoretical Foundations of Marxism-Leninism*. New York: International Publishers.

Agresti, Alan and Barbara Finlay. 1986. *Statistical Methods for the Social Sciences*. San Francisco: Dellen.

Albritton, Robert. 1999. *Dialectics and Deconstruction in Political Economy*. New York: St. Martin's Press.

Albritton, Robert, Shannon Bell, John R. Bell, and Richard Westa (eds.). 2004. *New Socialisms: Futures Beyond Globalization*. London: Routledge.

Alexander, Jeffery. 1982. *Theoretical Logic in Sociology*, Volumes 1–3. Berkeley and Los Angeles: University of California Press.

Allan, Kenneth. 2005. *Explorations in Classical Sociological Theory: Seeing the Social World*. Thousand Oaks, Calif.: Pine Forge Press.

Allen, James. 1969. Editor's Preface, in George Plekhanov, *Fundamental Problems of Marxism*, 1908. New York: International Publishers, pp. 7–17.

Althusser, Louis. 1969. *For Marx*. New York: Pantheon Books.

Althusser, Louis. 1970 [1968]. "From *Capital* to Marx's Philosophy", in Louis Althusser and Etienne Balidbar, *Reading* Capital. London: New Left Books, pp. 11–69.

Althusser, Louis. 1971. *Lenin and Philosophy*. New York: Monthly Review Press.

Althusser, Louis. 1977. "Introduction: Unfinished History", in Dominique Lecourt, *Proletarian Science? The Case of Lysenko*. London: New Left Books, pp. 7–16.

Anderson, Kevin. 1992. "Rubel's Marxology: A Critique." *Capital & Class* 47: 67–91.

Anderson, Kevin. 1998. "Uncovering Marx's Yet Unpublished Writings." *Critique (Glasgow)* 30–31: 179–187.

Antonio, Robert and Ronald Glassman (eds.). 1985. *A Weber-Marx Dialogue*. Lawrence: University Press of Kansas.

Appelbaum, Richard. 1988. *Karl Marx: Masters of Social Theory*. Newbury Park, Calif.: Sage.

Arthur, Christopher. 1979. "Dialectics and Labour", in John Mepham and David-Hillel Ruben, eds., *Issues in Marxist Philosophy*, Volume I. Atlantic Highlands, N.J.: Humanities Press, pp. 87–116.

Arthur, Christopher J. 2002. *The New Dialectic and Marx's* Capital. Leiden, The Netherlands: Brill.

Babbie, Earl. 1995. *The Practice of Social Research*. Sixth Edition. Belmont, Calif.: Wadsworth.

Ball, Terrence. 1979. "Marx and Darwin: A Reconsideration." *Political Theory* 7 (4): 469–483.

Ball, Terrence and Richard Dagger. 1999. *Political Ideologies and the Democratic Ideal*. New York: Longman.

Beamish, Rob. 1992. *Marx, Method, and the Division of Labor*. Urbana: University of Chicago Press.

Bellamy Foster, John. 1996. Introduction, in Ernst Fischer, *How to Read Marx*. New York: Monthly Review Press, pp. 7–30.

Bender, Frederic. 1975. *The Betrayal of Marx*. New York: Harper Torchbooks.

Berger, Peter and Thomas Luckmann. 1967. *The Social Construction of Reality*. Garden City, N.Y.: Anchor Books.

Bhaskar, Roy. 1989. *Reclaiming Reality: A Critical Introduction to Contemporary Philosophy*. New York: Verso.

Bhaskar, Roy. 1993. *Dialectic: The Pulse of Freedom*. New York: Verso.

Blakeley, Thomas J. 1961. *Soviet Scholasticism*. Dordrecht, The Netherlands: D. Reidel.

Blakeley, Thomas J. (ed.). 1975. *Themes in Soviet Marxist Philosophy*. Dordrecht, The Netherlands: D. Reidel.

Bochenski, Joseph M. 1963a. *The Dogmatic Principles of Soviet Philosophy*. Dordrecht, The Netherlands: D. Reidel.

Bochenski, Joseph M. 1963b. *Soviet Russian Dialectical Materialism (Diamat)*. Dordrecht, The Netherlands: D. Reidel.

Böhm-Bawerk, Eugen v. 1898. *Karl Marx and the Close of his System: A Criticism*. London: T. Fisher Unwin.

Bologh, Roslyn. 1979. *Dialectical Phenomenology*. Boston: Routledge & Kegan Paul.

Bosserman, Phillip. 1995. "The Twentieth Century's Saint-Simon: Georges Gurvitch's Dialectical Sociology and the New Physics." *Sociological Theory* 13 (1): 50–57.

Bottomore, Tom (ed.). 1973. *Karl Marx*. Engelwood Cliffs, N.J.: Prentice Hall.

Bottomore, Tom. 1975. *Marxist Sociology*. New York: Holmes and Meier.

Bottomore, Tom and Patrick Goode (eds.). 1983. *Readings in Marxist Sociology*. Oxford: Clarendon Press.

Bottomore, Tom et al. 1983. *A Dictionary of Marxist Thought*. Cambridge, Mass.: Harvard University Press.

Boyer, Richard and Herbert Morais. 1955. *Labor's Untold Story*. New York: United Electrical, Radio, and Machine Workers of America.

Brewer, Anthony. 1984. *A Guide to Marx's* Capital. Cambridge: Cambridge University Press.

Bukharin, Nikolai I. 1925. *Historical Materialism: A System of Sociology*. New York: International Publishers.

Bukharin, Nilokai I. 1931. *Science at the Cross Roads*. London: Kniga.

Bukharin, Nikolai I. and Evengii. Preobrazhensky. 1966. *The ABC of Communism: A Popular Explanation of the Program of the Communist Party of Russia*. Ann Arbor: University of Michigan Press.

Burawoy, Michael. 1982. "Introduction: The Resurgence of Marxism in American Sociology", in Michael Burawoy and Theda Skocpol, eds., *Marxist Inquires: Studies of Labor, Class, and States*. Chicago: University of Chicago Press, pp. 1–30.

Burawoy, Michael. 1990. "Marxism as Science: Historical Challenges and Theoretical Growth." *American Sociological Review* 55 (12): 775–793.

Callinicos, Alex. 1983. *Marxism and Philosophy*. Oxford: Clarendon Press.

Campbell, Donald and Julian Stanley. 1966 [1963]. *Experimental and Quasi-Experimental Designs for Research*. Chicago: Rand McNally.

Carver, Terrel. 1982. *Marx's Social Theory*. New York: Oxford University Press.

Carver, Terrel. 1983. *Marx & Engels: The Intellectual Relationship*. Bloomington: Indiana University Press.

Carver, Terrel. 1987. *A Marx Dictionary*. Totowa, N.J.: Barnes and Noble Books.

Coates, Zelda. 1945. *The Life and Teachings of Friedrich Engels*. London: Lawrence and Wishart.

Cohen, Gerald Allan. 1978. *Karl Marx's Theory of History: A Defense*. Princeton, N.J.: Princeton University Press.

Cohen, Gerald Allan. 1986. "Marxism and Functional Explanation", in John Roemer, ed., *Analytical Marxism*. Cambridge: Cambridge University Press, pp. 221–234.

Colletti, Lucio. 1972 [1969]. "Bernstein and the Marxism of the Second International", in Lucio Colletti, *From Rousseau to Lenin: Studies in Ideology and Society*. New York: Monthly Review Press, pp. 45–108.

Collier, Andrew. 1989. *Scientific Realism and Socialist Thought*. Boulder, Colo.: Lynne Rienner.

Comte, Auguste. 1974. *The Positive Philosophy*. Edited and translated by Harriet Martineau. New York: AMS Press.

Croce, Benedetto. 1982 [1914]. *Historical Materialism and the Economics of Karl Marx*. London: Transaction Books.

Davis, James A. 1985. *The Logic of Causal Order*. Newbury Park, Calif.: Sage Publications.

De Koster, Lester. 1964. *Vocabulary of Communism*. Grand Rapids, Mich.: William B. Eerdmans.

DeMartino, George. 1993. "The Necessity/Contingency Dualism in Marxian Crisis Theory: The Case of Long-Wave Theory." *Review of Radical Political Economics* 25 (3): 68–74.

Dietzgen, Joseph. 1906 [1869]. *The Positive Outcome of Philosophy*. Chicago: Charles Kerr.

Durkheim, Emile. 1968 [1915]. *Elementary Forms of the Religious Life*. New York: Free Press/Macmillan.

Durkheim, Emile. 1982a [1897]. "Review of Antonio Labriola, 'Essais sur la conception materialiste de l'histoire.'" *Revue philosophique* 44: 645–51. Reprinted as "Marxism and Sociology: The Materialist Conception of History", in Steven Lukes, ed., *Rules of the Sociological Method*. New York: Free Press, pp. 167–174.

Durkheim, Emile. 1982b [1908]. "The Method of Sociology", in Steven Lukes, ed., *Rules of the Sociological Method*. New York: Free Press, pp. 245–247.

Durkheim, Emile. 1982c. *Rules of the Sociological Method*. Edited by Steven Lukes. New York: Free Press.

Eldred, Michael and Mike Roth. 1978. *Guide to Marx's Capital*. London: Conference of Socialist Economists Books.

Elster, Jon 1982. "Marxism, Functionalism, and Game Theory." *Theory and Society* 11: 453–82.

Elster, Jon. 1985. *Making Sense of Marx*. Cambridge: Cambridge University Press.

Elster, Jon. 1986. "Further Thoughts on Marxism, Functionalism, and Game Theory", in John Roemer, ed., *Analytical Marxism*. Cambridge: Cambridge University Press, pp. 202–220.

Engels, Frederick. 1909 [1894]. Prefece. *Capital, Volume III*. Chicago: Charles H. Kerr.

Engels, Frederick. 1934. *The Dialectics of Nature*. Moscow, USSR: Progress Publishers.

Engels, Frederick. 1939. *Anti-Duhring*. New York: International Publishers.

Engels, Frederick. 1936a [1867]. Engels to Marx, June 16, 1867, in *The Correspondence of Karl Marx and Friedrich Engels*. New York: International Publishers, pp. 220–221.

Engels, Frederick. 1936b [1890]. Engels to Conrad Schmidt, August 5, 1890, in *The Correspondence of Karl Marx and Friedrich Engels*. New York: International Publishers, pp. 472–474.

Engels, Frederick. 1936c [1890]. Engels to J. Bloch, September 21, 1890, in *The Correspondence of Karl Marx and Friedrich Engels*. New York: International Publishers, pp. 475–477.

Engels, Frederick. 1936d [1890]. Engels to Conrad Schmidt, October 27, 1890, in *The Correspondence of Karl Marx and Friedrich Engels*. New York: International Publishers, pp. 477–482.

Engels, Frederick. 1941 [1888]. *Ludwig Feuerbach and the Outcome of Classical German Philosophy*. New York: International Publishers.

Engels, Frederick. 1971. *The Principles of Communism*. London: Pluto Press.

Engels, Frederick. 1975 [1844]. "Outlines of a Critique of Political Economy", in *Karl Marx and Frederick Engels: Collected Works*, Volume 3. New York: International Publishers, pp. 418–443.

Engels, Frederick. 1978a [1880/1892]. "Socialism: Utopian and Scientific", in Robert C. Tucker, ed., *The Marx-Engels Reader*. New York: W.W. Norton, pp. 683–717.

Engels, Friedrick. 1978b [1883]. Preface to the German Edition of 1883 – "The Communist Manifesto", in Robert c. Tucker, ed., *The Marx-Engels Reader*. New York: W.W. Norton, p. 472.

Engels, Frederick. 1978c [1883]. "Speech at the Graveside of Karl Marx", in Robert C. Tucker, ed., *The Marx-Engels Reader*. New York: W.W. Norton, pp. 681–682.

Engels, Frederick. 1980 [1859]. "Karl Marx, *A Contribution to the Critique of Political Economy*", in *Karl Marx and Frederick Engels: Collected Works*, Volume 16. New York: International Publishers, pp. 465–477.

Engels, Frederick. 1983 [1858]. Engels to Marx, April 8, 1858, in *Karl Marx and Frederick Engels: Collected Works*, Volume 40. New York: International Publishers, pp. 304–305.

Engels, Frederick. 1985 [1861]. Letter to Marx, December 2, 1861, in *Karl Marx and Frederick Engels: Collected Works*, Volume 41. New York: International Publishers, pp. 330–331.

Engels, Frederick. 1988 [1888]. Preface to the English Edition of 1888. "Manifesto of the Communist Party", in Martin Milligan, trans., *Economic and Philosophic Manuscripts of 1844*. Buffalo, N.Y.: Prometheus Books, pp. 203–208.

Engels, Frederick. 1989 [1875]. On Social Relations in Russia, in *Karl Marx and Frederick Engels: Collected Works*, Volume 24. New York: International Publishers, pp. 39–50.

Engels, Frederick. 1992 [1886]. Preface to the English Edition. *Capital, Volume I*. New York: International Publishers.

Engels, Frederick. 2001 [1889]. Engels to Karl Kautsky, February 20, 1889, in *Karl Marx and Frederick Engels: Collected Works*, Volume 48. New York: International Publishers, pp. 266–271.

Engels, Frederick. 2004 [1893]. Engels to Franz Mehring, July 14, 1893, in *Karl Marx and Frederick Engels: Collected Works*, Volume 50. New York: International Publishers, pp. 163–167.

Ferrarotti, Franco. 1985. "Weber, Marx, and the Spirit of Capitalism: Toward a Unitary Science of Man", in Robert J. Antonio and Ronald M. Glassman, eds., *A Weber-Marx Dialogue*. Lawrence: University Press of Kansas, pp. 262–272.

Feyerabend, Paul. 1975. *Against Method*. London: New Left Books.

Feyerabend, Paul. 1981. *Realism, Rationalism & Scientific Method*. New York: Cambridge University Press.

Fine, Ben and Laurence Harris. 1979. "Periodisation of Capitalism", in *Rereading Capital*. New York: Columbia University Press, pp. 104–119.

Fischer, Ernst. 1996. *How to Read Karl Marx*. New York: Monthly Review Press.

Foucault, Michel. 1972. *The Archaeology of Knowledge*. New York: Pantheon Books.

Foucault, Michel. 1977. *Discipline and Punish: The Birth of the Prison*. New York: Vintage Books.

Foucault, Michel. 1980. *Power/Knowledge: Selected Interviews and Other Writings*. Colin Gordon, ed. New York: Pantheon Books.

Foucault, Michel. 1984. "What is an Author?", in Paul Rabinow, ed., *The Foucault Reader*. New York: Pantheon Books, pp. 101–120.

Foucault, Michel. 1991. "Questions of Method", in Graham Burchell, Colin Gordon, and Peter Miller, eds., *The Foucault Effect: Studies in Governmentality*. Chicago: University of Chicago Press, pp. 73–86.

Fraser, Ian. 1997. "Two of a Kind: Hegel, Marx, Dialectic and Form." *Capital & Class* 61: 81–106.

Gandy, Daniel Ross. 1979. *Marx and History: From Primitive Society to the Communist Future*. Austin: University of Texas Press.

Garaudy, Roger 1967. *Karl Marx: The Evolution of His Thought*. New York: International Publishers.

Geras, Norman. 1971. "Essence and Appearance: Aspects of Fetishism in Marx's *Capital*." *New Left Review* 65: 69–85.

Gerdes, Paulus. 1985. *Marx Demystifies Calculus*. Minneapolis, Minn.: MEP Publications.

Gerratana, Valentino. 1973. "Marx and Darwin." *New Left Review* 82: 60–82.

Giddens, Anthony. 1971. *Capitalism and Modern Social Theory: Analysis of the Writings of Marx, Durkheim, and Max Weber*. Cambridge: Cambridge University Press.

Giddens, Anthony. 1995 [1981]. *A Contemporary Critique of Historical Materialism*. Second Edition. Stanford, Calif.: Stanford University Press.

Glazov, Yuri. 1985. *The Russian Mind Since Stalin's Death*. Dordrecht, The Netherlands: D. Reidel.

Godelier, Maurice. 2000. "The Disappearance of the 'Socialist System': Failure or Confirmation of Marx's Views on the Transition from One Form of Production and Society to Another?", in Werner Bonefeld and Kosmas Psychopedis, eds., *The Politics of Change: Globalization, Ideology and Critique*. New York: Palgrave, pp. 149–172.

Goldblatt, David, David Held, Anthony McGrew, and Jonathan Perraton. 1997. "Economic Globalization and the Nation-State: Shifting Balances of Power." *Alternatives* 22: 269–285.

Gottheil, Fred M. 1966. *Marx's Economic Predictions*. Evanston: Northwestern University Press.

Gould, Carol. 1978. *Marx's Social Ontology: Individuality and Community in Marx's Theory of Social Reality*. Cambridge, Mass.: MIT Press.

Gould, Stephen Jay. 1995. *Dinosaur in a Haystack: Reflections in Natural History*. New York: Harmony Books.

Graham, Loren R. 1972. *Science and Philosophy in the Soviet Union*. New York: Alfred A. Knopf.

Grier, Philip. 1978. *Marxist Ethical Theory in the Soviet Union*. Dordrecht, The Netherlands: D. Reidel.

Groff, Ruth. 2000. "The Truth of the Matter: Roy Bhaskar's Critical Realism and the Concept of Alethic Truth." *Philosophy of the Social Sciences* 30 (3): 407–435.

Gurvitch, George. 1983. "Marx's Sociology", in Tom Bottomore and Patrick Goode, eds., *Readings in Marxist Sociology*. Oxford: Clarendon Press, pp. 38–43.

Habermas, Jurgen. 1979a. "Historical Materialism and the Development of Normative Structures", in *Communication and the Evolution of Society*. Boston: Beacon Press, pp. 95–129.

Habermas, Jurgen. 1979b. "Toward a Reconstruction of Historical Materialism", in *Communication and the Evolution of Society*. Boston: Beacon Press, pp. 130–177.

Halle, Louis J. 1965. "Marx's Religious Drama." *Encounter* 4: 29–37.

Harris, Abram. 1950. "Utopian Elements in Marx's Thought." *Ethics* LX (2): 79–99.

Harstick, Hans-Peter (ed.). 1977. *Karl Marx: Uber Formen vorkapitalistischer Produktion*. Frankfurt, Germany: Campus Verlag.

Haydu, Jeffery. 1998. "Making Use of the Past: Time Periods and Cases to Compare and as Sequences of Problem Solving." *American Journal of Sociology* 104 (2): 339–71.

Hellman, Geoffrey. 1979. "Historical Materialism", in John Mepham and David-Hillel Ruben, eds., *Issues in Marxist Philosophy, Volume II*. Atlantic Highlands, N.J.: Humanities Press, pp. 143–166.

Hobsbawm, Eric. 1982a. "Marx, Engels, and Politics", in Eric Hobsbawm, ed., *The History of Marxism: Marxism in Marx's Day, Volumes I–IV*. Bloomington: Indiana University Press, pp. 227–264.

Hobsbawm, Eric. 1982b. "Marx, Engels and Pre-Marxian Socialism", in Eric Hobsbawm, ed., *The History of Marxism: Marxism in Marx's Day, Volumes I–IV*. Bloomington: Indiana University Press, pp. 1–28.

Hook, Ernest. 2002. "The Mutable Unmuted Sidney Hook (1902–1898) / Towards the Understanding of Karl Marx and Towards the Understanding of Sidney Hook", in Ernest Hook, ed., *Towards the Understanding of Karl Marx*. Amherst, N.Y.: Prometheus Books, pp. 9–33.

Horvath Ronald J. and Kenneth D. Gibson. 1984. "Abstraction in Marx's Method." *Antipode* 16 (1): 12–25.

Hughes, John, Peter Martin, and W.W. Sharrock. 1995. *Understanding Classical Sociology: Marx, Weber, Durkheim*. London: Sage Publications.

Isaac, Jeffery. 1987. *Power and Marxist Theory: A Realist View*. Ithaca, N.Y.: Cornell University Press.

Israel, Joachim. 1979. *The Language of Dialectics and the Dialectics of Language*. Atlantic Highlands, N.J.: Humanities Press.

Jakubowski, Franz. 1976. *Ideology and Superstructure in Historical Materialism*. New York: St. Martin's Press.

Jones, Gareth Stedman. 1982. "Engels and the History of Marxism", in Eric Hobsbawm, ed., in *The History of Marxism – Volume One: Marxism In Marx's Day*. Bloomington: Indiana University Press, pp. 290–326.

Jordan, Zbigniew A. 1967. *The Evolution of Dialectical Materialism: A Philosophical and Sociological Analysis*. New York: St. Martin's Press.

Jordan, Zbigniew A. 1971. *Karl Marx: Economy, Class and Social Revolution*. New York: Charles Scribner's Sons.

Kant, Immanuel. 1947. *Critique of Pure Reason*, in Saxe Commins and Robert Linscott, eds., *Man & Spirit: The Speculative Philosophers*. New York: Washington Square Press.

Karl Marx and his Contemporaries: Guide-book to the Permanent Exhibition in the Karl Marx House, Trier. 1994. Karl-Marx-Haus / Neu GmbH: Trier, Germany.

Kautsky, Karl. 1925. *The Economic Doctrines of Karl Marx*. J.H. Stenning, translator. London: A & C Black.

Kemple, Thomas. 1995. *Reading Marx Writing: Melodrama, the Market, and the 'Grundrisse.'* Stanford, Calif.: Stanford University Press.

Kline, George. 1988. "The Myth of Marx's Materialism – Appendix I: A Critical Examination of Engels Tendentious Editing of the First English Translation of *Das Kapital*, Volume 1; Appendix II: A Comparison of the First French Translation of *Das

Kapital, Volume 1 (in which Marx was heavily involved) with the Engels Edition", in Helmut Dahm, Thomas Blakeley, and George Kline, eds., *Philosophical Sovietology: The Pursuit of a Science*. Dordrecht, The Netherlands: D. Reidel, pp. 158–203.

Kolakowski, Leszek. 1978. *Main Currents of Marxism: Its Origins, Growth and Dissolution*. Volumes I–III. Oxford: Oxford University Press.

Kolko, Joyce. 1988. *Restructuring the World Economy*. New York: Pantheon Books.

Korsch, Karl. 1983. "Marxism and Sociology", in Tom Bottomore and Patrick Goode, eds., *Readings in Marxist Sociology*. Oxford: Clarendon Press, pp. 34–38.

Krader, Lawrence. 1982. "Theory of Evolution, Revolution and the State: The Critical Relation of Marx to His Contemporaries Darwin, Carlyle, Morgan, Maine and Kovalevsky", in Eric J. Hobsbawm, ed., *The History of Marxism, Volume One: Marxism in Marx's Day*. Bloomington: Indiana University Press, pp. 192–226.

Kuhn, Thomas. 1970. *The Structure of Scientific Revolutions*. Chicago: University of Chicago Press.

Labriola, Antonio. 1980 [1898]. *Socialism and Philosophy*. St. Louis, Mo.: Telos Press.

Laclau, Ernesto and Chantal Mouffe. 1982. "Recasting Marxism: Hegemony and New Political Movements." *Socialist Review* 12: 91–113.

Laclau, Ernesto and Chantal Mouffe. 1985. *Hegemony and Socialist Strategy*. London: Verso.

Lefebvre, Henri. 1968 [1940]. *Dialectical Materialism*. London: Jonathan Cape.

Leff, Gordon. 1969 [1961]. *The Tyranny of Concepts: A Critique of Marxism*. Tuscaloosa: University of Alabama Press.

Lenin, Vladimir I. 1929. *What is to be Done? Burning Questions for Our Movement*. New York: International Publishers.

Lenin, Vladimir I. 1952 [1908]. *Materialism and Empiro-Criticism*. Moscow, USSR: Foreign Language Publishing House.

Levine, Andrew, Elliot Sober, and Erik Olin Wright. 1987. "Marxism and Methodological Individualism." *New Left Review* 162: 67–84.

Liebknecht, Wilhelm. 1901. *Karl Marx: Biographical Memoirs*. E. Untermann, trans. Chicago: Charles H. Kerr.

Lindsay, Alexander Dunlop. 1973 [1925]. *Karl Marx's* Capital: *An Introductory Essay*. Westport, Conn.: Greenwood Publishers.

Ling, Trevor. 1980. *Karl Marx and Religion In Europe and India*. New York: Barnes and Noble Books.

Little, Daniel. 1986. *The Scientific Marx*. Minneapolis: University of Minnesota Press.

Löwith, Karl. 1982. *Max Weber and Karl Marx*. Tom Bottomore and William Outhwaite, eds. Hans Fantel, trans. Boston: George Allen & Unwin.

Lukács, Georg. 1971. *History and Class Consciousness*. Cambridge, Mass.: MIT Press.

Lukács, Georg. 1978. *Ontology of Social Being*. London: Merlin Press.

Maarek, Gerard. 1979. *An Introduction to Karl Marx's* Das Kapital: *A Study in Formalisation*. New York: Oxford University Press.

Macintyre, Stuart. 1980. *A Proletarian Science: Marxism in Britain 1917–1933*. Cambridge: Cambridge University Press.

Madan, G.R. 1979. *Western Sociologists on Indian Society: Marx, Spencer, Weber, Durkheim, Pareto*. London: Routledge & Kegan Paul.

Magill, Kevin. 1994. "Against Critical Realism." *Capital & Class* 54 (Autumn): 113–136.

Mandel, Ernest. 1968. *Marxist Economic Theory*. New York: Monthly Review Press.

Mandel, Ernest. 1970. "The Laws of Uneven Development." *New Left Review* 59: 19–37.

Mandel, Ernest. 1977. *From Class Society to Communism – An Introduction to Marxism*. London: Ink Links.

Mandel, Ernest. 1980. *Long Waves of Capitalist Development*. New York: Cambridge University Press.

Mandel, Ernest. 1990. "Karl Marx", in John Eatwell, Murray Milgate and Peter Newman, eds., *The New Palgrave Marxian Economics*. New York: W.W. Norton, pp. 1–38.

Marcuse, Herbert. 1954. *Reason and Revolution*. New York: Humanities Press.

Martin, Randy. 1998–1999. "Rereading Marx: A Critique of Recent Criticisms." *Science & Society* 62 (4): 513–536.

Marx, Karl. 1847. *The Poverty of Philosophy*. Moscow, USSR: Foreign Languages Publishing House.

Marx, Karl. 1909. *Capital, Volume III*. Chicago: Charles H. Kerr.

Marx, Karl. 1911 [1859]. Preface. *A Contribution to the Critique of Political Economy*. Chicago: Charles H. Kerr.

Marx, Karl. 1934. *Capital, Volume II*. New York: J.M. Dent & Sons/E.P. Dutton.

Marx, Karl. 1936a [1858]. Marx to Engels, January 14, 1858, in *The Correspondence of Karl Marx and Friedrich Engels*. New York: International Publishers, p. 102.

Marx, Karl. 1936b [1858]. Marx to Engels, April 2, 1858, in *The Correspondence of Karl Marx and Friedrich Engels*. New York: International Publishers, pp. 103–105.

Marx, Karl. 1936c [1858]. Marx to Engels, October 8, 1858, in *The Correspondence of Karl Marx and Friedrich Engels*. New York: International Publishers, pp. 117–118.

Marx, Karl. 1936d [1863]. Marx to Engels, January 28, 1863, in *The Correspondence of Karl Marx and Friedrich Engels*. New York: International Publishers, pp. 141–144.

Marx, Karl. 1936e [1865]. Marx to Engels, May 20, 1865, in *The Correspondence of Karl Marx and Friedrich Engels*. New York: International Publishers, pp. 202–203.

Marx, Karl. 1936f [1865]. Marx to Engels, July 31, 1865, in *The Correspondence of Karl Marx and Friedrich Engels*. New York: International Publishers, p. 204.

Marx, Karl. 1936g [1866]. Marx to Engels, February 13, 1866, in *The Correspondence of Karl Marx and Friedrich Engels*. New York: International Publishers, pp. 204–205.

Marx, Karl. 1936h [1866]. Marx to Engels, July 7, 1866, in *The Correspondence of Karl Marx and Friedrich Engels*. New York: International Publishers, pp. 209–210.

Marx, Karl. 1936i [1867]. Marx to Engels, June 22, 1867, in *The Correspondence of Karl Marx and Friedrich Engels*. New York: International Publishers, pp. 221–223.

Marx, Karl. 1936j [1868]. Marx to Engels, March 25, 1868, in *The Correspondence of Karl Marx and Friedrich Engels*. New York: International Publishers, pp. 235–237.

Marx, Karl. 1936k [1868]. Marx to Engels, April 22, 1868, in *The Correspondence of Karl Marx and Friedrich Engels*. New York: International Publishers, pp. 238–240.

Marx, Karl. 1936l [1868]. Marx to Engels, April 30, 1868, in *The Correspondence of Karl Marx and Friedrich Engels*. New York: International Publishers, pp. 240–245.

Marx, Karl. 1936m [1877]. Marx to Engels, July 18, 1877, in *The Correspondence of Karl Marx and Friedrich Engels*. New York: International Publishers, p. 346.

Marx, Karl. 1963. *Theories of Surplus-Value, Part I*. Moscow, USSR: Progress Publishers.

Marx, Karl. 1967. "The Centralization Question", in Loyd Easton and Kurt Guddat, eds., *Writings of the Young Marx on Philosophy and Society*. Garden City, N.Y.: Anchor/Doubleday, pp. 106–108.

Marx, Karl. 1968a [1870–1871]. *The Civil War in France*. New York: International Publishers.

Marx, Karl. 1968b. *Theories of Surplus-Value, Part II*. Moscow, USSR: Progress Publishers.

Marx, Karl. 1971. *Theories of Surplus-Value, Part III*. Moscow, USSR: Progress Publishers.

Marx, Karl. 1972. *The Ethnological Notebooks of Karl Marx (Studies of Morgan, Phear, Maine, Lubbock)*. Edited and translated by Lawrence Krader. Assen, The Netherlands: Van Gorcum.

Marx, Karl. 1973. *Grundrisse*. New York: Vintage Books.

Marx, Karl. 1975a [1835]. "Reflections of a Young Man on the Choice of a Profession", *Karl Marx and Frederick Engels: Collected Works*, Volume 1. New York: International Publishers, pp. 3–9.

Marx, Karl. 1975b [1837]. Letter From Marx To His Father, November 10[–11], 1837, in *Karl Marx and Frederick Engels: Collected Works*, Volume 1. New York: International Publishers, pp. 10–21.

Marx, Karl. 1975c [1840–1841]. "The Difference Between the Democritean and Epicurean Philosophy of Nature", in *Karl Marx and Frederick Engels: Collected Works*, Volume 1. New York: International Publishers, pp. 25–105.

Marx, Karl. 1975d [1842]. "Communism and the Ausburg *Allgemeine Zeitung*", in *Karl Marx and Frederick Engels: Collected Works*. Volume 1. New York: Progress Publishers, pp. 215–221.

Marx, Karl. 1975e [1842]. To Arnold Ruge, November 30, 1842, in *Karl Marx and Frederick Engels: Collected Works*, Volume 1. New York: Progress Publishers, pp. 393–395.

Marx, Karl. 1975f [1842]. To Dagobert Oppenheim, August 25, 1842, in *Karl Marx and Frederick Engels: Collected Works*, Volume 1. New York: Progress Publishers, pp. 391–393.

Marx, Karl. 1975g [1843]. "Justification of the Correspondent from the Mosel", in *Karl Marx and Frederick Engels: Collected Works*, Volume 1. New York: International Publishers, pp. 332–358.

Marx, Karl. 1975h [1843]. Passage from The Kreuznach Notebooks, July-August 1843, in *Karl Marx and Frederick Engels: Collected Works*, Volume 3. New York: Progress Publishers, p. 130.

Marx, Karl. 1975i [1844]. "Contribution to the Critique of Hegel's *Philosophy of Law*: Introduction", in *Karl Marx and Frederick Engels: Collected Works*, Volume 3. New York: Progress Publishers, pp. 175–187.

Marx, Karl. 1975j [1844]. Critical Marginal Notes on the Article 'The King of Prussia and Social Reform by a Prussian', in *Karl Marx and Frederick Engels: Collected Works*, Volume 3. New York: Progress Publishers, pp. 189–206.

Marx, Karl. 1975k [1844]. Letters from the *Deutsch-Franzosische Jahrbucher*, in *Karl Marx and Frederick Engels: Collected Works*, Volume 3. New York: Progress Publishers, pp. 133–145.

Marx, Karl. 1975l [1844]. To Ludwig Feuerbach, August 11, 1844, in *Karl Marx and Frederick Engels: Collected Works*, Volume 3. New York: Progress Publishers, pp. 354–357.

Marx, Karl. 1975m. *Karl Marx: Texts on Method*. Translated and edited by Terrell Carver. Oxford: Basil Blackwell.

Marx, Karl. 1976a [1843]. *Contribution to the Critique of Hegel's Philosophy of Law*, in *Karl Marx and Frederick Engels: Collected Works*, Volume 3. New York: Progress Publishers, pp. 3–129.

Marx, Karl. 1976b [1847]. "Moralising Criticism and Critical Morality", in *Karl Marx and Frederick Engels: Collected Works*, Volume 6. New York: Progress Publishers, pp. 312–340.

Marx, Karl. 1978a [1843]. "On the Jewish Question", in Robert C. Tucker, ed., *The Marx-Engels Reader*. Second Edition. New York: W.W. Norton, pp. 26–52.

Marx, Karl. 1978b [1845]. Theses on Feuerbach, in Robert C. Tucker, ed., *The Marx-Engels Reader*. Second Edition. New York: W.W. Norton, pp. 143–145.

Marx, Karl. 1978c [1850]. To the Editor of the *Neue Deutsche Zeitung*, June, 1850, in *Karl Marx and Frederick Engels: Collected Works*, Volume 10. New York: International Publishers, pp. 387–388.

Marx, Karl. 1978d [1851–1852]. *The Eighteenth Brumaire of Louis Bonaparte*, in Robert C. Tucker, ed., in *The Marx and Engels Reader*. Second Edition. New York: W.W. Norton, pp. 594–617.

Marx, Karl. 1978e [1872]. "The Possibility of Non-Violent Revolution", in Robert C. Tucker, ed., *The Marx-Engels Reader*. Second Edition. New York: W.W. Norton, pp. 522–524.

Marx, Karl. 1978f [1874–1875]. "After the Revolution: Marx Debates Bakunin", in Robert C. Tucker, ed., *The Marx-Engels Reader*. Second Edition. New York: W.W. Norton, pp. 542–548.

Marx, Karl. 1978g [1875]. "Critique of the Gotha Program", in Robert C. Tucker, ed., *The Marx-Engels Reader*. Second Edition. New York: W.W. Norton, pp. 525–541.

Marx, Karl. 1979a [1858]. Draft of Letter to Eduard Muller-Tellering, March 12, 1859, in Saul Padover, ed., *The Letters of Karl Marx*. Engelwood Cliffs, N.J.: Prentice-Hall, pp. 132–133.

Marx, Karl. 1979b [1868]. From Letter to Ludwig Kugelmann, March 6, 1868, in Saul Padover, ed., *The Letters of Karl Marx*. Engelwood Cliffs, N.J.: Prentice-Hall, pp. 243–244.

Marx, Karl. 1979c [1877]. From Letter to the Editor of the Petersberg Literary-Political Journal, *Otechestvennye Zapiski*, November 1877, in Saul Padover, ed., *The Letters of Karl Marx*. Engelwood Cliffs, N.J.: Prentice-Hall, pp. 321–322.

Marx, Karl. 1979d [1879]. To Maxim Maximovich Kovalevsky, April, 1879, in Saul Padover, ed., *The Letters of Karl Marx*. Engelwood Cliffs, N.J.: Prentice-Hall, pp. 324–325.

Marx, Karl. 1980 [1859]. "Population, Crime, and Pauperism", in *Karl Marx and Frederick Engels: Collected Works*, Volume 16. New York: International Publishers, pp. 487–491.

Marx, Karl. 1982a [1846]. Letter to Friedrich Julius Leske, August 1, 1846, in *Karl Marx and Frederick Engels: Collected Works*, Volume 38. New York: Progress Publishers, pp. 48–52.

Marx, Karl. 1982b [1846]. Letter to Pavel Vasilyevich Annenkov, December 28, 1846, in *Karl Marx and Frederick Engels: Collected Works*, Volume 38. New York: International Publishers, pp. 95–106.

Marx, Karl. 1982c [1851]. Marx to Joseph Weydemeyer, June 27, 1851, in *Karl Marx and Frederick Engels: Collected Works*, Volume 38. New York: International Publishers, pp. 375–377.

Marx, Karl. 1982d [1851]. Marx to Joseph Weydemeyer, August 2, 1851, in *Karl Marx and Frederick Engels: Collected Works*, Volume 38. New York: International Publishers, pp. 401–403.

Marx, Karl. 1983a [1847–1849]. *Wage-Labour and Capital / Value, Price and Profit*. New York: International Publishers.

Marx, Karl. 1983b [1852]. Marx to Joseph Weydemeyer, March 5, 1852, in *Karl Marx and Frederick Engels: Collected Works*, Volume 39. New York: International Publishers, pp. 60–66.

Marx, Karl. 1983c [1856]. Marx to Engels, March 5, 1856, in *Karl Marx and Frederick Engels: Collected Works*, Volume 40. New York: International Publishers, pp. 19–25.

Marx, Karl. 1983d [1856]. Marx to Engels, August 1, 1856, in *Karl Marx and Frederick Engels: Collected Works*, Volume 40. New York: International Publishers, pp. 61–64.

Marx, Karl. 1983e [1858]. Marx to Ferdinand Lassalle, February 22, 1858, in *Karl Marx and Frederick Engels: Collected Works*, Volume 40. New York: International Publishers, pp. 268–271.

Marx, Karl. 1983f [1858]. Marx to Ferdinand Lassalle, May 31, 1858, in *Karl Marx and Frederick Engels: Collected Works*, Volume 40. New York: International Publishers, pp. 315–316.

Marx, Karl. 1983g [1858]. Marx to Ferdinand Lassalle, November 12, 1858, in *Karl Marx and Frederick Engels: Collected Works*, Volume 40. New York: International Publishers, pp. 353–355.

Marx, Karl. 1983h [1859]. Marx to Joseph Weydemeyer, February 1, 1859, in *Karl Marx and Frederick Engels: Collected Works*, Volume 40. New York: International Publishers, pp. 374–378.

Marx, Karl. 1983i. *Mathematical Manuscripts of Karl Marx*. Translated by C. Aronson and M. Meo. London: New Park Publications.

Marx, Karl. 1984 [1861]. "The London *Times* on the Orleans Princes in America", in *Karl Marx and Frederick Engels: Collected Works*, Volume 19. New York: International Publishers, pp. 27–31.

Marx, Karl. 1985a [1860]. Marx to Ferdinand Freiligrath, February 29, 1860, in *Karl Marx and Frederick Engels: Collected Works*, Volume 41. New York: International Publishers, pp. 80–88.

Marx, Karl. 1985b [1860]. Marx to Engels, November 23, 1860, in *Karl Marx and Frederick Engels: Collected Works*, Volume 41. New York: International Publishers, pp. 216–217.

Marx, Karl. 1985c [1860]. Marx to Engels, December 19, 1860, in *Karl Marx and Frederick Engels: Collected Works*, Volume 41. New York: International Publishers, pp. 231–233.

Marx, Karl. 1985d [1861]. Marx to Ferdinand Lassalle, January 16, 1861, in *Karl Marx and Frederick Engels: Collected Works*, Volume 41. New York: International Publishers, pp. 245–247.

Marx, Karl. 1985e [1861]. Marx to Engels, December 9, 1861, in *Karl Marx and Frederick Engels: Collected Works*, Volume 41. New York: International Publishers, pp. 332–333.

Marx, Karl. 1985f [1862]. Marx to Ferdinand Lassalle, April 28, 1862, in *Karl Marx and Frederick Engels: Collected Works*, Volume 41. New York: International Publishers, pp. 355–358.

Marx, Karl. 1985g [1862]. Marx to Ludwig Kugelmann, December 28, 1862, in *Karl Marx and Frederick Engels: Collected Works*, Volume 41. New York: International Publishers, pp. 435–437.

Marx, Karl. 1985h [1863]. Marx to Engels, July 6, 1863, in *Karl Marx and Frederick Engels: Collected Works*, Volume 41. New York: International Publishers, pp. 483–487.

Marx, Karl. 1985i [1864]. Marx to Lion Philips, February 20, 1864, in *Karl Marx and Frederick Engels: Collected Works*, Volume 41. New York: International Publishers, pp. 508–510.

Marx, Karl. 1985j [1864]. Marx to Lion Philips, August 17, 1864, in *Karl Marx and Frederick Engels: Collected Works*, Volume 41. New York: International Publishers, pp. 550–551.

Marx, Karl. 1985k [1864]. Provisional Rules of the Association, in *Karl Marx and Frederick Engels: Collected Works*, Volume 20. New York: International Publishers, pp. 14–16.

Marx, Karl. 1985l [1865]. On Proudhon [Letter to J.B. Schweitzer], January 24, 1865, in *Karl Marx and Frederick Engels: Collected Works*, Volume 20. New York: International Publishers, pp. 26–33.

Marx, Karl. 1985m [1869]. Record of Marx's Speeches on General Education, August 10 and 17, 1869, in *Karl Marx and Frederick Engels: Collected Works*, Volume 21. New York: International Publishers, pp. 398–400.

Marx, Karl. 1986a [1857]. "The War Against Persia", in *Karl Marx and Frederick Engels: Collected Works*, Volume 15. New York: International Publishers, pp. 177–180.

Marx, Karl. 1986b [1871]. Record of Marx's Speech on the Political Action of the Working Class, in *Karl Marx and Frederick Engels: Collected Works*, Volume 22. New York: International Publishers, p. 618.

Marx, Karl. 1987a [1864]. Marx to Lion Philips, November 29, 1864, in *Karl Marx and Frederick Engels: Collected Works*, Volume 42. New York: International Publishers, pp. 46–48.

Marx, Karl. 1987b [1865]. Marx to Engels, August 19, 1865, in *Karl Marx and Frederick Engels: Collected Works*, Volume 42. New York: International Publishers, pp. 183–186.

Marx, Karl. 1987c [1867]. Marx to Sigfrid Meyer, April 30, 1867, in *Karl Marx and Frederick Engels: Collected Works*, Volume 42. New York: International Publishers, pp. 366–367.

Marx, Karl. 1987d [1867]. Marx to Engels, August 24, 1867, in *Karl Marx and Frederick Engels: Collected Works*, Volume 42. New York: Progress Publishers, pp. 407–408.

Marx, Karl. 1987e [1868]. Marx to Engels, January 8, 1868, in *Karl Marx and Frederick Engels: Collected Works*, Volume 42. New York: Progress Publishers, pp. 514–517.

Marx, Karl. 1988a [1844]. *Economic and Philosophic Manuscripts of 1844*. New York: Prometheus Books.

Marx, Karl. 1988b [1868]. Marx to Kugelmann, July 11, 1868, in *Karl Marx and Frederick Engels: Collected Works*, Volume 43. New York: Progress Publishers, pp. 67–70.

Marx, Karl. 1988c [1870]. Letter to Ludwig Kugelmann, June 27, 1870, in *Karl Marx and Frederick Engels: Collected Works*, Volume 43. New York: Progress Publishers, pp. 527–528.

Marx, Karl. 1988d [1872–1873]. "Political Indifferentism", in *Karl Marx and Frederick Engels: Collected Works*, Volume 23. New York: International Publishers, pp. 392–397.

Marx, Karl. 1989a [1871]. Marx to Ludwig Kugelmann, April 12, 1871, in *Karl Marx and Frederick Engels: Collected Works*, Volume 44. New York: International Publishers, pp. 131–132.

Marx, Karl. 1989b [1871]. Marx to Ludwig Kugelmann, April 17, 1871, in *Karl Marx and Frederick Engels: Collected Works*, Volume 44. New York: Progress Publishers, pp. 136–137.

Marx, Karl. 1989c [1871]. Marx to Ludwig Kugelmann, July 27, 1871, in *Karl Marx and Frederick Engels: Collected Works*, Volume 44. New York: International Publishers, pp. 176–177.

Marx, Karl. 1989d [1871]. Marx to Jenny Marx, August 25, 1871, in *Karl Marx and Frederick Engels: Collected Works*, Volume 44. New York: International Publishers, pp. 206–207.

Marx, Karl. 1989e [1878]. The Parliamentary Debate on the Anti-Socialist Law (Outline of an Article), in *Karl Marx and Frederick Engels: Collected Works*, Volume 24. New York: International Publishers, pp. 240–250.

Marx, Karl. 1989f [1880]. Workers' Questionnaire, in *Karl Marx and Frederick Engels: Collected Works*, Volume 24. New York: International Publishers, pp. 328–334.

Marx, Karl. 1991a [1877]. Marx to Sigmund Schott, November 3, 1877, in *Karl Marx and Frederick Engels: Collected Works*, Volume 45. New York: International Publishers, p. 287.

Marx, Karl. 1991b [1877]. Marx to Wilhelm Blos, November 10, 1877, in *Karl Marx and Frederick Engels: Collected Works*, Volume 45. New York: International Publishers, pp. 288–289.

Marx, Karl. 1991c [1879]. Marx to Nikolai Danielson, April 10, 1879, in *Karl Marx and Frederick Engels: Collected Works*, Volume 45. New York: International Publishers, pp. 353–358.

Marx, Karl. 1992a [1867]. Preface to First German Edition, in *Capital, Volume I: A Critical Analysis of Capitalist Production*. New York: International Publishers, pp. 18–21.

Marx, Karl. 1992b [1872]. Preface to the French Edition, in *Capital, Volume I: A Critical Analysis of Capitalist Production*. New York: International Publishers, p. 30.

Marx, Karl. 1992c [1873]. Afterword to Second German Edition, in *Capital, Volume I: A Critical Analysis of Capitalist Production*. New York: International Publishers, pp. 22–29.

Marx, Karl. 1992d* [1873]. "Das Kapital", in *European Messenger*, May, 1872, pp. 427–436. Reprinted in *Capital, Volume I: A Critical Analysis of Capitalist Production*. New York: International Publishers, pp. 27–28.

Marx, Karl. 1992e [1881]. Marx to Ferdinand Domela Nieuwenhuis, February 22, 1881, in *Karl Marx and Frederick Engels: Collected Works*, Volume 46. New York: International Publishers, pp. 65–67.

Marx, Karl. 1992f [1881]. Marx to Vera Zasulich, March 8, 1881, in *Karl Marx and Frederick Engels: Collected Works*, Volume 46. New York: International Publishers, pp. 71–72.

Marx, Karl. 1992g [1881]. Marx to Jenny Longuet, April 11, 1881, in *Karl Marx and Frederick Engels: Collected Works*, Volume 46. New York: International Publishers, pp. 81–85.

Marx, Karl. 1992h [1881]. Marx to Friedrich Adolph Sorge, June 20, 1881, in *Karl Marx and Frederick Engels: Collected Works*, Volume 46. New York: International Publishers, pp. 98–101.

Marx, Karl. 1992i [1881]. Marx to Friedrich Adolph Sorge, December 15, 1881, in *Karl Marx and Frederick Engels: Collected Works*, Volume 46. New York: International Publishers, pp. 161–163.

Marx, Karl. 1992j. *Capital, Volume I: A Critical Analysis of Capitalist Production*. New York: International Publishers.

Marx, Karl and Frederick Engels. 1956. *The Holy Family*. Moscow, USSR: Foreign Languages Publishing House.

Marx, Karl and Frederick Engels. 1976 [1846]. *The German Ideology. Karl Marx and Frederick Engels: Collected Works*, Volume 5. Moscow/New York: International Publishers.

Marx, Karl and Frederick Engels. 1978a [1848]. "The Manifesto of the Communist Party", in Robert C. Tucker, ed., *The Marx-Engels Reader*. Second edition. New York: W.W. Norton, pp. 469–500.

Marx, Karl and Frederick Engels. 1978b [1850]. Address of the General Authority to the League, March, 1850, in *Karl Marx and Frederick Engels: Collected Works*, Volume 10. New York: International Publishers, pp. 277–287.

Marx, Karl and Frederick Engels. 1988 [1872]. Fictitious Splits in the International: Private Circular from the General Council of the International Working Men's Association, in *Karl Marx and Frederick Engels: Collected Works*, Volume 23. New York: International Publishers, pp. 79–123.

Marx, Karl and Frederick Engels. 1989 [1879]. Circular Letter: To August Bebel, Wilhelm Liebknecht, Wilhelm Bracke and Others, in *Karl Marx and Frederick Engels: Collected Works*, Volume 24. New York: International Publishers, pp. 253–269.

Mayer, Tom. 2002. "The Collapse of Soviet Communism: A Class Dynamics Interpretation." *Social Forces* 80 (3): 759–811.

McIntyre, Lisa. 2006. *The Practical Skeptic: Core Concepts in Sociology*. Boston: McGrawHill.

McLellan, David. 1973. *Karl Marx: His Life and Thought*. New York: Harper and Row.

McLellan, David. 1975. *Karl Marx*. New York: Penguin.

McLellan, David (ed.). 1981. *Karl Marx: Interviews and Recollections*. Totowa, N.J.: Barnes & Noble Books.

McMichael, Philip. 1990. "Incorporating Comparison Within A World-Historical Perspective: An Alternative Comparative Method." *American Sociological Review* 55 (6): 385–397.

McMurtry, John. 1979. *The Structure of Marx's World-View*. Princeton, N.J.: Princeton University Press.

McMurtry, John. 1992. "The Crisis of Marxism: Is There a Marxian Explanation?" *Praxis International* 12: 302–321.

Meikle, Scott. 1979. "Dialectical Contradiction and Necessity", in John Mepham and David-Hillel Ruben, eds., *Issues in Marxist Philosophy, Volume I. Dialectics and Method*. Atlantic Highlands, N.J.: Humanities Press, pp. 5–35.

Merton, Robert. 1957. *Social Theory and Social Structure*. New York: Free Press.

Miliband, Ralph. 1983. *Class Power and State Power*. London: Verso.

Mills, Charles W. 1985–86. "Marxism and Naturalistic Mystification." *Science & Society* XLIX (4): 472–483.

Montano, Mario. 1971. "On the Methodology of Determinate Abstraction: Essay on Galvano Della Volpe." *Telos* 7 (Spring): 30–49.

Montano, Mario. 1972. "The 'Scientific Dialectics' of Galvano Della Volpe", in Dick Howard and Karl Klare, eds., *The Unknown Dimension: European Marxism Since Lenin*. New York: Basic Books, pp. 342–364.

Morrison, Ken. 1995. *Marx Durkheim Weber: Formations of Modern Social Thought*. London: Sage Publications.

Murray, Patrick. 1988. *Marx's Theory of Scientific Knowledge*. Atlantic Highlands, N.J.: Humanities Press International.

Nagel, Ernest. 1979. *Teleology Revisited and Other Essays in the Philosophy and History of Science*. New York: Columbia University Press.

Nicolaus, Martin. 1968. "The Unknown Marx." *New Left Review* 48: 41–61.

Nicolaus, Martin. 1973. "Foreword", in *Grundrisse*. New York: Vintage Books, pp. 7–63.

Norman, Richard. 1980. "On the Hegelian Origins", in Richard Norman and Sean Sayers, eds., *Hegel, Marx and Dialectic*. Atlantic Highlands, N.J.: Humanities Press pp. 25–46.

O'Malley, Joseph and Keith Algozin (eds.). 1981. *Rubel on Karl Marx: Five Essays*. Cambridge: Cambridge University Press.

O'Neill, John. 1996. "Engels Without Dogmatism", in Christopher J. Arthur, ed., *Engels Today: A Centenary Appreciation*. New York: St. Martin's Press, pp. 47–66.

Ollman, Bertell. 1976 [1971]. *Alienation: Marx's Conception of Man in Capitalist Society*. New York: Cambridge University Press.

Ollman, Bertell. 1979a. "Marxism and Political Science: Prolegomenon to a Debate on Marx's Method", in Bertell Ollman, ed., *Social and Sexual Revolution*. Boston: South End Press, pp. 99–123.

Ollman, Bertell. 1979b. "Marx's Vision of Communism", in Bertell Ollman, ed., *Social and Sexual Revolution*. Boston: South End Press, pp. 48–98.

Ollman, Bertell. 1993. *Dialectical Investigations*. New York: Routledge.

Ollman, Bertell. 2003. *Dance of the Dialectic: Steps in Marx's Method*. Urbana: University of Illinois Press.

Paolucci, Paul. 2000. "Questions of Method: Fundamental Problems Reading Dialectical Methodologies." *Critical Sociology* 26 (3): 301–328.

Paolucci, Paul. 2001a. "Assumptions of the Dialectical Method." *Critical Sociology* 27 (3): 116–146.

Paolucci, Paul. 2001b. "Classical Sociology Theory and Modern Social Problems: Marx's Concept of the Camera Obscura and the Fallacy of Individualistic Reductionism." *Critical Sociology* 27 (1): 77–210.

Paolucci, Paul. 2003a. "Foucault's Encounter with Marxism." *Current Perspectives in Social Theory* 22: 3–58.

Paolucci, Paul. 2003b. "The Scientific Method and the Dialectical Method." *Historical Materialism* 11 (1): 75–106.

Paolucci, Paul. 2004. "The Discursive Transformation of Marx's Communism into Soviet Diamat." *Critical Sociology* 30 (3): 617–667.

Paolucci, Paul. 2005. "Assumptions of the Dialectical Method: The Centrality of Labor for the Human Species, Its History, and Individuals." *Critical Sociology* 31 (3): 559–581.

Parsons, Howard. 1964. "The Prophetic Mission of Karl Marx." *Journal of Religion* XLIV: 1: 52–73.

Pilling, Geoffrey. 1980. *Marx's* Capital: *Philosophy and Political Economy*. London: Routledge & Kegan Paul.

Plaut, Eric A. and Kevin B. Anderson (eds.). 1999. *Marx on Suicide*. Evanston, Ill.: Northwestern University Press.

Plekhanov, George. 1972 [1895]. *The Development of the Monist View of History*. New York: International Publishers.

Plekhanov, George. 1969 [1908]. *Fundamental Problems of Marxism*. New York: International Publishers.

Poggi, Gianfranco. 1972. *Images of Society: Essays on the Sociological Theories of Tocqueville, Marx, and Durkheim*. Stanford, Calif.: Stanford University Press.

Popper, Karl. 1950. *The Open Society and Its Enemies: Volume I.* Princeton, N.J.: Princeton University Press.

Popper, Karl. 1966 [1962]. *The Open Society and Its Enemies: Volume II. The High Tide of Prophecy: Hegel, Marx, and the Aftermath.* Princeton, N.J.: Princeton University Press.

Popper, Karl. 1972 [1962]. "What is Dialectic?", in *Conjectures and Refutations: The Growth of Scientific Knowledge.* London: Routledge & Kegan Paul, pp. 312–335.

Popper, Karl. 1983 [1956]. *Realism and the Aim of Science.* Totowa, N.J.: Rowman & Littlefield.

Postone, Moishe. 1993. *Time, Labor, and Social Domination: A Reinterpretation of Marx's Critical Theory.* Cambridge: Cambridge University Press.

Rader, Melvin. 1979. *Marx's Interpretation of History.* New York: Oxford University Press.

Reuten, Geert. 2000. "The Interconnection of Systematic Dialectics and Historical Materialism." *Historical Materialism* 7 (1): 137–165.

Riazanov, David. 1927. *Karl Marx and Friedrich Engels.* New York: International Publishers.

Roberts, Marcus. 1996 [1986]. *Analytical Marxism: A Critique.* New York: Verso.

Robinson, Joan. 1966 [1942]. *An Essay on Marxian Economics.* New York: St. Martin's Press.

Roemer, John (ed.). 1986a. *Analytical Marxism.* Cambridge: Cambridge University Press.

Roemer, John. 1986b. "'Rational Choice Marxism: Some Issues of Method and Substance", in John Roemer, ed., *Analytical Marxism.* Cambridge: Cambridge University Press, pp. 191–201.

Rose, Margaret. 1984. *Marx's Lost Aesthetic: Karl Marx and the Visual Arts.* Cambridge: Cambridge University Press.

Rovatti, Pier Aldo. 1972. "Fetishism and Economic Categories." *Telos* 14 (Winter): 87–105.

Rubel, Maximilien. 1980 [1965]. *Marx: Life and Works.* New York: Facts on File.

Rubel, Maximilien. 1981. *Rubel on Karl Marx.* Edited by Joseph O'Malley and Keith Algozin. Cambridge: Cambridge University Press.

Ruben, David-Hillel. 1979. "Marxism and Dialectics", in John Mepham and David-Hillel Ruben, eds., *Issues in Marxist Philosophy, Volume I.* Atlantic Highlands, N.J.: Humanities Press, pp. 37–85.

Sawer, Marian. 1977. *Marxism and the Question of the Asiatic Mode of Production.* The Hague, The Netherlands: Martinus Nijhoff.

Sayer, Andrew. 1984. *Method in Social Science: A Realist Approach.* London: Hutchinson.

Sayer, Andrew. 2000. *Realism and Social Science.* London: Sage.

Sayer, Derek. 1979. *Marx's Method: Ideology, Science and Critique in* Capital. Atlantic Highlands, N.J.: Humanities Press.

Sayer, Derek. 1987. *The Violence of Abstraction: The Analytical Foundations of Historical Materialism*. Oxford: Basil Blackwell.

Sayers, Sean. 1980. "Dualism, Materialism and Dialectics", in Richard Norman and Sean Sayers, eds., *Hegel, Marx, and Dialectic*. Atlantic Highlands, N.J.: Harvester, pp. 67–143.

Sayers, Sean. 1998. *Marxism and Human Nature*. London: Routledge.

Schmidt, Alfred. 1981. *History and Structure*. Translated by Jeffrey Herf. Cambridge, Mass.: MIT Press.

Schumpeter, Joseph. 1954. *Capitalism, Socialism, Democracy*. London: George Allen & Unwin.

Sekine, Thomas. 1998. "The Dialectic of Capital: An Unoist Interpretation." *Science & Society* 62 (3): 434–445.

Seve, Lucien. 1978. *Man in Marxist Theory and the Psychology of Personality*. Atlantic Highlands, N.J.: Humanities Press.

Sherman, Howard. 1995. *Reinventing Marxism*. Baltimore, Md.: Johns Hopkins University Press.

Simmel, Georg. 1964. *Sociology of George Simmel*. Translated by Kurt H. Wolf. New York: Free Press.

Simmel, Georg. 1977. *The Problems of the Philosophy of History: An Epistemological Essay*. Translated & edited, with an Introduction, by Guy Oakes. New York: Free Press.

Simmel, Georg. 1980. *Essays on Interpretation in Social Science*. Edited and translated by Guy Oakes. Totowa, N.J.: Rowman & Littlefield.

Smith, Cyril. 1996. *Marx at the Millennium*. London: Pluto Press.

Solzhenitsyn, Aleksandr. 1974 [1973]. *The Gulag Archipelago, 1918–1956: An Experiment in Literary Criticism*, Volumes I–II. New York: Harper and Row.

Solzhenitsyn, Aleksandr. 1975 [1974]. *The Gulag Archipelago, 1918–1956: An Experiment in Literary Criticism*, Volumes III–IV. New York: Harper and Row.

Solzhenitsyn, Aleksandr. 1978 [1976]. *The Gulag Archipelago, 1918–1956: An Experiment in Literary Criticism*, Volumes V–VIII. New York: Harper and Row.

Soper, Kate. 1979. "Marxism, Materialism and Biology", in John Mepham and David-Hillel Ruben, eds., *Issues in Marxist Philosophy, Volume II*. Atlantic Highlands, N.J.: Humanities, pp. 61–99.

Sorel, Georges. 1983. "Marxism in Sociological Theory", in Tom Bottomore and Patrick Goode, eds., *Readings in Marxist Sociology*. Oxford: Clarendon Press, pp. 23–29.

Sprigge, C.J.S. 1962 [1938]. *Karl Marx*. New York: Collier.

Stalin, Joseph. 1942. *Leninism: Selected Writings*. New York: International Publishers.

Steila, Daniela. 1991. *Genesis and Development of Plekhanov's Theory of Knowledge*. Dordrecht, The Netherlands: Kluwer Academic Publishers.

Stockhammer, Morris (ed.). 1965. *Karl Marx Dictionary*. New York: Philosophical Library.

Struik, Dirk. 1948. "Marx and Mathematics." *Science & Society* XII (1): 181–196.

Sweezy, Paul. 1964 [1942]. *The Theory of Capitalist Development: Principles in Marxian Political Economy*. New York: Monthly Review Press.

Sweezy, Paul (with Paul Baran). 1966. *Monopoly Capital: An Essay on the American Economic System*. New York: Monthly Review Press.

Swinton, John. 1983. "What Karl Marx Himself Said About the Translating of 'Capital'," *John Swinton's Paper*. November 29, 1885, in Philip Foner, ed., *Karl Marx Remembered*. San Francisco: Synthesis Publications, pp. 266–267.

Therborn, Goran. 1976. *Science, Class and Society: On the Formation of Sociology and Historical Materialism*. London: New Left Books.

Thomas, Paul. 1976. "Marx and Science." *Political Studies* 24 (3): 1–23.

Tilly, Charles. 1984. *Big Structures, Large Processes, Huge Comparisons*. New York: Russell Sage Foundation.

Timasheff, Nicholas. 1957 [1955]. *Sociological Theory: Its Nature and Growth*. New York: Random House.

Tucker, Robert. 1957–1958. "Marxism – Is it Religion?" *Ethics* lxviii: 125–130.

Tucker, Robert. 1964. *Philosophy and Myth in Karl Marx*. New York: Cambridge University Press.

Turner, Jonathan H. 1993a. "Marx and Simmel: Reassessing the Foundations of Conflict Theory", in *Classical Sociological Theory: A Positivist's Perspective*. Chicago: Nelson Hall, pp. 87–100.

Turner, Jonathan H. 1993b. "Where Marx Went Wrong", in *Classical Sociological Theory: A Positivist's Perspective*. Chicago: Nelson Hall, pp. 101–108.

Van Den Braembussche, Antoon. 1990. "Comparison, Causality and Understanding: The Historical Explanation of Capitalism by Marx and Weber." *Cultural Dynamics* 3: 190–223.

Walker, Angus. 1989 [1978]. *Marx: His Theory and its Context*. London: Rivers Oram Press.

Wallerstein, Immanuel. 1974. "The Rise and Future Demise of the World Capitalist System: Concepts for Comparative Analysis." *Comparative Studies in Society and History* 16: 387–425.

Wallerstein, Immanuel. 1982. "Crisis as Transition", in Samir Amin, Giovanni Arrighi, Andre Gunder Frank, and Immanuel Wallerstein, eds., *Dynamics of Global Crisis*. New York: Monthly Review Press, pp. 11–54.

Wallerstein, Immanuel. 1991. *Unthinking Social Science: Limits of Nineteenth Century Paradigms*. London: Polity Press.

Wallerstein, Immanuel. 1999 [1983]. *Historical Capitalism*. New York: Verso.

Weeks, John. 1985–1986. "Epochs of Capitalism and the Progressiveness of Capital's Expansion." *Science & Society* XLIX, 4 (Winter): 414–436.

Wessell, Leonard. 1984. *Prometheus Bound: The Mythic Structure of Karl Marx's Scientific Thinking*. Baton Rouge: Louisiana State University Press.

Wetter, Gustav. 1963 [1958]. *Dialectical Materialism: A Historical and Systematic Study of Philosophy in the Soviet Union*. New York: Frederick A. Praeger.

Whitehead, Alfred North. 1960 [1925]. *Science and the Modern World*. New York: Mentor.

Wiley, Norbert (ed.). 1987. *The Marx-Weber Debate*. Newbury Park, Calif.: Sage.

Williams, Raymond. 1973. "Base and Superstructure in Marxist Cultural Theory." *New Left Review* 82: 3–16.

Williams, Raymond. 1978. "Problems of Materialism." *New Left Review* 109: 3–17.

Wolff, Robert Paul. 1984. *Understanding Marx: A Reconstruction and Critique of Capital*. Princeton, N.J.: Princeton University Press.

Wolff, Robert Paul. 1988. *Moneybags Must Be So Lucky: On the Literary Structure of Capital*. Amherst: University of Massachusetts Press.

Wolton, Suke (ed.). 1996. *Marxism, Mysticism, and Modern Theory*. New York: St. Martin's Press.

Wright, Erik O. 1978. *Class, Crisis, and the State*. London: New Left Books.

Wright, Erik O. 1995 [1989]. "What is Analytical Marxism?", in Terrell Carver and Paul Thomas, eds., *Rational Choice Marxism*. State College: Pennsylvania University Press, pp. 11–30.

Wright, Erik O., Andrew Levine, and Elliot Sober. 1992. *Reconstructing Marxism*. New York: Verso.

Yuille, Judith. No Date. *Karl Marx: from Trier to Highgate*. Friends of Highgate Cemetery. Charity No. 1058392. VAT No. 544 5652 34. Highgate Cemetery. Swains Lane, London N6.

Zeleny, Jindrich. 1980. *The Logic of Marx*. London: Blackwell.

Zimmerman, Marc. 1976. "Polarities and Contradictions: Theoretical Bases of the Marxist-Structural Encounter." *New German Critique* 7: 69–90.

Zirkle, Conway. 1949. *Death of a Science in Russia*. Philadelphia: University of Pennsylvania Press.

Index

a priori (concepts and approaches), 16, 25, 26n10, 27, 31–32, 35, 54–55, 92, 96, 99, 106–107, 120–122, 160, 175, 222n17, 251, 268n4
  See also critique (of Kant); Kant; philosophy (speculative philosophy)
abstract (the), 20, 32, 57, 76, 77, 85, 115, 143–144, 161, 164, 177, 193, 200, 270
  abstract thought, 108, 115
  categories, concepts, and frameworks, 32, 61, 71, 91, 113, 116, 133, 140, 142–143, 159, 160–162, 164–168, 176, 182–185, 190–191, 193, 195, 202, 207, 212–213, 217–222, 263–264, 275
    See also conceptual doublets: moments of abstraction; determinism (abstractions of determination); labor (abstract labor); laws; successive abstractions (general abstract; specific abstract)
abstract (to), 38, 40, 46, 57–59, 70–72, 74, 76–78, 87, 93, 96, 100, 112, 116, 118, 124–125, 130, 131, 146, 151–153, 156–158, 161–162, 164–169, 174–178, 181–182, 187–188, 191, 204–205, 218
  See also abstraction/method of; conceptual doublets: moments of abstraction; successive abstractions
abstract (too), as alienation, 90, 95, 174
  as distortion, 61, 80, 108, 187, 264
abstraction (the method of), 58, 59, 65, 103, 111–114, 147, 150, 157, 183–185, 263
  abstracting flexibly and conceptualizing precisely, 151–153
  and category construction, 126, 162–163, 176
  and the search for sociological laws, 58–59
  naming abstractions, 156
  See also conceptual doublets: moments of abstraction; conceptual doublets: moments of conceptualization; historical and structural analysis/relationships; levels of generality; structure; successive abstractions
abstractions, 37–39, 46, 59, 61, 72, 87, 123, 127, 148, 187
  as historically specific, 122, 157
  as variables, 121–142
  general abstractions, 160, 164, 165

history as an abstraction, 86
of determination/determinant abstractions, 161–173, 174, 178, 193
of extension, 111–113, 117, 135
of identity/difference, 153–156
of precondition, 116–118
of presupposition, 117–118
of vantage point, 111. 127, 131, 135
within historical and structural analysis, 115–118, 129, 134, 152, 156
  See also concepts; conceptual doublets: moments of abstraction; determinism (abstractions of determination); essence/essentialism; extension; identity/difference; preconditions; presuppositions; successive abstractions; vantage point
abstractions (inappropriate), 100, 105–107
  abstract system building, 162
  carved on appearances, 174
  carved too narrowly, 140
  "humanity" and/or "man" as an abstraction, 86, 106, 123
  hypostatised abstractions, 122
  "Marxism" as an abstraction, 243
  operational definitions, 121–122
  "society" in general, 9, 59–60, 62–63, 64, 69–70, 82, 89, 120–124, 139–140, 160, 160n3, 163, 163n4, 165, 204, 271
    See also appearance; critique; fallacy of misplaced concreteness; individualism
accidents, 26n10, 34, 38, 91, 180–182
agency, 31, 70–71, 80, 246, 247, 276
  intentional action, 35, 40
    See also consciousness
agrarian societies, 5, 167, 196, 238, 241
  See also production (mode/s of)
algebra
  See mathematics
alienation, 23–24, 32, 74, 90, 93–95, 202, 230, 235–237, 250, 255, 264
Althusser, Louis, 22, 71, 111, 245
analysis, 9, 14, 19–20, 22, 32–33, 35, 37, 42, 43, 51–53, 58, 60, 71, 77–79, 82, 99–101, 104, 110–111, 117–118, 121, 124–125, 129, 133–136, 145–146, 148, 151–152, 155, 159, 165, 169–175, 175n6,

176–179, 182–183, 187, 193, 204, 216, 252, 264, 273, 277
analytical procedures (Marx's), 103–146
and interpretation, 156–159, 160–173, 187
functional, 82
historical, 70
historical materialist, 57
multivariate, 126–135
political-economic, 7, 97, 101, 110, 178
structural, 70
unit of, 40, 60, 70, 168, 179
See also class (analysis); conceptual doublets: epistemological moments; critique (as method of analysis); function(al analysis); historical analysis; historical materialism; historical and structural analysis/relationships; induction; political economy; structure
Analytical Marxism, 35, 39–41, 187
See also philosophy (analytical)
analytical procedures (Marx's)
See analysis (analytical procedures/Marx's)
antagonism (antagonistic relationships), 70, 80
See also class (antagonisms)
anthropological studies, Marx's, 17, 98
antithesis, 38, 39, 50, 110, 176, 236, 251
See also thesis-antithesis-synthesis myth
appearance, 37, 49–53, 62–63, 65, 90, 96–97, 99–101, 119, 129, 145, 174, 190, 196, 209, 218, 262, 278
and essence, 49, 54, 65, 99, 173–175
disjuncture of appearance and essence, 49–53
See also conceptual doublets: moments of conceptualization; essence/essentialism
appropriation (method/mode of), 88, 94, 129–130, 167, 170, 175, 179, 180, 182, 183, 184, 185, 196, 198, 209, 220, 273
of surplus-value, 180, 197, 202n6
of wealth, 202, 230
archaeology, x–xi
See also Foucault
art(s), 39, 77, 78, 268
Asiatic mode of production, 7, 9, 30, 89, 130, 160, 169, 171, 196, 219, 220
See also class; production (mode/s of)

assemblages, 205
See also ensemble
assumptions (Marx's), 15, 38, 50, 63, 67, 77–78, 123–124, 126, 130, 134, 151, 157, 161, 187, 190, 245, 269
centrality of labor, 84–96
essentialism, 96–97
evolution, 98–100
naturalism, 72–74
necessity of a dialectical science, 100–101
onto-epistemological, 69–101
organicism, 97–90
relation of human individuals to their species, 74–76
social universals, 76–84
totality, 71–72
See also base/superstructure; essence/essentialism; evolution; naturalism; organicism; totality
atomism, 106, 107, 109, 110, 273
See also individualism; abstractions/inappropriate

backward, reading concepts, 60–62
reading Marx, 19–23, 67
study of history, 111, 114–118
study of political economy, 142–143
See also teleology
barbarism (fascism), 219, 279
base/superstructure (relations), 76–77, 78n3, 79–84, 85, 90, 113–114, 123, 191–192, 275
asymmetrical relationships, 83
contradictory relationships, 82–83
functional interdependence, 81–82
inverting relationships, 83–84
necessary and contingent relationships, 79–80
power of ideological influences, 80
primacy of material conditions, 78–79
reciprocal relationships, 80–81
See also art(s); education; ideology; law; materialism; morals; politics; production (forces of; means of; mode/s of); religion; scientific dialectics; state/the
Bernstein, Eduard, 242, 246
biology (biological), 26n10, 47, 73, 74, 97, 98
bourgeois, culture, 241
economists, 29n11, 37
historians, 29n11

ideology, 74, 149, 220
production, 20, 107, 108, 114, 215
property relations, 163n4, 230, 236–237
revolution/counterrevolution, 231, 233
society, 19, 30, 52, 54, 65, 74, 160, 168, 196, 212, 222n17, 230, 235
states 235
See also capitalism; society (bourgeois/capitalist/modern); state/the
bourgeoisie (the), 74, 185, 186, 198, 202, 212, 220, 229–230, 234, 235, 250, 259, 268, 280
petit, 181, 185, 230, 231
Bukharin, Nicolai, 10, 249–250
bureaucracy, 109, 175, 177, 198, 216n16, 220, 233n4, 235, 253–254, 271

camera obscura, 83
See also inversion(s)
Capital, as experimental model, 126–143
how Marx put *Capital* together, 20–22
interpreting its volumes, 22–23
Marx's presentation, 142–145
capital, 54, 82, 90, 97, 115, 127n8, 128n9, 131, 133, 145, 149, 179, 180, 197–201, 206, 206nn11–12, 207, 209–214, 218, 256, 266, 276, 278–280
accumulation, 82, 133, 137, 180, 185, 203, 220, 265
as a class relation, 131, 137, 155, 200, 203, 208, 208n13, 210, 214, 266, 273
capital-state relation, 216, 221
constant and variable, 82, 179, 201n3, 202
individuals as incarnations/capital personified, 40, 207, 209
in general, 80, 211, 266
international, 81
in the abstract, 88, 266
logic of, 39
organic composition of, 123, 185, 198, 203, 208n13, 217, 220
primitive accumulation of, 185, 214, 217, 221
social capital, 127n8, 211
stocks of, 265
See also capitalism; political economy
capitalism, as a *differentia specifica*, 133–134
as a system/in general, 23, 40, 113, 126, 130, 168, 170, 182, 185, 215–216, 275

capitalism versus society, 59–62, 64, 69–70, 73–77, 82, 87–96, 101, 110, 113–115, 120–122, 124, 129–130, 130n10, 138–139, 159–160, 160n3, 163n4, 165, 222, 267, 271, 275
capitalist material relations, 97, 124, 175, 185, 199
capitalist mode of production, 7, 9, 10, 30, 39, 44, 57, 80, 88, 90–91, 94, 97, 99, 114, 116, 121, 130, 130n10, 139, 166, 167, 168, 168n5, 169, 171, 175, 176, 182, 185, 191, 196–199, 203n7, 211, 214, 217, 219, 220, 221, 226
central tendencies/laws of motion, 31, 40, 49, 63, 97–99, 114, 130, 138, 139–141, 170, 173, 196–199, 206–209, 211, 213, 215–220, 275
circuit of capital, 97, 205, 208
development, 8, 9, 92, 94, 138, 212, 216, 218, 222, 241
elementary parts/structural components, 21, 147, 168, 198
historical origins, 57, 143
subforms/submodes of, 168–169, 171–173, 186, 207
agricultural, 21, 168–169, 179, 186, 207, 216–217
entrepreneurial/merchant, 168, 216–217, 221, 166
global, 169, 215, 216–217, 221, 222, 266, 276, 278
industrial, 128n9, 169, 170, 179, 208, 210, 216–217, 221, 234, 265, 279
monopoly, 169, 216, 217, 221
transnational, 169, 216–217, 221
See also counteracting forces; laws; political economy; production (mode/s of)
categories, 19, 32, 39, 51, 53, 59–61, 67, 76, 87, 100, 110, 157, 161, 164, 173
arbitrariness, 27, 58, 109, 162
construction of, 28, 99, 111–115, 121–126, 129–130, 143, 146, 150–156, 159–173
economic, 32–33, 37, 101, 107, 148, 160, 163–173, 196, 216, 218–219
ideal, 38
logic and, 38, 40, 123–124, 152, 156
See also conceptual doublets; knowledge (categories of); successive abstractions; typifying
cause (causal forces/properties, causality), 7, 26n10, 34–36, 37, 39–42,

44, 47, 55, 60–61, 80–83, 86–89, 92, 96, 99–101, 104, 110, 116–118, 121–122, 124, 126–128, 131–135, 137–140, 158–159, 161, 176–179, 182, 187, 197, 203–204, 208, 214
central tendencies
  See capitalism (central tendencies/ laws of motion); counteracting forces; laws
centrality of labor
  See labor (centrality of)
change within and between systems
  See models (change within and between systems)
Christianity
  See religion
civil society, 71, 154, 176, 180, 235
clarification, 4, 11, 39, 108, 118, 127, 185, 188
  self-clarification, 105, 110–111, 143, 264
class, 8, 29, 29n11, 34, 48, 78, 91, 92, 150, 168, 185, 210, 233–236, 236n6, 252, 264, 268
  analysis, 6, 7, 62, 91–100, 110, 114–116, 125, 129, 167, 170–172, 178–188
  antagonisms, 30, 53, 63, 90, 94, 207–208, 232
  class-state relationship, 9, 184, 185, 203n7, 261, 278
  dynamics, 77, 80, 88–89, 94, 180, 192–195, 219
  exploitation, 95, 180, 236, 261
  history/societies/systems, 6, 7, 10, 32, 40, 63, 77, 82–83, 88, 90, 94, 95, 113–115, 125, 129–130, 156, 163, 166–173, 182, 196, 198, 214, 261, 273–275
  interests, 58, 94, 109, 230
  relations, 7, 40–41, 58, 73, 83, 86, 90–92, 94–95, 129, 133, 150, 180, 214, 234, 265–266, 273
  structure, 74, 86, 129, 192, 227–228, 246, 265–66
  struggle, 13, 29, 29n11, 31, 34, 53, 62–64, 77, 88, 94, 97–98, 113, 129, 179–183, 187, 194–195, 202, 216–217, 220–221, 227, 230, 232, 239, 252, 266–267, 273, 279
  See also Asiatic mode of production; capitalism; classes; feudalism; exploitation; production (mode/s of); tribal society; slavery; state/the
classes, capitalist, 80, 214, 234, 241, 265, 266, 268
  intellectual, 241, 280

  managerial, 96, 181, 217, 221, 266
  middle, 231, 233
  owning, 86, 96
  ruling, 5n2, 180, 220, 233, 250, 261, 277, 280
  working, 8, 12n8, 22, 23, 80, 120, 203, 228, 229, 232–233, 252, 262, 263, 265–267, 279–280
  See also bourgeoisie; capital; production (mode/s of); proletariat
commensurable (incommensurable), 5, 13, 62, 64, 113, 150, 173, 187, 204, 251, 265
commodity (commodification), 57, 81–82, 90, 92, 94–97, 100, 122–123, 129, 131–132, 134, 137–139, 155, 158, 172, 175, 177–180, 182, 191, 192, 196, 198–200, 202, 204–205, 207–208, 210, 213, 217, 220–221, 273
  fetishism, 97, 175, 273
communism, 5, 6, 10, 20, 63, 67, 99, 219, 221, 227
  as a metaphysical inevitability, 31–32, 42–43, 238, 238n7, 240–249
  communal ownership/primitive communism, 77, 89, 94, 98–99, 160, 171, 196, 219, 239
  communalism in scientific values, 48–49
  communalism in the human animal, 273
  See also communist project; Diamat; Soviet Union
communist project (Marx's), 8–11, 13–15, 22, 91, 109, 114, 150, 215, 227–257
  and spatial relations, 267–268
  dictatorship of the proletariat, 219, 233–234, 240, 248, 250
  early communist society, 228, 235–236
  flexibility and precision and, 228–237, 252, 256, 266–267, 280
  idealism and, 268
  inequality in communism, 236n6
  intellectuals' role, 231–232
  long-term communist development, 14, 188, 228–229, 236–237
  necessary and contingent relationships in, 228–229, 230, 232–233, 236, 239, 266–267, 272, 275–276, 278
  non-elitist nature of revolutionary change, 231–232, 232n3
  preconditions of, 227, 239, 241
  presuppositions of, 227, 239
  revolutionary program, 228–232
  science and, 235, 256, 263, 269–270, 280–281

Index • 309

socialist transitionary period, 188, 228, 232–235
   socialism and the transition from capitalism, 230–232, 277–281
   socialism and the transition to communism, 219, 229, 232–235, 255, 275
   transformation of Marx's communism into Soviet Diamat, 237–256
   See also revolution (worker/proletarian); socialism
comparative method (the), 47, 124, 126–135, 143, 146, 165–173, 177, 182–186, 195–199, 202, 215–216, 273
   experimental model in structural analysis, 130–134
   in historical analysis, 128–130
   in the synthesis of historical and structural analysis, 134–135
   See also experiment
compelling (forces, external coercion), 34–35, 40, 51, 56–58, 73, 79, 92–96, 98–99, 176, 198, 217
Comte, Auguste, 46, 63, 120
concepts, 4–5, 7, 25, 26n10, 27, 39, 51, 54–55, 58, 60, 67, 100, 105, 106, 111, 121, 133, 136, 207–209, 247, 248, 264, 268
   conceptualizing precisely (and abstracting flexibly), 151–153
   construction and development, 3, 15, 23, 36, 40, 53, 56, 60–62, 64, 67, 71–72, 75–77, 79, 89, 100–101, 107, 111–118, 121–126, 131, 133–135, 142, 189–190, 193, 196, 199–200, 209, 218, 223, 272
   material basis of, 28, 79, 84, 87–89, 92, 96, 108–109, 116, 122, 125–126, 128–129, 138
   through realist and internal relations philosophies of science, 148–156
   See also a priori; abstraction/method of; abstractions; conceptual doublets; language (conversion terms); Dietzgen; duality; flexibility; historical materialism; identity/difference; language; precision; typifying
conceptual doublets: epistemological moments, 156–159
   See also analysis; description; epistemology (Marx's); explanation; inquiry; moment(s) (of inquiry); observation; presentation

conceptual doublets: moments of abstraction, 156, 159–173
   See also abstract/the; abstraction/method of; concepts (construction and development); concrete; duality; flexibility; general; identity/difference; language; precision; specific/specificity; successive abstractions; typifying
conceptual doublets: moments of conceptualization, 156, 173–178
   See also appearance; content; contingency; essence/essentialism; form(s); history; necessity; processes; quantity and quality; relations; structure; time and space; wholes and parts
concrete, 26, 27, 39, 42, 53, 57, 60, 71, 76, 80, 84, 87, 88, 99, 106, 110, 114–116, 118, 122, 124, 125, 130–133, 138, 140, 142, 143–144, 148, 150, 152–153, 157, 159, 160–173, 176–177, 183, 188, 192, 193, 197, 199, 200, 202, 207, 209, 210, 211, 218, 220, 222, 241n9, 264, 266, 274
   See also abstract/the; successive abstractions (the general concrete; the specific concrete)
conflate (conflating/conflation), 60, 64, 75, 113, 134, 160
conflation of Marx's moments of inquiry, 5–14, 44, 240, 243, 252
congeal (congelation), 52, 205, 206
congealed expression, 87
consciousness, 34, 40, 49, 50, 57–58, 62, 72, 79, 246
   class consciousness, 267
   false consciousness, 5n2
   See also freedom; ideology
content, 26, 27, 43, 51, 62, 75, 76, 113, 153, 156, 163, 173–174, 177, 177n7, 178, 215, 236n6, 237, 262, 265, 267
   See also conceptual doublets: moments of conceptualization; form(s)
contingency (contingent), 8, 13–14, 19, 26n10, 59, 63, 70, 78, 79, 80, 123, 144, 174, 178, 179–186, 191, 197, 210, 218, 228, 272, 275, 279
   historical events, 79, 116–117, 140, 193, 180–186, 193, 195, 228, 275, 279
   See also conceptual doublets: moments of conceptualization; relations (necessary/contingent)
contradiction (contradictory), 6, 266

310 • Index

between productive forces and
  relations, 7, 12, 13, 30, 97, 195
  in capitalism, 23, 95, 100, 138, 180,
    208–209, 214–215, 276
  in dialectical thought, 21, 28, 38,
    42n14, 108, 149, 152, 191, 273
  in material life, 78, 78n18, 82–84, 176,
    194–195, 227, 253
  in moment of critique, 28, 105, 121
  in moment of analysis, 191
control variable
  See variables
  See also comparative method
conversion terms
  See language
corporations, 32, 216n16, 216–217, 278
counteracting forces (countervailing
  tendencies), 128, 140, 197–199,
  213–215, 222
crises, 13, 128n9, 195, 208, 213–217, 219,
  220, 227, 230, 233, 263, 273, 279
critical realism
  See philosophy (realism)
critical theory, 39
  See also Frankfurt School
critique (Marx's), 15, 21–22, 35, 49,
  96–97, 105–111, 112, 123, 138, 144, 148,
  157, 159, 174, 261, 273
  as method of analysis, 104–111
  as self-clarification, 105, 110, 111, 143
  negative critique, 104, 110, 143, 273
  of Feuerbach, 106–108, 110
  of Hegel, 9, 20, 24, 37–38, 50, 83,
    105–108, 110, 122, 177n7, 182
  of idealism, 6, 7, 25, 27, 32, 34–35,
    37, 38, 40, 52, 72, 79, 80, 84–85, 97,
    98–99n5, 104, 106–110, 122, 148,
    273
  of Kant, 20, 27–28, 37–38, 105–107, 110
  of metaphysics, 5, 6, 15, 27, 31, 37–39,
    41–43, 86–87, 98, 98–99n5, 107–109,
    110, 158, 245, 273
  of mysticism/mystification, 5, 15,
    37–39, 107, 177n7
  of obscurantism, 5, 15, 39–41, 43–44
  of political economy, 19, 21–22, 28,
    37–38, 74, 107, 109, 110, 138, 143, 149
  of popular knowledge, 52, 83–84
  of positivist sociology, 120–123
  of Proudhon, 32–33, 38, 87, 106–107
  of Smith and Ricardo, 107, 110
  of speculative philosophy, 5, 6, 15,
    37–39, 41, 72, 79, 103, 106–107, 109,
    110, 122, 270, 273
  political critique, 228, 231, 239n8, 255

positive critique, 38, 107–109, 110,
  143, 273
religious, 76
self-critique, 39, 109
true criticism, 104
vulgar criticism, 104
See also individualism (Marx's
  critique of)
crystallizations (crystallized), 48, 55, 59,
  61, 88, 153, 195, 207, 210, 216

Darwin, Charles, 29, 98, 98n4, 119, 139,
  243, 248, 260
Darwinian principles, 98–99
See also evolution
deduction (deduce/deductive), 8, 16,
  24n9, 54, 114, 120–121, 122, 152, 158,
  190, 244, 249
  in analysis, reasoning, and formulating
    laws, 114, 132, 136, 138–139, 142,
    161, 204
  See also induction
democracy, 83, 192, 233, 241, 248, 250,
  252, 267, 272, 275–276, 280
demystification (demystified/demystify),
  41, 90, 96, 105
dependent variable
  See variables
  See also comparative method
description (described/describing), 35,
  41, 54, 71, 73n2, 114, 120, 122, 124,
  126, 136–137, 143, 145, 162, 178, 189,
  204, 206, 207, 215, 218, 269
  and explanation, 15, 67, 55, 121, 124, 141,
    148, 156, 158–159, 164, 173–174, 178,
    183–188, 190–191, 195, 197–198, 216–222
  See also concepts; conceptual
    doublets: epistemological moments;
    language; models
determinism (determination), 8, 29–33,
  34–45, 56–58, 63, 70, 77, 80–81, 83–84,
  88–89, 91, 97, 98–99n5, 99, 117, 122,
  125, 135, 137–140, 148, 157, 159, 161,
  175, 177, 187, 191, 194, 201n3, 202, 205,
  237–238, 239, 244, 245, 266, 267, 271
  abstractions of determination/
    determinate abstractions, 161–173,
    178, 193
  forms of, 178–186, 193–195
  modes of, 192–195
  rigid, 11, 29, 31–35, 223
    economic, 4, 34, 62–63, 80, 240, 247, 268
    materialist, 33–34
    technological, 34
  See also successive abstractions

Index • 311

development, 19–20, 25, 27, 33, 34, 48, 60–61, 74, 75–76, 78n3, 82–83, 88, 91–92, 93, 97, 99, 116, 125n6, 132, 143, 154, 157, 159, 177, 195, 197, 199, 238n7, 249, 252
  and concepts and methods, 15, 19–20, 22, 33, 37, 44, 53, 67, 88, 131–132, 135, 144–145, 158, 160, 163, 166, 170, 173, 176, 180, 187–188, 189, 215–216, 265
  capitalist, 8, 9, 59, 81, 83, 92, 94, 113, 138, 143, 175, 176, 179, 182, 190, 196, 210, 212, 216–218, 222, 239, 240, 241
  historical, 10, 19–20, 32, 34, 57, 69, 74, 77, 81, 84, 86–88, 99–100, 108, 117–118, 127–128, 130–132, 142, 170, 176, 181, 190, 191, 196, 202, 218, 243, 245, 248, 273
  of material/productive forces, 29–30, 37, 75, 78–79, 84, 87, 92, 107, 139, 188, 214
  spiral form of, 211
  uneven, 77, 92, 93, 100, 229
  See also time and space
"dialectical materialism"/Marxism as world philosophy, 5, 7, 10, 14, 29, 41, 42, 240, 242–243, 247, 249–254
  See also Diamat; Lenin; Plekhanov; Soviet Union
dialectical method (the), 6n3, 6–11, 13, 14, 24–25, 83, 103–105, 111, 123, 142, 145–146, 151, 225, 240, 245, 273–274
  and scientific method, 118–145
  Marx's scientific dialectic, 67–223
    onto-epistemological assumptions, 69–101
    analytical procedures, 103–146
    conceptual doublets, 147–188
    models, 189–223
  necessity of a dialectical science, 100–101
  See also dialectics; scientific dialectics
dialectics (dialectic/dialectical), 3–4, 7, 8n4, 13, 15, 20–21, 23, 24–25, 28, 29, 33, 34–36, 38–39, 41–44, 52, 67, 78, 82, 84, 85, 87, 96, 98, 105–106, 110, 145, 147, 149, 149n1, 151, 153, 158, 187, 205, 211, 211n14, 227, 239, 240, 243–246, 248–255, 261, 265, 268–271
  Hegelian/idealist dialectic, 6–7, 19, 20, 25, 31, 32, 35, 37–38, 41–43 (42, n14), 43, n15, 50, 52, 75–76, 83, 105–108, 110, 122, 126–127, 145, 154, 158, 176, 178, 191, 211, 211n14, 244, 251, 269
  movements of negativity and qualitative change, 36, 108, 152, 207, 211–215, 273
  "the dialectic", 35, 41–42, 185, 240, 249
  See also dialectical method; Diamat; Engels; scientific dialectics
Diamat, 237, 253
  See also "dialectical materialism"; Soviet Union
dictatorship of the proletariat
  See communist project
Dietzgen, Joseph, 28, 151
division of labor, 28, 32–33, 47, 60–61, 87–91, 92–94, 108, 113, 122, 130n10, 175, 179, 201n4, 217, 273
  See also alienation; class; exploitation; labor; labor (centrality of)
dogma (doctrine/doctrinaire), 32, 34, 36–37 (36, n13), 53–54, 56, 98n4, 104, 108, 109, 118, 151, 158, 234, 243, 246–247, 240–241, 250–255, 259, 261, 268, 269, 272, 279
  See also critique
duality (dualism), 129, 154–156, 159, 162, 188, 263
  abstract dualism, 154
  Cartesian dualism, 251
  double character, 155
  double movement/process, 94, 153, 155, 176, 179
  false dualism, 106
  mystical dualism, 38
  ontological duality, 153
  See also concepts; conceptual doublets; historical and structural analysis/relationships; identity/difference; quantity and quality; successive abstractions
Durkheim, Emile, 31, 45–56, 58, 60–65

economic determinism
  See determinism (rigid)
education, 22, 31, 75, 77, 82–83, 96, 231, 234, 261, 263, 266, 268, 280
embodiment, 37, 58, 97, 207
embryonic, form, 61
  growth, 97
emergence (emergent), 7, 63, 70, 168, 181, 188, 194–196, 216–219, 227, 229, 252, 273, 275
empirical, reality (data, referents, regularities, representations), 26–27, 57, 60, 69, 73, 73n2, 96, 116, 119, 122, 124–125, 131, 134, 140, 162–163, 165–168, 179, 222, 267

research (demonstrations, inquiry), 22, 32, 39, 51, 55, 70, 79, 98, 104, 107, 110, 111, 116, 120, 122–123, 132, 138–139, 142, 151, 157–159 (158, n2), 164–169, 188, 191, 193, 197, 216, 222, 263, 267, 272
  See also inquiry; observation
Engels, Frederick, 4, 7, 16, 19, 28, 31, 37, 44, 80, 98, 105, 111, 136, 138, 145, 181, 189, 205, 228, 231, 240, 246, 255, 262, 267, 268
  and dialectic, 23–25, 41, 239–242, 242 nn10–11, 243–245, 245n13, 247, 249–253
  and historical materialism, 7, 15, 15n7, 16, 24n9, 31–32, 34–35, 39, 60, 63–64, 71, 79–81, 83, 86, 179–180
  and the invention of "Marxism", 242–244, 255
  as editor and influence on interpretations of Marx, 5, 15, 21, 23–25
  on Marx's terms, 124, 149–150, 153
ensemble (of social relations),
  individuals as, 74
  social wholes as, 153
  See also assemblages
epistemology (epistemological), 67, 136, 251
  See also ontology; epistemology/Marx's
epistemology (Marx's), 3, 8, 28, 36, 42, 69–73, 75, 86, 91, 130, 147, 148, 151, 156–159, 190, 193, 223, 272
  and classical theory, 47–65
  dialectical method and scientific method, 118–146
  the order and logic of investigation, 103–111
  See also abstraction/method of; analysis; backward (study of history); comparative method; concepts; conceptual doublets: epistemological moments; critique; deduction; determinism; dialectical method; evolution; experiment; historical analysis; historical and structural analysis/relationships; laws; models; moments(s); onto-epistemology; statistics; structure; successive abstractions; typifying
epoch(s)
  See periodization
equality
  See communism
era(s)
  See periodization

essence/essentialism (characteristics, parts, relations, structures), 37, 64, 70, 74, 76, 77, 89, 96–97, 113–116, 121, 125–126, 129–130, 138–142, 149, 150, 151, 154, 158, 160–162, 174–175, 181, 190, 196, 198–199, 206n12, 209, 219, 233n4), 242, 265, 267
  and appearance, 49, 50–53, 65, 96–97, 99, 173–175
  dual-essence, 154
  See also appearance; conceptual doublets: moments of conceptualization
evolution, social, 52, 58
  Engels's dialectic and, 244
  Marx's theoretical frameworks/models, 29, 49, 53, 60, 73n2, 88, 96–100, 116, 122, 166, 181, 194–196, 218, 239, 273
  misinterpreting Marx's as a linear evolutionist theory, 29–33, 42n4, 63, 238, 238n7, 239, 240, 276
  natural, 29, 60
  Soviet use of, 248, 250
  See also Darwin; metaphysics; models (change within and between systems); teleology
exchange-value
  See value
excluded middle (the), 152, 156
  See also philosophy (analytical; of science)
experiment (experimental model/experiments), 47, 49, 126–135, 273
  from the vantage point of structural analysis, 130–134
  See also comparative method
explanation (explain/explanatory models), 15, 26, 28, 29, 30, 32, 34–35, 40–41, 51, 56, 59–61, 67, 71–72, 74–76, 79, 82, 94, 96–98, 101, 111, 114–117, 120–146, 147–149, 149n12), 156–158, 161–162, 164, 167, 173, 174, 176, 178–181, 185, 187–193, 203, 210, 218, 222, 263
  See also analysis (multivariate); comparative method; conceptual doublets: epistemological moments; description (and explanation); determinism; experiment; models
exploitation (class, of labor, of labor-power), 64, 88, 93, 95, 115, 126, 133, 137, 168, 180, 183, 186, 189–190, 197–198, 200, 201n3, 201–202, 209, 214, 220, 230, 236, 237, 261, 265, 267, 273

express (expresses/expressions), 27, 30, 39, 49, 53, 56, 58, 75, 77–78, 78n3, 87, 88, 90, 93, 96, 98–99n5, 99, 104, 105, 106, 107, 115, 125, 129, 130, 133, 139, 148, 149, 149n1, 153, 154, 155, 160, 166, 168, 175, 178, 179, 180, 187, 190, 207, 208, 211, 215, 230, 237, 238
expropriation, 185, 186, 214–215, 215n15, 234
extension (extend/extends), 13, 57, 63, 71, 80, 83, 100, 101, 107, 112, 118, 121, 124, 126, 128–130, 140–142, 148–149, 149n1, 156, 163, 166–167, 176, 177, 181, 192, 203, 204, 206, 208, 208n13, 214, 228, 253, 266, 267, 272, 281
See also abstractions (of extension)

fallacy of misplaced concreteness, 122
See also philosophy (of science)
falling rate of profit
See profit (tendency for rate of profit to fall)
false consciousness
See consciousness false
family, the, 48, 74, 99, 177, 181, 192, 194
fetish form
See commodity (fetishism)
feudalism, 7, 30, 32, 57, 63, 74, 88, 89, 107, 118, 129, 130, 160, 167, 169, 171, 180, 182, 184, 192, 196, 214, 217, 219, 220, 229, 238n7
See also class; production (mode/s of)
Feuerbach, Ludwig, 7, 15, 20, 25, 31, 64, 84–85, 106–108, 110
See also critique (of Feuerbach)
flexibility (flexible), 31, 33, 127, 151, 163, 166, 191, 222, 267
   in abstraction and concepts, 150–153, 156, 164, 166, 191
   flexible precision/precise flexibility, 151–153, 164, 178, 185, 188, 228, 280
   See also abstraction/method of; communist project; precision; successive abstractions
forces of production
See production (forces of)
form(s), 7, 27, 33, 40, 43, 46, 50, 51, 53, 55, 62, 63, 67, 71, 74, 75, 76, 77, 78, 79, 81, 83, 84, 85, 88, 89, 90, 93, 95, 96, 97, 98, 99, 100, 104, 107, 108, 109, 110, 113, 115, 116, 117, 118, 122, 123n5, 132, 141, 151, 153, 155, 157, 160, 163, 164, 166, 168, 171, 172, 173, 175, 177, 177n2, 178, 180, 183, 185, 189, 191, 192, 202, 203, 205, 206, 211, 215, 216, 218, 219, 222, 239, 261, 267, 271
   antagonistic forms, 30
   economic form(s), 127, 129
   embryonic form(s), 61
   emergent form(s), 194
   emergent forms of capitalism, 216
   historical form(s), 9, 19, 28, 61, 114, 129, 167, 200
   institutional form(s), 73, 113, 123, 177, 192, 194
      commodity, equivalent, and money forms, 81, 97, 100, 129, 138, 177, 178, 196, 204
   of being/existence, 98n5, 160
   of development, 30
   of discourse/ideology/knowledge/reasoning/thought, 26, 26n10, 30, 39, 52, 60, 73n2, 80, 84, 90, 92, 101, 106, 111, 128, 129, 148, 160, 190, 220–221, 229, 277–278
   of governance/state-form, 92, 115, 192, 220–221
   of individuality, 94
   of inquiry, 22, 84, 101, 119
   political forms of class struggle, 34, 37
   See also conceptual doublets: moments of conceptualization; content; determinism (forms of); development; production (mode/s of); property; value
Foucault, Michel, x–xi, 9n5, 148, 158n2, 175n6, 243n12
Fourier, Charles, 5, 227, 232, 234
Frankfurt School, 39
freedom, 23, 74, 93–95
   and bourgeois ideology, 272
   and communism, 231, 236–238, 241, 255, 276
   and dialectical method, 151, 187–188
French Revolution, 39, 229, 233n4
function (functions/functional analysis), 47, 52, 60–63, 70, 75, 78, 81–82, 95, 101, 133, 140, 141–142, 152, 155, 169, 178, 179–180, 185, 191–194, 203, 206, 208, 209, 210–211, 213, 214, 229, 236, 249
   differences between Marx's and traditional functional analysis, 82
   See also determinism (modes of determination); limits (of functional compatibility)

general abstract
    See successive abstractions (the general abstract)
general concrete
    See successive abstractions (the general concrete)
general, 98, 99, 99n5, 144, 155, 157, 159, 161, 164, 166–170, 180, 182–183, 207, 215
    categories/types, 26, 125, 130, 160
    conditions, 58
    rule(s)/law(s), 25, 59, 80, 90, 118, 120, 140, 156, 162, 187, 197, 202, 202n6, 210, 213, 214, 243–244
    systemic relationships, 207
    theories, 251
        and the materialist conception of history, 29–33, 42, 45, 63, 73, 128
        and dialectic, 25, 35, 247–248
        Marx's general theory, 272–275
        and political economy, 21, 218
    See also generality; generalization(s); laws; levels of generality; models; production ("production in general"); successive abstractions (the general abstract; the general concrete)
generality (generalities), abstract 106, 241
    and explanation, 75, 135, 245
    See also general; generalization(s); levels of generality; society
generalization(s), 62, 76, 118, 128–130, 139, 140–142, 162, 204
    See also extension
god(s), 27, 37, 90, 104, 105
Gould, Carol, 71, 88–91, 122

Hegel, Georg, 6, 7, 19, 20, 24–25, 31–32, 35, 37–38, 41–42, 42n14, 43n15, 50, 52, 75, 83, 98–99n5, 105–108, 110, 120, 122, 126, 145, 154, 158, 176, 177n7, 178, 182, 191, 200n2, 211, 211n14, 223, 244, 249, 250–251, 254–255, 269
    See also critique (of Hegel)
historical analysis, 11, 12n6, 19–20, 21, 23, 25, 31, 33–35, 40, 48, 49, 53, 57, 58, 60–63, 84, 85, 90, 98–99n5, 101, 106, 108, 110, 114–115, 119, 122–123, 124–125, 126–129, 138–139, 143, 148, 157, 159–160, 168, 176, 183–186, 202, 215–216, 222, 265, 273
    ahistoricism, 7, 31, 60, 107–107, 110, 120, 134, 140, 150, 165, 269, 273
    historical change/development/evolution/movement/processes, 3, 10, 19–20, 29, 29n11, 53, 55, 74, 76, 86, 89, 91–92, 93–96, 97–99, 127–128, 170, 176, 190–191, 202, 243, 244
    history as a variable, 86–89
    See also analysis; backward (study of history); class (history); comparative method; conceptual doublets: moments of conceptualization; contingency; evolution; form(s) (historical); historical and structural analysis/relationships; historical materialism; labor; labor (centrality of); laws (and history); levels of generality; necessity; periodization; processes; progress; relations; structure; teleology
historical and structural analysis/relationships, 22, 33, 35, 56–64, 69–70, 84, 97, 101, 103, 111–118, 123–131, 134–135, 141–143, 148, 151–153, 156–157, 158n2, 159–160, 162–165, 168–170, 174, 176–177, 178–186, 188, 189–199, 202, 204, 209–223, 273–274
    See also analysis; comparative method; conceptual doublets: moments of conceptualization; contingency; internal relations; levels of generality; models (change within and between systems); necessity; structure; successive abstractions; typifying
historical materialism (principles, variables), 6–9, 10, 11, 12, 13–15, 16, 22, 24, 29–30, 34, 39, 45, 52, 71–73, 73n2, 74–78, 78n3, 79–101, 105, 106–107, 113, 114, 121, 227, 271, 273, 275
    "materialist conception of history", 7, 15, 16, 24n9, 31, 34, 45, 242, 271
    See also analysis; base/superstructure; comparative method; determinism; labor; labor (centrality of); levels of generality; political economy; precision (over- and under-precise readings of historical materialist principles); Preface; production; revolution; typifying
history, 7, 8, 10, 11, 12, 14, 19, 20, 29, 38, 41, 42, 55, 56, 67, 79, 91–93, 100, 150, 168, 223, 240, 243, 244, 260, 272, 275, 279
    economic history, 33, 153, 160
    teleological view of, 4
    in Soviet society, 247–251, 276

world history, 222
See also class; historical analysis; historical and structural analysis/relationships; historical materialism; levels of generality; Preface (traditional readings); structure; teleology
human nature (human essence), 73–76, 86, 89, 90, 107, 134, 154, 174, 237, 263, 278
See also individual(s)
humanism, 228, 255
hypothesis (hypotheses), 7, 32, 72, 121, 138–139, 216, 271

idealism, 19, 24–28, 38, 40–44, 50–51, 54–57, 62, 77–80, 97, 98–99n5, 101, 110, 114, 124, 132, 136, 148, 162, 250
See also critique (of Hegel; of idealism; of Kant); Hegel; philosophy (speculative philosophy)
identity (identities), 27, 131, 154–156, 175, 188, 204, 206, 208, 211n14, 238, 265, 266
false identity, 106, 154, 156, 160
See also identity/difference
identity/difference, 131, 153–156, 160, 166–169, 171, 173, 177, 188, 199, 265
and concept construction, 162–164
essential identity/difference, 27, 129, 154–155, 161, 166–168, 206
See also determinism; duality; quantity and quality; successive abstractions
ideology (ideological, ideological superstructure), 5n2, 7, 12, 14, 30, 35, 36, 36n13, 54, 73, 73n2, 74, 77, 78, 80–81, 83–84, 92, 114, 149, 220, 221, 235, 237, 239, 242n11, 245, 250–251, 253–254, 262–263, 271, 276, 277
See also base/superstructure
incarnation, 37, 207
independent variable
See variables
individual(s), 30, 35, 65, 72–73, 73n2, 74–75, 82, 84–87, 89–91, 101, 113, 114, 133, 159, 160n3, 183, 201
individuals and the human species, 74–76
individuals in communism, 235, 236n6, 237
See also agency; consciousness; freedom; individualism; labor; labor (centrality of); levels of generality; supra-individual reality

individualism (abstract individualism, individualistic reductionism, methodological individualism), 40–41, 72–76, 84–85, 107, 157, 266
Durkheim's critique of, 55–56, 62
ideology of, 220
Marx's critique of, 40–41, 54, 57–58, 72, 74–76, 79, 85–86, 146, 148, 150, 154, 157
Simmel's critique of, 50, 56–57
See also critique (of idealism); human nature; reduction(ism)
induction (inductive analysis/method), 54–58, 65, 114, 142, 152
See also analysis; concepts; deduction; successive abstractions
industry, 32, 57, 133, 165, 169, 176, 186, 203n7, 206, 208, 208n13, 234, 246, 278, 280
inner-connections ("inner connexion"/interconnection), 34, 50, 118, 123, 139, 152, 157, 161–162, 166, 175, 188, 200
between history and system, 84, 117, 127, 134–135, 163, 222, 222n6
between Marx's various moments of inquiry, 7–9, 272–275
inner-structure, 135, 168, 199
See also historical and structural analysis/relationships; internal relations; successive abstractions
inquiry, 4, 25, 28, 38–39, 48–49, 52, 53–59, 65, 84, 86, 93, 96, 101, 104, 111–113, 119, 120, 122, 127, 138, 142, 187, 204, 228, 241, 251, 256, 265
and analysis, 22, 32, 121, 153, 159, 170–173, 175, 179, 182, 193, 205
and observation, 162, 170, 175, 179, 187
and presentation, 143, 156–157, 159, 184
moments of, 5, 6–15, 22, 25, 44, 91, 100, 155, 179, 218, 272–281
order of, 118, 143, 159
See also analysis; backward (study of history); categories; clarification; conceptual doublets; epistemological moments; critique; dialectical method; historical analysis; historical and structural analysis/relationships; historical materialism; moment(s) (of inquiry); observation; political economy; structure
institution(s) (institutional frameworks, forms), 39, 50, 55, 58, 61–62, 64, 70, 73–74, 82, 86, 108, 112–113, 123, 133,

141–142, 157, 162, 177, 188, 191–192, 194, 210, 216, 238, 262, 263
See also structure; wholes and parts
interiorizing (interiorize), 5, 49, 69, 70, 77, 85, 150
internal relations, 36, 63, 81, 86, 94, 96, 99, 112, 125n6, 138–139, 150, 162, 180, 187, 205, 222
  and concept formation, 148–156, 162–164
  and the force of abstraction, 69–72
  between Marx's science and his politics, 149
  between the specific abstract and the general concrete, 165–166
  in society and the disjuncture of appearance and essence, 49–54, 141–142
  ontology and philosophy of science, 49, 71, 101, 121–122, 124–125, 135, 147, 150–152, 223
  See also abstraction/method of; epistemology (Marx's); flexibility; Ollman; onto-epistemology; ontology; philosophy (realism); precision; processes; relations; scientific dialectics; successive abstractions; totality; wholes and parts
interpenetration, 234
  of opposites, 25, 244
interpretation (interpreting), 54, 55, 64, 77, 78, 104, 133, 151, 153, 158–159, 164, 170–173, 176, 182, 193, 215, 216
  analysis and, 133, 156–157, 159–162, 169, 171, 173, 174, 182, 187, 222
  See also analysis; conceptual doublets: epistemological moments; Marx (reading/problems reading); successive abstractions
interrelations (interrelationships), 71–72, 79, 81, 111–116, 124, 127–128, 130, 141–142, 147, 148, 151–152, 162, 175, 223
  See also historical and structural analysis/relationships; inner-connections; internal relations; moment(s) (of inquiry); onto-epistemology; wholes and parts
inversion(s) (invert/inverted), 35, 37, 61, 78, 83–84, 114, 149, 184, 207–208, 273
  Hegel's inversion, 83, 106, 158
  See also critique (of Hegel; of idealism; of Proudhon); idealism

Judaism
  See religion

Kant, Immanuel (Kantian), 20, 25–26, 26n10, 27–29, 31, 37–38, 41, 56, 92, 105–107, 122, 185, 244, 246, 251, 260
  See also a priori; critique (of Kant); knowledge (categories of); philosophy (speculative philosophy)
Kautsky, Karl, 39, 242, 245, 246, 247, 251, 255
knowledge, 25–28, 39, 46, 48, 53, 69, 70, 75, 77–78, 81, 85, 86, 90, 92, 95–96, 123, 160–161, 229, 237, 244, 249, 250
  categories of, 26n10, 26–27
  common/conventional/everyday/ popular/practical, 51–53, 75, 76, 83–84, 90, 92, 100–101, 106, 128, 134, 147–149, 149n1, 153, 154, 185, 263, 278
  critical/scientific, 51–52, 55–56, 90, 108, 119–120, 123, 138, 147–150, 153, 187, 196, 220, 235, 243, 265, 269
  See also a priori; appearance; ideology; inversion(s); Kant; science

labor (in general), 42, 50, 60–61, 62, 73, 75, 76, 78, 82, 89, 117, 125, 126, 126n7, 180, 183, 186, 198, 199, 204, 210, 217, 220, 280
  abstract labor, 87–89, 92–93, 97, 160, 198, 207, 220
  as the origin of time, 91–93
  capital and, 131, 137, 155, 180–181, 185, 200–201, 201n3, 202–203, 203n8, 205, 205n9, 206, 206n11, 208n13, 209, 213–214, 215n15, 230, 231, 265–267, 273
  centrality of, 84–96, 97, 101, 113, 187
  for human history, 91–92
  for the history of capitalism, 92–93
  for the individual, 93–94
  for the individual under capitalism, 94–96
  for the species, 89–90
  for the species under capitalism, 90–91
  the variability of labor's centrality, 84–89
  in communism, 235–236, 236n6, 237
laborer(s), 117, 129, 132, 265
  powers of determination, 88–89
labor-process, 82, 210
labor-product, 122
slave-labor, 168, 168n5

See also alienation; capital; class; class (struggle); division of labor; exploitation; freedom; labor theory of value; materialism; production; proletariat; value; wage(s)

labor-power, 80, 82, 84, 88, 90, 94, 132, 133, 137, 155, 179, 180, 181, 198, 201, 201n3, 202, 204–206, 207, 209, 214, 220, 267

See also exploitation; labor; labor theory of value

labor theory of value, 29, 109, 110, 264, 273
labor-time, 92, 97, 132, 133
labor-value relation, 87–89, 92–93, 200, 207, 209
socially necessary labor-time, 93, 97, 129, 132–133
unpaid labour/surplus-labor, 88, 91, 93, 126n7, 129, 155, 167–168, 168n5, 169, 198, 203, 210, 220

See also capitalism; labor; labor-power; political economy; value

language, 33, 44, 75, 79, 129, 148
dialectical reason and mathematics, 190–214
analytical terms: qualitative and qualitative comparisons, 203–205
conversion terms, 205–211
dialectical movements of negativity and qualitative change, 211–215
quantitative terms: algebra and statistics, 201–203
quantitative terms: basic mathematical functions, 200–201
linguistic-terminological development and strategy, 22, 25, 39, 56, 124, 137, 146, 148–149, 149n1, 150–156, 163, 179, 182, 187–188, 189–190, 270

See also clarification; concepts; internal relations; philosophy (realism); precision

Lassalle, Ferdinand, 24, 105
law (legal, juridical), 24n9, 27, 29–30, 34, 39, 77, 78n3, 86, 90, 92, 109, 165, 186, 220, 260, 265, 268

See also base/superstructure

laws (scientific, sociological), 25, 31, 46, 48, 50, 58–59, 62, 65, 72, 110, 113, 120, 175, 191, 275
abstract laws, 141, 197, 209, 244
and capitalism, 7, 31, 40, 59, 60, 63, 90, 97, 109, 114, 132, 134, 139, 158–159, 190, 197–199, 203, 203n7, 207, 209–215, 219, 275

and dialectics, 25, 151, 200, 200n2, 211, 211n14, 245
and history, 20, 37, 89–90, 113, 187, 244–246
"eternal laws"/laws of nature, 32, 37, 38, 53, 90, 104, 158, 163n4, 228, 243
"iron laws", 31, 139
laws of motion (in general), 25
Marx's formulation of laws, 138–143, 146, 170–173, 197, 199

See also capitalism (central tendencies/laws of motion); contingency; history; interpenetration (of opposites); metaphysics; models; necessity; negation (of the negation); quantity and quality (transformation of quantity into quality); value (general law of value)

Lenin, Vladimir, 5, 241, 245, 248–254
Leninism (Leninist), 240, 252–254, 272
Leninist-Stalinism, 5, 245, 262

See also Diamat; Soviet Union; vanguard party

levels of generality (historical generality), 9, 12, 40–41, 49, 52, 57, 89, 92, 111–118, 123–130, 132–133, 134–135, 139, 153, 156, 159, 163, 181, 187, 197, 198, 215–216, 272–275
capital/capitalism in general, 80, 130, 141, 168, 177, 180, 185, 188, 211, 215–216, 266
class systems/history in general, 77, 166, 167
society in general, 10, 59–60, 64, 76–77, 89, 120, 122, 130, 139, 149n1, 163, 165, 181, 182, 183

See also backward (study of history); capitalism; class; individual(s); labor (centrality of); nature; society; successive abstractions

Liebknecht, Wilhem, 119, 153, 269
limit(s) (limitations), 26, 26n10, 53, 70, 73, 79–80, 82, 127, 138, 139, 157, 161, 177, 191, 192, 194, 197, 203n7, 213, 215, 218, 239, 265, 266, 273, 278
of functional compatibility, 192–193, 211
structural limitation, 192–193

See also determinism (modes of)

manifestation (manifests), 37, 53, 58, 62, 73n2, 75, 90, 94, 97, 130, 139, 140, 159, 166, 178, 191, 207, 209, 243n12
manufacture, 32, 118, 130n10, 169, 201n4, 205n9, 205–206, 211

318 • Index

Marx, and sociology, 10, 31, 45–59, 59–65, 87, 113, 116, 218, 238, 268, 271, 275
  as prophet, 36, 269
  reading/problems reading, 3–44, 158, 263–265, 272–277
    conflation of his moments of inquiry, 5, 8–9
    Engels's influence, 5, 23–25
    intellectual and scientific roots, 5, 25–31
    moments of inquiry and their interrelations, 6–9
    presentation of his work, 4, 5, 15–23, 142–146, 156–157, 264–265
    political and analytical in moments of inquiry, 11–14
    reading forward/backward, 19–23
    reading historically or logically, 22–23
    under-/over-precision, 5, 31–36
    unpublished work, 15–18
  See also analysis; assumptions/Marx's; Capital; class; communist project; concepts; conceptual doublets; critique; determinism; dialectical method; Engels; epistemology (Marx's); historical materialism; metaphysics; models; mystification; obscurantism; onto-epistemology; political economy; precision; Preface; philosophy (speculative philosophy)
Marx (works), Capital, Volume I, 1, 3, 6, 6–7n3, 16, 18, 20, 21, 22, 23, 24, 31, 35–36, 41, 43, 45, 46, 52, 53, 57, 97, 105, 109, 111, 112, 119, 121, 122, 123, 124, 125n6, 131, 134, 136, 138, 139, 144, 145, 149n1, 158, 162, 165, 182, 187, 189, 190, 215, 260, 265, 269
  Capital, Volume II, 18, 21, 22, 23, 24, 105, 112, 134, 145, 155
  Capital, Volume III, 18, 21, 22, 23, 24, 112, 134
  Civil War in France, 18
  Class Struggles in France, 18, 124
  Contribution to the Critique of Hegel's Philosophy of Law, 17, 105, 110
  Critique of Hegel's Philosophy of Law: Introduction, 17
  Contribution to the Critique of Political Economy, 3, 7, 16, 18, 19, 21, 21n8, 29, 33, 50, 110, 112, 114
  Critique of the Gotha Programme, 18, 255
  doctoral thesis, 17, 24, 72
  Economic and Philosophical Manuscripts of 1844, 15, 17, 22, 23, 24, 109, 110
  Eighteenth Brumaire of Louise Bonaparte, 18, 233n4
  The German Ideology, 15, 17, 23, 45, 83, 110
  Grundrisse, 3, 15, 18, 21, 23, 24, 33, 46, 110, 112
  The Holy Family, 17, 23, 37, 110
  The Communist Manifesto, 17, 23, 29, 31, 35, 45, 181, 231, 234, 234n5, 281
  The Poverty of Philosophy, 17, 33, 38, 106, 110
  Theories of Surplus-Value (generally), 18, 21, 22, 110, 215
  Theses on Feuerbach, 15, 15n7, 17, 84–85, 110
  See also Preface
Marxism, 4–5, 11, 13–14, 36, 41, 46, 187, 238, 242, 242n11, 262, 271, 279
  as religion, 36, 36n13
  in Russia/Soviet Union, 237–256
materialism (materialist), 5, 11, 20, 24, 24n9, 25–31, 35, 38, 53, 57–58, 59, 61–63, 65, 67, 71–74, 77–89, 96–97, 100–101, 107, 124–126, 128, 136, 139, 148, 150, 152–153, 159–160, 162–163, 174–175, 185, 187–188, 192, 199, 206, 214, 223, 227, 238, 264, 265, 268, 269, 271
  and the communist project, 228, 232–233, 239, 264, 273
  mechanistic materialism, 240, 249–254
  primacy of material relationships, 7, 78–80, 101
  See also base/superstructure; critique (of Feuerbach; of Hegel; of idealism; of Kant); "dialectical materialism"; Diamat; historical materialism; idealism; labor; labor (centrality of); models; moment(s) (of inquiry); precision; production; Marx/works (The German Ideology; Theses on Feuerbach)
mathematics (mathematical), 17, 29n12, 84, 109, 119, 119n4, 136–139, 189–191, 199, 205, 213, 223, 273
  addition and subtraction, 200
  algebra, 119, 123n5, 136, 201
  average(s), 140–141, 203, 203n8, 222
  calculus, 29, 29n12, 119, 119n4, 136
  derived functions, 29n12
  multiplication, 200
  mathematical reason and language, 190, 199–200, 201n4, 202–203, 203n8
  See also language (dialectical reason and mathematics); statistics; variables

means of production
  See production (means of)
mediation, 10, 27, 39, 61, 78, 86, 89, 124, 151, 153, 192–193, 198, 204, 210
metamorphosis (metamorphoses), 137, 149, 191, 200, 210
metaphysics, 4, 5, 6, 10, 15, 25, 27, 31, 34, 37–44, 56, 65, 72, 84, 86, 90, 98, 98–99n5, 100, 105, 107, 109–110, 149, 158, 185, 240, 243, 245, 268–269, 272–273
  communism as non-metaphysical potential, 232, 238
  transcendental/transhistorical forces, 38, 56, 72, 99, 107, 240
  See also critique (of metaphysics); Diamat; Soviet Union; thesis-antithesis-synthesis myth
mode of production
  See production (mode/s of)
mode (modes), 149
  of abstraction, 122
  of determination, 192–195
  of existence/being, 33, 84, 85, 243n12
  of explanation, 75
  of inquiry, 4, 256
  of knowing, 84, 131
  of presentation, 134, 154
  See also appropriation; moment(s) (of inquiry); production (mode/s of)
models (model building), 7, 15, 16, 29, 31–32, 47, 60, 67, 76–77, 82, 91, 97–101, 107, 113, 118, 122, 126–135, 136–142, 143, 146, 159, 165–173, 182, 188, 189–223
  abstract and concrete in historical materialism and political economy, 220–221
  capitalism and its central tendencies, 196–199
  change within and between systems, 29–30, 34, 190, 192–197, 215
  change within the capitalist mode of production, 199–215, 217–218
  model building within dialectical and mathematical reason and language, 190–215
  model of capitalism and its historical development, 190
  models of history and system from the general and abstract to the specific and concrete, 215–223
  modes of production and historical change, 218–222

  See also base/superstructure; comparative method (experimental model in structural analysis); deduction; description; experiment; explanation; evolution; language; laws; predictions; structure; successive abstractions
modernity, 57, 60, 95–96, 101, 124, 158n2, 175, 218, 259, 270, 279
  See also society (bourgeois/capitalist/modern)
moment(s), 20, 51, 74, 78n3, 116, 128, 143, 159, 161, 164, 166, 170, 178, 181, 183, 197, 204, 211n14, 214
  of indefiniteness, finiteness, and periodization, 209–210
  of inquiry (in Marx's work and their interrelationships), 6–15, 22, 25, 67, 74, 91, 100, 111, 118, 125, 131, 142, 143, 155–156, 156–188, 197, 204–205, 218, 272–275
  three moments of determination, 193–194
  See also conceptual doublets; determinism; dialectical method; epistemology/Marx's; historical and structural analysis/relationships; historical materialism; inquiry; political economy; successive abstractions
money, 54, 81, 97, 109, 115, 119n4, 125, 126n7, 133, 138, 149, 177, 178, 179, 182, 188, 199, 201n5, 202, 205, 207, 208, 209, 213, 220
morals (morality, moralism, moralist), 33, 71, 90, 269, 270
  See also base/superstructure
movement, 34–35, 47, 55, 73, 78, 78n3, 111, 117, 127n8, 151–152, 159, 162, 165, 168, 171, 181, 189, 204, 207, 209, 228, 239, 240, 243, 246, 260
  double movement, 94, 117, 129, 153, 176, 211–215
  See also contradiction; development; duality; history; metamorphosis; negation; processes; structure; successive abstractions (movement of); transformation; wholes and parts
mystification, (mystify), 5, 37–41, 44, 63, 79, 84, 101, 106, 149n1, 152, 163, 189–190, 279
  See also demystification; naturalism (naturalistic mystification)

320 • Index

natural, history/science, 23, 25, 29–31, 37, 47, 49, 72–73, 96, 97, 98, 98n4, 111, 113–114, 119, 120–121, 139, 147, 158, 185, 200n2, 240, 243, 244
   laws, 163n4
   processes, 73n2, 93, 243
   resources, 202, 213
   selection, 98n4, 98–99, 243, 273
   See also Darwin; evolution
naturalism, 72–74, 100, 113–114, 273
   naturalistic mystification, 73
   predictive-theory naturalism, 120
nature, 9, 25–27, 47, 53, 72–73, 75, 81, 84–86, 89–91, 93, 95, 113, 160n3, 183, 200, 205, 219, 237, 240, 243–246, 249, 253
   necessity (necessary relations/conditions/qualities), 8–9, 26, 26n10, 28, 30, 34, 59, 62, 63, 64, 70, 76, 78n3, 79, 83, 85, 89–90, 99, 105, 117, 123, 130, 138, 190, 203, 209, 210–211, 211n14, 213, 214, 222n17, 227, 271
   See also conceptual doublets: moments of conceptualization; contingency; laws; processes; relations (necessary/contingent)
needs, 86, 89, 237, 260
   and capacities, 74, 76, 237
   and desires, 75
   need for labor, 86
   of capital, 82–83, 216
   social needs, 82
negation (negate/s), 26n10, 28, 33, 56, 72, 95, 100, 140, 252
   dialectical negativity, 36, 42n14, 43n15, 108, 152, 191, 207, 211–215, 273
   of the negation, 25, 244, 245
   See also dialectics

objectivity, 48–49, 54–55, 252
   objective conditions/laws/reality, 49, 85, 90, 91, 240, 242, 248
obscurantism (obscure), 36, 37, 39–41, 43–44, 60, 63, 70, 73, 107, 121–122, 140, 163, 187
   See also abstractions/inappropriate ("society" in general); critique (of Hegel; of obscurantism); individualism (Marx's critique of)
observation (observe, observable reality), 4, 20, 28, 46, 48, 50, 53, 54, 56, 60, 61, 64, 72, 80, 87, 96, 111, 114, 118, 120–127, 128n9, 131–132, 133, 135, 138–139, 141, 143, 153, 154, 158, 159–160, 165, 173, 179, 182, 201, 206, 222, 263, 264, 269
   and conceptualization, 156–157, 159, 160–170, 175, 183–186, 187, 188, 191, 199–200
   See also conceptual doublets: epistemological moments; concrete; determinism (abstractions of determination/determinate abstractions); successive abstractions
Ollman, Bertell, 11, 71, 111, 117, 135, 150
   See also internal relations
onto-epistemology, 71, 73, 76, 101, 200, 275
   Marx's assumptions, 65, 69–101
   See also abstraction/method of; dialectical method (necessity of a dialectical science); duality; epistemology/Marx's; essence/essentialism; evolution; identity/difference; internal relations; labor (centrality of); naturalism; organicism
ontology (ontological), 62, 67, 71, 105, 153, 188, 245
   Marx's, 9, 36, 41, 49–50, 69–84, 116, 121, 130, 150, 151, 153, 156
   See also epistemology; epistemology/Marx's; internal relations; labor (centrality of); onto-epistemology
operational definitions, 121–122, 124–125, 128, 156, 244
organic composition of capital
   See capital (organic composition of)
organicism, 96, 97–99, 164, 218, 273
   organic totality, 77
   organic whole, 70, 97, 145, 159, 193
   See also evolution; totality
over-precision
   See precision
Owen, Robert, 5, 227, 232

Paris Commune, 109, 234
part(s)
   See wholes and parts
periodization, 130, 169, 215
   era(s), 92, 114, 127–131, 216, 235, 266
   of capitalist development, 216–217, 218, 220
   epoch(s), 32–33, 61, 62, 76, 88, 96, 108, 116, 128–129, 159, 163–164, 210, 216

"epochs in the progress of the economic formation of society", 30, 32–33
moments of periodization, 210
period of history, 35, 50, 72, 78, 78n3, 81, 98–99n5, 116, 129, 131, 137, 157, 173, 177, 202, 205, 216, 221
periods of development/maturation/ transformation/decline, 20, 30, 61, 70, 84, 88, 140, 180, 199, 202, 202n6, 206, 211–216
stage(s), 20, 32, 51, 52, 55, 71
prehistoric stage of human society, 30
stage of development/historical development, 20, 22, 29, 30, 33, 88, 107, 129, 132, 153, 163, 177, 188, 214, 216–218
stage of production, 91, 107, 209
See also capitalism (subforms/ submodes of); class; comparative method; historical analysis; historical and structural analysis/ relationships; Preface
philosophy, 10, 20, 24, 24n9, 27, 30, 34, 38, 39, 58, 71, 72, 76, 84, 105, 106, 187, 238, 240, 243, 244, 245, 247, 249, 250, 252
analytical, 11, 40–39, 152, 191
of science, 4, 16, 101, 120, 122, 147, 148–156, 185, 187, 223, 264
philosophers, 261, 264
realism, 20
critical realism, 9
Marxist-realism, 151
realist philosophy of science/ scientific realism, 147–150, 272
representative realism, 244
speculative philosophy, 5, 6, 37–38, 39, 41, 96, 103, 106–107, 270
See also concepts (though realist and internal relations philosophies of science); critique (of Hegel; of idealism; of Kant; of speculative philosophy); flexibility; internal relations; language; precision; science; scientific dialectics
Plekhanov, Georgi, 242, 245, 246–247, 249, 250–251, 254
political economy, 3, 6, 7–9, 10, 11, 13–14, 15, 16, 19–23, 28–29, 31, 39, 43, 48, 50, 53–54, 58, 59–60, 65, 67, 74, 83, 86, 91, 96, 98, 102, 105–108, 112, 113–116, 121–122, 124–125, 136–146, 147–149, 155, 158, 159, 170–173, 174, 180, 182, 185–186, 190–192, 193, 196–199, 199–215, 215–222, 227, 229, 239, 255, 264–265, 270, 272, 273–275
See also capitalism; class (analysis); conflate; dialectical method; evolution; historical and structural analysis/relationships; historical materialism; models (capitalism and its central tendencies); moment(s) (of inquiry); organicism; structure; successive abstractions
politics (political relations), 10, 12, 34, 57, 77, 78n3, 79, 80, 81, 83, 86, 89, 106, 108, 109, 121, 149, 154, 180–182, 184, 214, 220, 221, 225, 230, 232, 238, 247, 253, 262–263, 266–268
knowledge and practice/praxis, 11–15, 22, 31, 34, 37, 52–54, 63–64, 84, 107–108, 149, 181, 215, 227–228, 235, 244, 259, 263, 265–281
"revolutionizing practice", 85
See also base/superstructure; communist project: law; revolution
positivism
See science (positivism)
possibility, conditions and ranges of, 14, 63, 73, 82, 92, 94, 114, 148, 177, 187, 188, 194–195, 206, 214, 223, 237, 239
conditions of possibility and socialism/ communism, 5, 13, 23, 188, 227–229, 255, 259, 273
potential (individual abilities/ capacities/powers/traits), 8, 74–76, 88, 93, 237
potential (social), 82, 115, 129, 178, 213–215, 230
precision, 15, 30, 50, 92, 123, 125, 139, 150, 162, 182, 185, 187–188
and flexibility, 150–153, 155–156, 162–164, 166, 178, 188
imprecision, 5, 9, 184
over- and under-precise readings of historical materialist principles, 31–36
over-precise readings, 31–35, 185, 240
under-precise readings, 35–36
See also communist project (flexibility and precision); concepts (conceptualizing precisely/and abstracting flexibly); flexibility (flexible precision/precise flexibility); successive abstractions; typifying

preconditions, 116–118, 125, 135, 160n3, 178–179, 182, 194–195, 205–206, 216–219, 227, 239
  See also backward (study of history); communist project (preconditions of); historical and structural analysis/relationships; presuppositions
predictions, 11, 12–14, 31, 98, 129, 139, 176, 181, 197, 228, 236
  history as predictable stages, 33, 238, 238n7, 244
  predictive-theory and sociology, 120–122, 139
  See also a priori, critique (of positivist sociology); laws; naturalism (predictive-theory naturalism); Preface (traditional readings)
Preface (to *A Contribution to the Critique of Political Economy*), 16, 18, 29–30
  traditional readings/critique of traditional readings, 31–35, 40–41, 276
  See also determinism (rigid); evolution; precision (over- and under-precise readings of historical materialist principles); progress; universal(s) (universality and Marxian theory)
presentation, 16–17, 44, 51–53, 82, 103, 105, 109, 111, 117–118, 124–125, 132–135, 137–138, 149n1, 153, 155, 162, 173, 179, 182, 184, 190, 193, 200, 263, 270
  as epistemological moment, 156–159
  of data and findings, 142–146
  See also conceptual doublets; epistemological moments; Marx (presentation of his work); models
presuppositions, 12, 115, 117, 125–126, 135, 144, 178–179, 182, 188, 194–195, 216–219
  See also abstractions; backward (study of history); communist project (presuppositions of); historical and structural analysis/relationships; preconditions
price(s), 134, 136, 155, 177, 200, 201n5
processes, 5n2, 6, 30, 35, 41, 46, 51, 64, 71, 77, 89, 93, 100, 115, 117, 149, 152, 155, 158, 161, 175n6, 200, 211
  and relations(hips) as conceptual pair, 89, 141, 173–175, 177, 179–186, 193, 199, 202, 222, 210, 266, 267

emergent historical-structural processes in capitalism, 217
historical, 3, 69, 73, 88, 91, 98, 100, 115, 124, 141, 150, 181–186, 199, 214–215, 215n15, 230, 239
necessary/contingent historical/structural processes, 181–186
processes of formation, organizing, transformation, renewal, 201, 210, 215
structural/institutional, 70, 82–83, 92, 96, 100, 132, 139, 141, 176, 179–180, 192–193, 194, 196, 201, 201n4, 202–203, 204, 207, 209–210, 213, 214, 217, 265–266
  See also conceptual doublets: moments of conceptualization; contingency; determinism (forms of); historical and structural analysis/relationships; internal relations; language; models (change within and between systems); necessity; production (production process); relations
production, 29, 34, 53, 62, 71, 73, 76, 78–79, 81, 84–86, 89–92, 94, 95, 107, 108, 122, 123, 131, 134, 150, 159–161, 179, 180, 202, 206, 212, 219, 271, 280
bourgeois, 108, 215
capitalist, 12, 40, 83, 138, 153, 189, 190, 202, 203, 209, 239n8
forces of/productive forces, 7, 30, 34, 37, 75, 76–77, 79, 84, 97, 107, 183, 194–195, 249, 273
means of, 76, 90, 94, 133, 137, 181, 184, 185, 198, 203, 206, 207, 211, 213, 219, 220, 232, 234, 261, 265, 266, 275
mode/s of, 7, 9–10, 29–30, 32–34, 39, 44, 57, 71, 76–77, 80, 82–84, 86, 87–90, 93, 94, 99, 114, 121, 123–124, 128, 129, 139, 148, 165, 167, 171–172, 176, 181, 182, 184–185, 188, 192–199, 218
"production in general", 129, 155, 159–160, 165, 166, 169–171, 180, 182, 183, 188, 220
production process/process of production, 30, 71, 91, 92, 114, 145, 155, 204, 210
productive powers, 12, 39, 206n11
relations of, 13, 29–30, 34, 77, 78n3, 99, 107, 114–115, 155, 181, 194–195, 220, 233–234

spheres of, 210
See also Asiatic mode of production; capitalism (capitalist mode of production); class; communism (communal ownership/primitive communism); feudalism; labor; materialism; relations; slavery; tribal society

profit(s), 88, 90, 93, 131, 133, 134, 136, 139, 155, 168, 182, 191, 198, 199, 200, 204, 205, 210, 212, 220, 236, 266
tendency for rate of profit to fall, 134, 197, 198, 217, 220, 264
See also capitalism (central tendencies/ laws of motion); laws

progress (progressive), in history/ society/capitalism, 20, 30, 32–33, 59, 63, 81, 98, 100, 118, 176, 180, 207, 211, 213, 273
in revolutionary tactics/values, 229, 231, 256, 261, 266, 267, 278, 279
in science, 48, 52, 81, 119

proletariat (proletarian), 21, 29, 63, 77, 185, 198, 202, 214, 229–230, 232–235, 251, 255, 261, 265–267, 268, 268n4, 274
lumpenproletariat, 181
proletarianization, 92
See also bourgeoisie; capital; class; communist project (dictatorship of the proletariat); labor (laborers); revolution (worker/proletarian)

property (forms, relations), 30, 50, 94, 117, 126n7, 160n3, 168, 177, 196, 211n14, 215n15, 230, 234, 236, 239, 261, 277
private property, 50, 42n14, 94, 123, 160n3, 174, 194, 198, 205, 214, 215n15, 220, 232, 234, 236, 239, 274, 275

psychology, 40, 48

quality (qualitative phenomena), 7, 26–27, 56–57, 70, 90, 112, 125–126, 129, 135, 137, 152, 155, 164, 165, 166, 170, 187, 188, 200, 210
qualitative comparisons/equality, 115, 131, 201n4, 204, 236
qualitative relations/changes/ transformations, 14, 70, 88, 92–93, 113, 131, 135, 152, 153, 163, 173, 175, 181, 200, 209, 211–215, 216
See also commensurable; quantity and quality

quantity (quantitative phenomena/ relations), 26n10, 89, 93, 155, 189, 199–201, 201n5, 202, 205, 211, 213
terms/measures, 136–138, 155, 146, 191, 199, 200–203
See also language; mathematics

quantity and quality (qualitative and quantitative relations), 173–174, 178, 191, 199–200, 209, 236
qualitative and quantitative comparisons, 203–205
transformation of quantity into quality, 25, 168n5, 178, 191, 199, 200, 200n2, 205–207, 244, 245
See also conceptual doublets: moments of conceptualization; dialectics; Hegel; language (conversion terms); mathematics

questionnaire (Marx's), 18, 120, 136, 231

radical(ism), in Marx's thought and politics, 12, 64, 86, 101, 175, 176, 229–230, 232n3, 239n8, 230, 254, 263, 267, 269–270

rational (rationalism/rationality), 26n10, 27, 38, 40–41, 85, 98, 105, 220, 235

realism
See philosophy (realism)

reciprocity (reciprocal relationships), 35, 72, 73, 78, 85, 135, 175
See also base/superstructure

reduction(ism) (reductive frameworks), 25, 33, 40, 56, 58, 62, 64, 79, 107, 120, 139, 185, 199, 242, 243
reduction of dialectics, 244–245
See also determinism (rigid); individualism (Marx's critique of)

reflexivity, creative, 93
See also freedom

reify (reifying/reification), 70, 163
See also fallacy of misplaced concreteness

relational (qualities, analysis, truth), 27, 85, 121–128, 150–152, 160, 162, 187–188

relations (inter/relationships), 19, 25, 29, 35, 37, 39, 56, 73, 73n2, 78, 97, 105, 115, 154, 281
as political-economic unit of analysis, 28, 39, 40–41, 50, 57–58, 70, 87, 99, 101, 107, 109, 111–113, 116, 130–131, 133, 137–139, 148–149, 157–158, 168, 199–215
general systemic relations, 207–209

relations of composition, 205–206
relationships of negativity and
  change, 152, 207, 211–215
time-space relations, 209–211
as subject matter for social-scientific
  thinking, 62–64, 72–77, 84, 86–90,
  96–97, 99, 101, 111–113, 116, 119,
  121, 132, 135, 139, 147, 151–152,
  162–164, 191–192, 264
exchange relations, 222n17
necessary/contingent (historical/
  structural relations), 181, 183–186
relations and processes as conceptual
  doublet, 89, 141, 173–175, 177,
  179–186, 193, 196, 199, 202, 222, 210,
  266, 267
See also abstraction/method of; base/
  superstructure; capital (as a class
  relation); capitalism (capitalist
  material relations); class (analysis;
  class-state relationship; relations);
  comparative method; conceptual
  doublets: moments of abstraction;
  conceptual doublets: moments of
  conceptualization; contingency;
  contradiction; determinism (forms
  of); experiment; generalization(s);
  historical and structural analysis/
  relationships; identity/difference;
  internal relations; inversion(s);
  labor (centrality of); labor theory of
  value (labor-value relations); laws;
  materialism; models; necessity;
  politics; processes; processes (and
  relationships as conceptual pair);
  production (relations of); property;
  quality; quantity; quantity and
  quality; reciprocity; statistics
  (correlations); successive
  abstractions; time and space;
  totality; wholes and parts
religion (religious), 19–20, 24n9, 30, 34,
  36, 37, 56, 57, 60, 62, 71, 76, 77, 78, 80,
  83, 85, 89, 90, 96, 98–99n5, 108, 129,
  175, 192, 220, 261, 263, 268, 268n4,
  269, 279, 280
  Christianity, 36n13, 95, 105
  Judaism, 95
reproduction, 34, 114, 123, 155, 180,
  192–193, 197, 202n6, 210, 211
See also determinism (modes of)
revolution, 4, 12–14, 31, 186, 263, 265,
  266–268, 280–281, 281n5
  and the communist project, 215, 219,
    227, 228–234, 238–239, 275

and historical materialist principles,
  30, 78n3, 85, 100, 194, 238, 273
economic, 202
industrial, 84, 221
technological revolutions in capitalism,
  198, 217, 220
theoretical/scientific, 108, 144, 149
worker/proletarian, 8, 11–14, 42, 181,
  188, 220, 238–240, 255, 259, 275
See also communist project; French
  Revolution; historical materialism;
  politics; Soviet Union

Sayer, Andrew, 96
Sayer, Derek, 11, 152, 263
science (methods, principles, and values),
  3–4, 5, 7, 9–11, 16, 22, 25, 39, 41, 46,
  47–49, 55–56, 65, 77, 108, 111, 119–121,
  141, 147, 182, 199, 228, 241, 251, 259,
  263
  critical conceptions of/in science, 49,
    50–54, 56, 59, 87, 100–101, 105, 108,
    116, 144, 148–150, 160, 187–188,
    222n17, 261, 263, 281
  positive science, 37, 41, 120
  positivism, 4, 103, 118–121
  scientific abstraction/
    conceptualization, 71–72, 79, 96,
    100–101, 112, 118, 162–164
  scientific thought and/in the
    capitalist era, 82, 96, 220
  social science, 11, 23, 47, 48, 52, 73,
    119–123, 126, 131, 134, 137, 148, 160,
    185, 271
See also a priori; analysis; comparative
  method; concepts; critique;
  deduction; evolution; experiment;
  induction; laws; mathematics;
  natural (history/science); objectivity;
  philosophy; precision; prediction(s);
  progress (in science); scientific
  dialectics; statistics; typifying;
  variables
scientific dialectics, 3, 22, 41–42, 103, 105,
  110, 136–141, 146, 150–153, 189–190,
  222, 223, 225, 228, 251, 263, 264,
  274–275
  dialectical method and scientific
    method, 118–146
  Marx's moments of inquiry and
    scientific methods, 14–15, 44, 67, 74,
    222–223
See also communist project (science
  and); critique; dialectical method;
  dialectics; essence/essentialism;

evolution; experiment; flexibility; induction; internal relations; language; laws; Marx (intellectual and scientific roots); naturalism; precision; prediction(s); science; typifying; vantage point
self-clarification
See clarification; critique
Simmel, Georg, 45, 46, 47, 48, 49–50, 52, 54, 55, 56–57, 59, 60, 62–63, 64, 65, 116
slavery, 7, 88, 89, 126, 126n7, 129–130, 160, 167, 168n5, 169, 171, 179, 196, 219, 220, 238n7, 279, 281, 281n5
See also class; production (mode/s of)
social, action, 27, 54, 84, 174, 256
   facts, 46, 48, 50–51, 55–56, 58, 60, 62, 120, 137, 187
   formation, 30, 77, 96, 97, 99, 112, 117, 130, 159, 160, 177, 179, 181, 192, 194
   organism, 48
socialism (socialist), 12, 23, 109, 181, 219, 221, 238–241, 246, 248, 251, 254, 260, 272, 277
   in one country, 253
   "scientific socialism", 240, 243, 252
   utopian socialism, 234
   vulgar socialism, 235
   See also communist project (socialist transitionary period); revolution; Soviet Union
socially necessary labor-time
   See labor theory of value (socially necessary labor-time)
society (in general), 7, 9–10, 12, 24n9, 26, 32, 34, 41, 51, 53, 57, 59–62, 67, 69–70, 71–74, 57, 76–83, 85, 87, 89, 91, 94, 108, 113–116, 120, 122, 130, 130n10, 165, 175, 176, 188, 214, 222, 229, 233, 236n6, 261, 262, 266, 276, 278
   bourgeois/capitalist/modern, 7, 13, 19, 29n11, 54, 61, 65, 70, 74, 76, 87, 90, 94, 97, 101, 110, 116, 122, 129, 130n10, 132, 138, 149n1, 154, 160, 168, 174, 182, 190, 196, 199, 202, 207, 212, 213, 214, 219, 222n17, 230, 235, 252, 263, 264, 267, 275
   civil, 71, 154, 176, 180
   class/alienated, 32, 93, 130, 166, 170, 171
   See also abstractions/inappropriate ("society" in general); modernity; production (mode/s of); universal(s) (social/sociological)
sociology (sociological/sociologists), 1, 10, 44, 45, 46–49, 52, 54–55, 57, 59, 62, 64, 65, 116, 120, 238, 264, 265, 268, 271

Marx and classical sociology, 45–67
Marx's critique and alternative to positivist sociology, 120–142
sociological laws/models/variables, 31, 57, 58–59, 63–65, 113, 218, 241, 275
unit of analysis, 60–62
See also abstractions/inappropriate ("society" in general); base/superstructure; laws; social (facts); universal(s) (social/sociological)
Soviet Union (Soviets), 5, 12–13, 237–256, 262
   and revolutionary ideas, 237–244, 246–249, 250, 252, 254
   Bolsheviks/Russian Revolution, 12, 238, 239n8, 241, 241n9, 243, 246–248, 250, 253–255, 261
   metaphysics, dialectical laws, and historical materialism in, 227, 238, 238n7, 239–240, 243–245, 245n13, 246–249, 250–55
   See also Bukharin; Diamat; ideology; Kautsky; Lenin; Plekhanov; Stalin
space
   See time and space
species-being, 74–75, 89, 114, 219, 274
   fixed and relative features, 75
specific abstract
   See successive abstractions (the specific abstract)
specific concrete
   See successive abstractions (the specific concrete)
specific/specificity (analysis, cases, systems), 6–7, 9, 11, 15, 26, 29, 31–32, 37, 40, 59, 60–61, 76, 80, 82, 87–89, 94, 106, 112–117, 121, 122, 123, 130, 132, 140, 141, 144, 155–156, 159, 160, 160n3, 161, 163–173, 178, 180–182, 184–188, 191, 193, 196, 197, 215–216, 218, 220, 231, 237, 271
differentia specifica, 75, 133, 265
See also conceptual doublets: moments of abstraction; general; generality; successive abstractions (the specific abstract; the specific concrete)
speculative philosophy
   See critique (of Hegel; of Kant; of speculative philosophy); Kant; philosophy (speculative philosophy)
stage(s)
   See periodization
Stalin, Joseph, 245, 249, 253, 254
Stalinism, 5, 240, 245, 253, 254, 255, 262
   See also Leninism (Leninist-Stalinism)

starting-point, 20, 53, 55, 57, 108, 125, 196, 204
stasis (static frameworks), 27, 70, 75, 77, 121–122, 175, 176, 194
  See also abstractions/inappropriate ("society" in general); historical analysis (ahistoricism)
state (the), 7, 8, 12, 27, 48, 56, 71, 77, 83, 92, 98–99n5, 114, 115, 117, 125, 154, 170, 175, 177, 180, 184, 185, 192, 194, 198, 203n7, 213, 214, 216, 216n16, 217, 219, 220, 221, 230, 233n4, 261, 266–268, 273, 278–279
  See also base/superstructure; communist project (dictatorship of the proletariat)
statistics, 29, 43, 48, 49, 111, 127, 136–138, 146, 201, 203, 246, 271, 273
  correlation(s), 136–137, 137n11, 140–141, 203
  multiple regression, 137–138
  regression to the mean, 203
  standard errors, 203n8
  See also language (dialectical reason and mathematics); mathematics variables
structure (relations/hips and dynamics, structural analysis), 7, 12, 13, 33, 34–35, 40–41, 46, 47, 48, 50, 57, 60, 63, 65, 69–72, 74, 75–84, 85–86, 88–90, 93–94, 96–97, 99–101, 103, 111–118, 121, 123–145, 146, 150, 151, 153, 155–156, 157, 162–173, 174–188, 189–199, 200, 202, 204–222, 223, 227, 236n6, 262, 264–267, 271, 272–275
  See also abstraction/method of; base/superstructure; conceptual doublets: moments of conceptualization; historical analysis; historical and structural analysis/relationships; historical materialism; levels of generality; models; Preface; processes; relations; successive abstractions; teleology; universal(s) (social/sociological); wholes and parts
successive abstractions (the method of), 161–173, 174, 178, 188
  movement of, 171–172
  the general abstract, 160, 164, 165, 166, 167–171, 183, 193
  the general concrete, 164, 165, 166, 167–168, 169–172, 173, 182, 184, 185, 193, 221, 275

the specific abstract, 164, 165–166, 167, 168, 169–172, 173, 182, 184, 185, 193, 220
the specific concrete, 164, 165–166, 167, 168, 169–173, 178, 186, 191, 193, 221
  See also successive approximations
successive approximations (method of), 161–162
  See also successive abstractions
superstructure
  See base/superstructure
  See also art(s); family/the; ideology; knowledge; morals; politics; state
supra-individual reality, 54, 56–58, 65, 71
  See also sociology (unit of analysis)
surplus-labor
  See labor theory of value; value
surplus-value
  See appropriation; labor theory of value; value
synthetic unities
  See successive abstractions

tautology (tautological), 100–101, 105–106, 110, 160n3, 204, 253
technology (machinery, technological complexity, tools), 76, 80, 84, 86, 93, 96, 118, 119, 133, 137, 149, 155, 197, 198, 202, 206, 209, 217, 220, 227, 228, 236, 239, 261, 267, 268
  See also determinism (rigid)
teleology (teleological), 35, 59–60, 88, 99, 194, 277
  and communism, 248, 268n4, 272, 276, 277
  and Durkheim, 61–62
  and the metaphysical view of history, 4, 31, 99–100, 218
  Marx on Darwin and non-teleological evolution, 98
  See also Preface (traditional readings)
temporality
  See time and space
tendency (tendencies)
  See capitalism (central tendencies/laws of motion); laws
theory of history
  See class; historical analysis; historical materialism ("materialist conception of history"); history; Preface (traditional readings); teleology (and the metaphysical view of history)

thesis-antithesis-synthesis myth, 42, 42n14, 43, 43n15, 110, 268n4
See also metaphysics
time and space (time/space, temporal and spatial relations), 6, 22, 26n10, 27, 55, 57, 72, 87, 93, 96, 107, 110, 112–113, 115–116, 118, 122, 124–126, 126n7, 128, 132, 134, 140, 163, 166, 173, 176–178, 180, 194, 196, 199, 201, 203, 206, 209, 211–215, 216–222
conversion terms and, 209–211
space/spatial relations and qualities, 47, 92, 95, 112, 116, 176, 177
time/temporal relations and qualities, 7, 51, 52, 61, 70, 76, 78, 82, 87–94, 97, 114–115, 116, 118, 127–131, 160, 175–177, 191, 201n4, 202–203, 203n7, 209–210, 222, 278
See also conceptual doublets: moments of conceptualization; development (uneven) historical and structural analysis/relationships; history; labor (abstract labor; as the origin of time); levels of generality; periodization
totality (totalities), 14, 26n36, 59, 70–72, 77, 124, 153, 163, 175, 196, 207, 218
transcendental forces
See metaphysics
transformation (transforms), 13, 25, 30, 60, 63, 70, 75, 76, 84, 86, 88, 91, 94, 95, 97, 117, 124, 140, 152, 175, 185, 192–194, 196, 199, 200, 203, 205, 207, 211, 214–217, 227, 229, 230, 232, 235, 267, 273, 278
See also communist project (transformation of Marx's communist project into Soviet Diamat); metamorphosis; models (change within and between systems; change within the capitalist mode of production); quantity and quality (transformation of quantity into quality)
transhistorical forces
See metaphysics
tribal society (ancient society), 7, 30, 32, 88, 89, 129, 160, 164, 169, 171, 219, 220
See also class; production (mode/s of)
typifying (typification), 123–126, 143
See also concepts (construction and development); successive abstractions

under-precision
See precision
uneven development
See development (uneven)
unit of analysis
See analysis
United States, 167, 168, 170, 229, 260, 262, 279
universal(s), abstract universality (Marx's critique of), 42, 72–73, 273
social/sociological, 76–77
sociological laws versus universal truth, 58, 62
universal equivalent, 81, 138
universality and Marxian theory, 75–76, 87–89, 99, 100, 124, 128, 158, 160, 165, 218, 238–239, 245
See also abstractions/inappropriate ("society" in general); base/superstructure; critique; Diamat; Engels (and dialectic); freedom; human nature; individualism; Kant; labor (centrality of); Preface (traditional readings); science (positivism); species-being
use-value
See value
utopianism (utopia; utopian), 5, 8, 36, 227, 234, 234n5, 236, 238, 243, 261, 269, 272, 273, 276, 277, 281

value, 41, 82, 88, 90–91, 92, 94, 97, 100, 122, 131, 134, 137, 138, 145, 155, 158, 177, 178, 179, 190, 196, 199, 200, 202n6, 205, 206n11, 207, 208n13, 209, 213
exchange-value, 155
general law of value, 140, 199
of labor-power, 132
rate of surplus-value, 132–133
realization of value, 182
social value, 40, 88, 131
social value in communism, 236
surplus-value, 90, 92, 93, 133–134, 155, 179, 180, 185, 197, 198, 201, 201n3, 202, 202n6, 205, 206n11, 209, 210–211, 217, 220, 236, 265
absolute and relative, 132, 198
use-value, 90, 97, 155, 205, 207
See also labor; labor theory of value; wage(s)
vanguard party, 251–252
See also Leninism; Soviet Union
vantage point, 87, 111, 155, 178
and backward study of history, 117

and controlled comparison, 127–128
and model building, 191, 202, 204, 218–219
controlled comparison, and historical analysis, 128–130
experimental model, and structural analysis, 130–134
in the synthesis of historical and structural analysis, 135, 182–186
See also abstractions (of vantage point)
variables, 27, 35, 55, 57, 79, 88, 113, 121–122, 126–137, 139, 140–142, 144, 146, 157, 163, 191, 194, 199, 200, 201n5, 202–205, 208, 210, 222, 228, 264, 273
control variables (constants), 131–135, 137, 139, 146, 191, 201, 204–205
dependent variables (caused), 130, 132, 137, 204
independent variables (causal), 130, 132, 137, 142, 204
multivariate analysis, 127, 132–133, 137, 141–142
operational variables, 121–122
See also analysis; capital (constant and variable); comparative method; concepts; experiment; historical analysis (history as a variable); models; typifying

wage(s), 88, 109, 125, 136, 137, 149, 155, 181, 200, 203, 205, 206, 208, 220, 236, 266, 267
labor-wage market, 209
wage-bill, 180
wage-labor(er)/work(er), 115, 126, 126n7, 129, 198, 235, 236n8, 267
wage-system, 90, 123, 134, 139, 170
See also labor; proletariat
wholes and parts (whole/part relations), 21, 36, 40, 49–50, 57, 59, 60–61, 70–72, 77, 81–82, 89, 97, 112, 114, 116, 118, 123, 125n6, 127, 139, 141–142, 147, 150, 153, 158–159, 162, 173, 175–176, 187, 191, 193, 196–197, 201n4, 208, 211, 268n4
parts, 96, 113, 114, 125–126, 127–129, 131–132, 139, 141–143, 175–176, 177, 191–192, 197, 205, 213
wholes, 9, 156, 191
See also conceptual doublets: moments of conceptualization; models; successive abstractions; totality
working-class(es)
See class; proletariat
working-day, 109, 126n7, 132, 133, 149, 170, 197, 201n3, 208, 210, 213, 231
Wright, Erik Olin, 192

STUDIES IN CRITICAL SOCIAL SCIENCES

The Studies in Critical Social Science *book series, through the publication of original manuscripts and edited volumes, offers insights into the current reality by exploring the content and consequence of power relationships under capitalism, by considering the spaces of opposition and resistance to these changes, and by articulating capitalism with other systems of power and domination – for example race, gender, culture – that have been defining our new age.*

ISSN 1537-4234

1. LEVINE, Rhonda F. (ed.) *Enriching the Social Imagination.* How Radical Sociology Changed the Discipline. 2004. ISBN 90 04 13992 3
2. COATES, Rodney D. (ed.) *Race and Ethnicity.* Across Time, Space and Discipline. 2004. ISBN 90 04 13991 5
3. PODOBNIK, B. & T. REIFER (eds.) *Transforming Globalization.* Challenges and Opportunities in the Post 9/11 Era. 2005. ISBN 90 04 14583 4
4. PFOHL, S., A. VAN WAGENEN, P. AREND, A. BROOKS & D. LECKENBY (eds.) *Culture, Power, and History.* Studies in Critical Sociology. 2005. ISBN 90 04 14659 8
5. JORGENSON, Andrew & Edward KICK (eds.) *Globalization and the Environment.* 2006. ISBN 90 04 15132 X
6. GOLDSTEIN, Warren (ed.) *Marx, Critical Theory, and Religion.* A Critique of Rational Choice. 2006. ISBN 90 04 15238 5
7. DELLO BUONO, Richard A. & José BELL LARA (eds.) *Imperialism, Neoliberalism and Social Struggles in Latin America.* 2007. ISBN 90 04 15365 9
8. PAOLUCCI, Paul B. *Marx's Scientific Dialectics.* A Methodological Treatise for a New Century. 2007. ISBN 978 90 04 15860 3
9. OTT, Michael R. (ed.) *The Future of Religion.* Toward a Reconciled Society. 2007. ISBN 978 90 04 16014 9
10. ZAFIROVSKI, Milan. *Liberal Modernity and Its Adversaries.* Freedom, Liberalism and Anti-Liberalism in the 21st Century. 2007. ISBN 978 90 04 16052 1

www.ingramcontent.com/pod-product-compliance
Lightning Source LLC
Chambersburg PA
CBHW071148070526
44584CB00019B/2703